OXFORD MEDICAL PUBLICATIONS

Fast and Slow Chemical Signalling
in the Nervous System

IVERSEN, L.L. and GOODMAN,
E.C. (eds.)

Fast and Slow Chemical Signalling in the Nervous System

EDITED BY

L. L. IVERSEN AND E. C. GOODMAN

Neuroscience Research Centre,
Merck Sharp & Dohme Research Laboratories,
Harlow, Essex

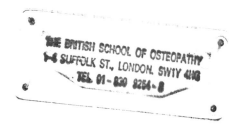
Oxford New York Tokyo

OXFORD UNIVERSITY PRESS

1986

Oxford University Press, Walton Street, Oxford OX2 6DP

Oxford New York Toronto
Delhi Bombay Calcutta Madras Karachi
Kuala Lumpur Singapore Hong Kong Tokyo
Nairobi Das es Salaam Cape Town
Melbourne Auckland

and associated companies in
Beirut Berlin Ibadan Nicosia

Oxford is a trade mark of Oxford University Press

Published in the United States
by Oxford University Press, New York

British Library Cataloguing in Publication Data

Fast and slow chemical signalling in the nervous system.
1. Neurotransmitters
I. Iversen, L. L.   II. Goodman, E. C.
599.01'88   QP364.7
ISBN 0-19-857216-6

Library of Congress Cataloging in Publication Data

Fast and slow chemical signalling in the nervous system.
(Oxford medical publications)
Includes bibliographies and index.
1. Neural transmission.   2. Neurotransmitters.
3. Neuropeptides.   4. Amino acids.   5. Neuropharmacology.
I. Iversen, Leslie L.   II. Goodman, E. C.   III. Series.
[DNLM: 1. Brain Chemistry.   2. Neural Transmission.
3. Neurons.   4. Neuroregulators. WL 102.7 F251]
QP364.5.F37   1986   599'.01'88   86-2379
ISBN 0-19-857216-6

Set by Spire Print Services Ltd, Salisbury
Printed in Great Britain by
St Edmundsbury Press,
Bury St Edmunds, Suffolk.

Contents

Contributors

J. L. BARKER, *Laboratory of Neurophysiology, National Institute of Neurological and Communicative Disorders and Stroke, National Institute of Health, Bethesda, Maryland 20205, USA.*

M. J. BERRIDGE, *AFRC Unit of Invertebrate Neurophysiology and Pharmacology, Department of Zoology, Cambridge CB2 3EJ, UK.*

F. E. BLOOM, *Division of Preclinical Neuroscience and Endocrinology, Scripps Clinic and Research Foundation, 10666 North Torrey Pines Road, La Jolla, CA 92037, USA.*

N. G. BOWERY, *Neuroscience Research Centre, Merck Sharp & Dohme Research Laboratories, Terlings Park, Eastwick Road, Harlow, Essex CM20 2QR, UK.*

D. A. BROWN, *MRC Neuropharmacology Research Group, Department of Pharmacology, School of Pharmacy, University of London, 29–39 Brunswick Square, London WC1N 1AX, UK.*

V. CRUNELLI, *Department of Pharmacology and Clinical Pharmacology, St. Georges Hospital Medical School, Cranmer Terrace, London SW17 0RE, UK.*

B. DUFY, *Laboratoire de Neurophysiologie, Université de Bordeaux II, Bordeaux, France 33076.*

B. EVERITT, *Department of Histology, Karolinska Institute, PO Box 60400, S-104 01 Stockholm, Sweden.*

S. FORDA, *Department of Pharmacology and Clinical Pharmacology, St. Georges Hospital Medical School, Cranmer Terrace, London SW17 0RE, UK.*

S. FREEDMAN, *Neuroscience Research Centre, Merck Sharp & Dohme Research Laboratories, Terlings Park, Eastwick Road, Harlow, Essex CM20 2QR, UK.*

P. GREENGARD, *Laboratory of Molecular and Cellular Neuroscience, The Rockefeller University, 1230 York Avenue, New York 10021, USA.*

T. HÖKFELT, *Department of Histology, Karolinska Institute, PO Box 60400, S-104 01 Stockholm, Sweden.*

V. R. HOLETS, *Department of Histology, Karolinska Institute, PO Box 60400, S-104 01 Stockholm, Sweden.*

R. F. IRVINE, *Department of Biochemistry, AFRC Institute of Animal Physiology, Babraham, Cambridge CB2 4AT, UK.*

C. E. JAHR, *Section of Molecular Neurobiology, Yale University School of Medicine, New Haven, CT 06510, USA.*

T. M. JESSELL, *Department of Biochemistry and Molecular Biophysics, and Howard Hughes Medical Institute, Columbia University, College of Physicians and Surgeons, New York, NY 10032, USA.*

J. S. KELLY, *Department of Pharmacology, University of Edinburgh, 1 George Square, Edinburgh EH8 9JZ, UK.*

J. KEMP, *Neuroscience Research Centre, Merck Sharp & Dohme Research Laboratories, Terlings Park, Eastwick Road, Harlow, Essex CM20 2QR, UK.*

T. KREINER, *Department of Biological Sciences, Stanford University, CA 94305, USA.*

D. E. KOSHLAND, JR, *Department of Biochemistry, University of California, Berkeley, CA 94720, USA.*

E. A. KRAVITZ, *Department of Neurobiology, Harvard Medical School, 25 Shattuck Street, Boston MA 02115, USA.*

K. KRNJEVIĆ, *Room 1208, McIntyre Building, McGill University, 3655 Drummond Street, Montréal, Québec, Canada H3G 1Y6.*

P. KROGSGAARD-LARSEN, *Department of Chemistry BC, The Royal Danish School of Pharmacy, DK-2100, Copenhagen Ø, Denmark.*

N. LERESCHE, *Department of Pharmacology and Clinical Pharmacology, St. Georges Hospital Medical School, Cranmer Terrace, London SW17 0RE, UK.*

J. M. LUNDBERG, *Department of Pharmacology, Karolinska Institute, PO Box 60400, S-104 01 Stockholm, Sweden.*

R. N. MCBURNEY, *MRC Neuroendocrinology Unit, Newcastle General Hospital, Newcastle-upon-Tyne NE4 68, UK.*

U. MADSEN, *Department of Chemistry BC, The Royal Danish School of Pharmacy, DK-2100, Copenhagen Ø, Denmark.*

B. MEISTER, *Department of Histology, Karolinska Institute, PO Box 60400, S-104 01 Stockholm, Sweden.*

T. MELANDER, *Department of Histology, Karolinska Institute, PO Box 60400, S-104 01 Stockholm, Sweden.*

E. Ø. NIELSEN, *Department of Chemistry BC, The Royal Danish School of Pharmacy, DK-2100, Copenhagen Ø, Denmark.*

L. NIELSEN, *Department of Chemistry BC, The Royal Danish School of Pharmacy, DK-2100, Copenhagen Ø, Denmark.*

M. PIRCHIO, *Department of Pharmacology and Clinical Pharmacology, St. Georges Hospital Medical School, Cranmer Terrace, London SW17 0RE, UK.*

J. D. SALAMONE, *Neuroscience Research Centre, Merck Sharp & Dohme Research Laboratories, Terlings Park, Eastwick Road, Harlow, Essex CM20 2QR, UK.*

M. SCHALLING, *Department of Histology, Karolinska Institue, PO Box 60400, S-104 01 Stockholm, Sweden.*

R. H. SCHELLER, *Department of Biological Sciences, Stanford University, CA 94305, USA.*

F. O. SCHMITT, *Department of Biology, Massachusetts Institute of Technology, Cambridge, Massachusetts, USA.*

A. M. SILLITO, *Department of Physiology, University College, Cardiff, UK.*

S. H. SNYDER, *Departments of Neuroscience, Pharmacology and Experimental Therapeutics and Psychiatry and Behavioural Sciences, John Hopkins University School of Medicine, 725 North Wolfe Street, Baltimore, MD 21205, USA.*

W. SOSSIN, *Department of Biological Sciences, Stanford University, CA 94305, USA.*

W. STAINES, *Department of Histology, Karolinska Institute, PO Box 60400, S-104 01, Stockholm, Sweden.*

S. I. WALAAS, *Laboratory of Molecular and Cellular Neuroscience, The Rockefeller University, 1230 York Avenue, New York, NY 10021, USA.*

J. C. WATKINS, *Department of Pharmacology, The Medical School, Bristol BS8 1TD, UK.*

K. WATLING, *Neuroscience Centre, Merck Sharp & Dohme Research Laboratories, Terlings Park, Eastwick Road, Harlow, Essex CM20 2QR, UK.*

J. K.-T. WANG, *Laboratory of Molecular and Cellular Neuroscience, The Rockefeller University, 1230 York Avenue, New York, NY 10021, USA.*

E. H. F. WONG, *Neuroscience Research Centre, Merck Sharp & Dohme Research Laboratories, Terlings Park, Eastwick Road, Harlow, Essex CM20 2QR, UK.*

G. N. WOODRUFF, *Neuroscience Research Centre, Merck Sharp & Dohme Research Laboratories, Terlings Park, Eastwick Road, Harlow, Essex CM20 2QR, UK.*

Introduction

L . L . I V E R S E N

The chapters in this volume arise from the proceedings of a symposium held in May 1985 to celebrate the inauguration of the Neuroscience Research Centre established by Merck Sharp & Dohme Research Laboratories in Harlow, England. The new laboratories will be entirely devoted to applying new knowlege about brain chemistry to the discovery and development of novel treatments for CNS disorders.

There have been remarkable advances in our understanding of chemical mediators in the nervous system during the past few decades. The traditional concepts of chemical transmission in the nervous system developed largely from detailed studies of the actions of acetylcholine as the fast chemical signal used at the neuromuscular junction. Fast chemical signalling, in which the neurotransmitter released at specialized synaptic junctions stimulates the opening of receptor-controlled ion channels in the post-synaptic cell within a millisecond time frame, does occur in the mammalian CNS. Indeed the amino acids glutamate and GABA may represent the principal fast signals used by most of the 'main line' fast conducting circuits. However, many chemical transmitters in the CNS do not operate in this classical manner. The actions of monoamines and neuropeptides are slow (acting over periods of seconds or minutes) and rather than directly excitatory or inhibitory they are modulatory in character (Iversen 1984).

Furthermore, the 'slow' modulators may not always be released at morphologically specialised synapses, but can sometimes act at a distance from their sites of release. This has given rise to the concept of 'chemically addressed' chemical transmission, in which information is transmitted by the use of a wide range of different chemical signals acting diffusely, but achieving selectivity by the uneven distribution of suitable receptors on target cells to recognise these signals (Fig. 1). Slow mediators also act largely by triggering persistent metabolic responses in target cells, rather than by controlling ion channels directly.

The chapters in Part I of this volume focus on the fast transmitters, GABA and glutamate. In recent years there has been a rapid development of new pharmacological tools for the selective stimulation or antagonism of

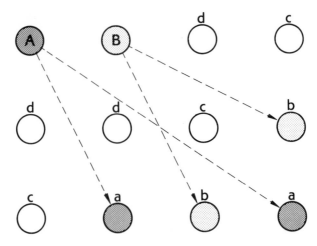

FIG. 1. The concept of 'chemically addressed' transfer of information between neurones. Cells A and B release neurotransmitters A and B diffusely, but these can be recognized by cells that possess the appropriate surface receptors (a and b).

the various receptors for GABA and glutamate that are now known to exist. Indeed, as so often in the past, the discovery of novel pharmacological agents has revealed the diversity of the receptors involved both for GABA and for glutamate. Although these new discoveries have not yet led to new therapeutic discoveries, the field holds considerable promise.

Part II deals entirely with slow transmitters, the monoamines and peptides, and illustrates the complexity of their actions, both neurophysiologically and biochemically, where two major second messenger systems are involved: the cyclic nucleotides and the breakdown products of phosphatidyl inositides. The repertoire of chemical signalling is also greatly enhanced by the use of cosecreted mixtures of bioactive substance in individual neurones, as richly illustrated by T. Hökfelt and colleagues.

Part III focuses on the concept of 'chemically addressed' information transfer, and the wide variety of neurobiological and biological mechanisms that this involves. The use of chemical signals by micro-organisms (D. Koshland) illustrates the subtlety of information transfer and the associated biochemical machinery that have evolved in bacteria, and this has many lessons for students of higher brain function in mammals.

I am very grateful to all who participated in the present survey of emerging concepts in this field, who gave their time to attend the symposium and later to prepare the chapters for this volume. Thanks also to Elisabeth Goodman and to Oxford University Press for their help in editing the manuscripts for publication.

References

Iversen, L. L. (1984). Amino acids and peptides—fast and slow chemical signals in the nervous system? (The Ferrier Lecture). *Proc. Roy. Soc. Lond. Ser. B—Biol. Sci.* **221**, 245–60.

Part I

AMINO ACIDS AS FAST SIGNALS

1

Amino acid transmitters: 30 years' progress in research

K. KRNJEVIĆ

HISTORY

Research on excitatory and inhibitory amino acids as possible synaptic transmitters has extended over some 30 years. As summarized in a very simplified way in Table 1.1, its history can be broken up in decades, starting with the 1950's.

TABLE 1.1. *History of amino acids as transmitters (in decades)*

Pre-history (till 1950)
Neither synaptic involvement (glutamate), nor even presence in brain (GABA) suspected.

Ancient history (1950's)
1. Discovery of GABA in brain
2. Excitatory and depressant actions first observed
3. Transmitter role postulated for GABA

Dark ages (1960's)
Amino acids are generally viewed as non-transmitters in spite of increasing positive evidence.

Renaissance (1970's)
Following discovery of some specific antagonists, amino acids become widely accepted as probable CNS transmitters

Baroque era (1980's)
By now 'classical' transmitters; topic of uninhibited, exuberant activity in all directions.

Pre-history (before 1950)

These amino acids either were not known even to exist in the brain—as in the case of γ-aminobutyrate (GABA)—or, though well known as important constituents of brain tissue, had clearly assigned roles in cerebral metabolism: thus, glutamate had been studied extensively in the context of brain respiration, the Krebs cycle, and transamination reactions (Weil-Malherbe 1950; Braunstein 1947). Albeit less prominent biochemically, aspartate was also hardly a chemical in search of a function. After the success of transmitter studies at peripheral junctions (Loewi 1945; Dale 1938; Feldberg 1951), the general expectation was that ACh and other peripheral transmitters would also be the predominant transmitters in the CNS.

Ancient history (1950's)

The decade was ushered in with the discovery of GABA in the brain (Roberts and Frankel 1950; Awapara *et al*. 1950), which would later prove to be of the greatest importance for CNS studies; although its significance at first could hardly be appreciated, since GABA was not known to have any clear function, whether biochemical or physiological, anywhere in the biosphere. Unlike most other amino acids, GABA is not a constituent of protein. Its existence as a free compound, therefore, could not be explained in terms of protein manufacture and breakdown. Being such a close relative of glutamate, however, it seemed likely to take part in metabolic processes linked to the Krebs cycle.

Some kind of involvement in excitation and inhibition became apparent after the middle 1950's. Firstly, because a correlation was found between seizure activity and low GABA levels caused, for example, by agents which interfere with GABA synthesis (Killam and Bain 1957). Secondly, and more directly, when GABA proved to have a strong inhibitory action on crustacean nerve cells (Bazemore *et al*. 1957), that fully explained the inhibitory potency of brain extracts ('Factor I'). GABA thus came to be considered as a possible inhibitory transmitter, especially in crustacea (Kuffler and Edwards 1958).

Though proposed by Hayashi (1956) as a significant excitatory agent, glutamate was taken less seriously, until the demonstration of its strong excitatory action on spinal neurons (Curtis and Watkins 1960).

Dark ages (1960's)

The new microiontophoretic technique had initiated a real breakthrough by providing conclusive evidence of the powerful excitatory or depressant actions of glutamate and aspartate or GABA. However, the strong conviction of the pioneers in these studies (Curtis and Watkins 1960) that neither GABA nor glutamate could be physiological transmitters did much to dampen the early enthusiasm. In any case, the amino acids did not belong to the category of 'accepted' transmitters—such as acetylcholine, noradrenaline, dopamine, 5HT, and histamine. As I pointed out at the time (Krnjević 1965), they would perhaps not have been ignored for so long if they had a recognized action on one of the classical smooth muscle preparations. The search therefore went on for the 'real' transmitters in the brain.

The situation was different in the crustacean world, where amino acids—at first GABA (Kuffler 1960; Kravitz *et al*. 1963) and later glutamate (Takeuchi and Takeuchi 1964)—continued to be seriously considered as putative transmitters, especially as increasingly compelling evidence became available. As far as the vertebrate CNS was concerned, with the exception of some early enthusiasts (Krnjević 1965; Werman *et al*. 1968), the consensus was principally against amino acids.

Renaissance (1970's)

The tide began turning in the late 1960's, when the combination of intracellular recording and iontophoresis made evident the similarity between natural inhibition and the action of GABA in the cerebral cortex (Krnjević and Schwartz 1967), or between inhibition and the action of glycine in the spinal cord (Werman *et al*. 1968). It was reinforced by the failure to isolate any other potent neuroactive-transmitter-like agents from the brain. Also very influential was the finding that the well-known convulsant drug strychnine is a potent and selective glycine antagonist (Curtis *et al*. 1968). This set the stage for a systematic search for a GABA antagonist, which culminated in the rediscovery of bicuculline (Curtis *et al*. 1970)—a hitherto obscure convulsant alkaloid extracted from the bulbous roots of dicentra cucullaria by Manske (1932).

Throughout the 1970's, neuroscientists therefore turned to GABA with renewed interest. It was investigated as a putative transmitter at numerous sites throughout the brain and spinal cord (and even in peripheral ganglia). Much of the new evidence came from the application of more-or-less novel neurochemical techniques: radio-labelled GABA to demonstrate GABA release and uptake (Srinivasan *et al*. 1969; Iversen 1971); the isolation of pure glutamic acid decarboxylase (GAD) from the brain, which was then used to produce specific antibodies (Roberts 1976)—this inaugurated a new era of GABA-related immuno-histochemistry; far quicker and more sensitive techniques for measuring endogenous amino acids (Bradford and Richards 1976; Enna and Snyder 1976).

Inhibitory cells, pathways and synapses were gradually recognized as of the utmost importance in the functional organization of the brain, and GABA as the transmitter at the great majority of inhibitory synapses (Eccles 1969; Krnjević 1974; Ito 1976). An unexpected new development was that GABA-mediated inhibition proved to be readily modulated by some of the most widely used CNS depressants: benzodiazepine tranquillizers (Choi *et al*. 1977), acting through a specific benzodiazepine (BZD) receptor, and barbiturates (Nicoll *et al*. 1975) acting at another site, but also part of the GABA-receptor complex. This enormously amplified the scope of GABA-related studies, notably with regard to convulsive disorders and anticonvulsant therapy.

The highly productive research on GABA led to a marked increase of interest in the closely related excitatory amino acids, which were surprisingly quickly accepted as the most likely excitatory transmitters at fast-acting synapses throughout the CNS. A rapid advance in real knowledge, however, was hampered by great technical difficulties in establishing their precise mode of action and in comparing them rigorously with synaptic potentials. A further obstacle was the multiplicity of receptors on which glutamate appeared to be acting. At least three types were identified, according to the amino acids of relatively rigid molecular conformation for which they had a particularly high affinity: N-methyl-D-aspartate (NMDA), kainate (KA), and quisqualate (QUIS) (Watkins 1978).

PRESENT ERA (1980's)

As a result of this rapid progress, we have reached a point where the amino acids are generally viewed as 'classical' or 'canonical' transmitters. Whether this is fully justified by the available evidence might be questioned—especially with respect to the excitatory amino acids—nevertheless, the widespread belief is there. Partly because of this strong conviction, and partly because of the exuberant, uninhibited activity which this has generated, the 1980's are described in Table 1 as a Baroque era. For the role of the principal quick-acting transmitters in the brain, the amino acids have apparently won the battle over other contenders (ACh, monoamines, and neuropeptides), which are increasingly seen as predominantly involved in slower and longer lasting modulatory or 'plastic' mechanisms.

In recent years, investigations have expanded in all directions: on the one hand towards defining with ever greater precision the membrane, or even the molecular, mechanisms involved in their actions; on the other, in defining their role in the *organization* of brain function, from the cellular level, to systems of cells and pathways, and even to the animal's behaviour.

Only a few examples can be mentioned in this brief overview, enough, however, to give an idea of the extremely wide range of activities and some of the principal foci of endeavour.

Inhibitory amino acids

Unitary channel studies

The properties of Cl^- channels activated by GABA and glycine have been examined by two kinds of approaches. First by 'noise analysis', from the spectra of which it is possible to deduce the characteristics of the elementary units of anionic current activated by the interaction between single (or whatever minimum number is necessary) molecules of amino acid and the appropriate membrane receptor (Barker *et al.* 1982). An interesting finding was a consistent smaller, but longer-lasting elementary conductance increase evoked by GABA, which is in keeping with the typically longer time course of GABA-mediated as opposed to glycine-mediated IPSPs. In other respects, including their anion selectivity, the GABA- and glycine-sensitive events were identical, but they were mediated by distinct receptors (blocked, respectively, by bicuculline and strychnine).

The analysis of literally single channels has become possible thanks to the development of the patch-clamp technique (Hamill *et al.* 1983). The main new finding has been that the opening of Cl^- channels is not an all-or-none event, several conductance substates (degrees of opening) being evident. Moreover, in these experiments—which confirmed their separability—the GABA- or glycine-sensitive channels differed not by the duration of opening, but by the predominant conductance state. The results of such experiments are described in much greater detail in the next chapter of the present volume (J. L. Barker).

Unconventional GABA receptors

In addition to its now 'classical' Cl^- mediated inhibitory action, GABA appears to have some other significant effects. Dunlap and Fischbach (1978) found that GABA can depress the Ca^{2+} component of the action potential of dorsal root ganglion cells, by a bicuculline-insensitive action. Extensive studies of the bicuculline-insensitive block of transmitter release at a variety of synapses led to the recognition of a special class of 'type B' GABA-receptors (Bowery *et al.* 1980, 1984)—which curiously enough, may be linked to adenylate cyclase (Karbon *et al.* 1984). These receptors appear to be the sites of action of the phenylated GABA compound baclofen. This drug—which is a useful antispastic, because it readily crosses the blood–brain barrier and so can be given systemically—has little if any action on Cl^- channels, but very effectively blocks synaptic transmission at a presynaptic site. To what extent GABA-B receptors are activated under physiological conditions, and by which endogenous agents, remains to be established. There is now some evidence that they may also be responsible for a particularly prolonged form of *post-synaptic* inhibition (Newberry and Nicoll 1984).

Neurochemical and pharmacological studies on the GABA-receptor complex

The properties of the GABA-binding complex can readily be studied *in vitro*. This has been an area of intense activity, with numerous, systematic investigations of the binding affinity to various agonists and antagonists of GABA itself (Olsen *et al*. 1984). An early discovery, that the affinity of isolated membranes could be greatly enhanced by freezing and thawing (and some other procedures) suggested the presence of an endogenous regulator molecule (dubbed 'GABA modulin' by Costa and Guidotti 1979). The precise identity of this modulator, however, has so far remained elusive.

Related *in vitro* studies have continued to examine the function of the benzodiazepine (BZD) receptor (Costa and Guidotti 1979; Haefely *et al*. 1983; Squires 1984). Of particular interest has been the identification of benzodiazepine antagonists and reverse agonists and, of course, the intriguing possibility that endogenous substances (analogous to endogenous opiates) interact selectively with BZD receptors as agonists, antagonists, etc. Such agents might play an important role in the physiological or pathological regulation of anxiety.

There has been much progress also in somewhat comparable *in vitro* studies on the mechanisms of action of barbiturates, which also influence GABA binding, but independently of the BZD receptors (Willow and Johnston 1983; Olsen *et al*. 1984). Other related studies are discussed separately by P. Krogsgaard-Larsen (see later).

GABA-receptor induction

A new field of studies has been opened up by Sumikawa *et al*.'s (1984) demonstration that intracellular injections of appropriate messenger RNA (extracted from rat brain) render *Xenopus* oocytes capable of giving Cl^--mediated, selective responses to GABA or glycine. This greatly enhances prospects of understanding the mechanisms of gene expression and translation that are necessary for the synthesis and selective distribution of neurotransmitter-related macromolecules.

Excitatory amino acids

A comparable wide range of investigations is in progress.

How does glutamate excite?

After a long period of uncertainty, the mechanism of excitation is becoming much clearer, thanks to more stringent electrophysiological investigations, first by conventional voltage-clamping (MacDonald *et al.* 1982; Flatman *et al.* 1983) and most recently by patch clamping of single channels (Nowak *et al.* 1984). The apparent increase in membrane resistance caused by glutamate, aspartate, etc.—especially when acting via NMDA receptors—is really a manifestation of a voltage-dependent inward current. Rather surprisingly, the voltage dependence disappears in the absence of external Mg^{2+}, and therefore can be explained by a potential-sensitive block of inward current channels by Mg^{2+} (Nowak *et al.* 1984).

Cytotoxic action

That glutamate might cause cell necrosis in the brain was first suggested by Olney (1969). Originally, the concern was especially about the damaging effect of circulating glutamate at sites where there is no blood brain barrier (e.g. in the ventromedian hypothalamus), but according to very recent experiments, a massive release of glutamate and aspartate within the brain, under abnormal conditions such as seizure activity and anoxia, may result in necrosis (e.g. Rothman 1984; Simon *et al.* 1984), because depolarization allows an excessive influx of either Ca^{2+} or Cl^- (or both).

Excitatory amino acid receptors

Numerous binding studies (Roberts 1981) have been performed *in vivo* and *in vitro*. Receptors are being induced *de novo* in 'naive' cells by injections of foreign mRNA (Sumikawa *et al.* 1984). However, the most active—and potentially most important—field of investigation is the search for reliable antagonists at the three main types of receptors (for NMDA, KA, and QUIS). This topic is covered systematically by J. Watkins in the present volume (Chapter 6). A number of antagonists are beginning to prove useful in identifying probable glutamate- and perhaps aspartate-releasing fibres, at various sites, from the hippocampus (cf. J. S. Kelly, later) to the primary afferents (cf. T. Jessel, also later).

Role of amino acids in organization of brain function

While a vast number of investigations are going on at the cellular, membrane and even molecular level, increasing awareness of the fundamental importance of synaptic signalling is encouraging more and more studies in the opposite direction, which aim towards a better understanding of the role of transmitters in the organization of behaviour, whether reflected in the activity of small groups of cells, more elaborate neural systems, or even the integrated behaviour of the whole animal.

On small groups of neurons

Particularly illuminating has been Sillito's (1984—see also Chapter 4) use of a GABA antagonist to demonstrate the essential role played by local inhibitory inputs in determining such functional characteristics as the directional and orientation selectivity or the ocular dominance of central visual neurons. Comparable experiments are now being performed to elucidate the role of inhibitory connections in the functional organization of other parts of the CNS such as the somatosensory cortex (Dykes *et al.* 1984), and the superior colliculus (Hikosaka and Wurtz 1985). Complementary investigations with glutamate antagonists are now becoming increasingly useful in identifying excitatory synapses (as described elsewhere in this volume by T. Jessell, J. S. Kelly, as well as J. Watkins).

On larger systems

The regulatory function of inhibition is self-evident in such major components of the brain as the cerebellum and the basal ganglia, where inhibitory (mainly GABAergic) neurons are not only very numerous as internal connectors, but may even make up the principal—if not indeed the only—system of output neurons (Eccles *et al.* 1967; Ito 1976; Roberts 1980). GABAergic cells are found in all regions of the brain, as well as the rest of the CNS, including the retina. Throughout the vertebrate CNS, information transfer is probably mediated by various degrees of selective disinhibition.

Glutamate and specific antagonists are increasingly being used to investigate the role of excitatory amino acids at higher levels of organization, for example in relatively long term plastic 'phenomena' such as LTP in the hippocampus (Dolphin 1983) or prolonged modifications in the efficacy of cerebellar synapses (Ito *et al.* 1982).

Behavioural states

The close relationship between GABA and benzodiazepine receptors on the one hand, and states of anxiety on the other is receiving much attention (Haefely *et al.* 1983), although no doubt other transmitters and/or modulators may well also play a significant role in anxiety (Gray 1982). There is some evidence that potent psychoactive drugs related to phencyclidine are effective NMDA antagonists (Anis *et al.* 1982), and therefore may act by disturbing synaptic mechanisms mediated by an excitatory amino acid.

Involvement in pathological states

A loss of GABAergic inhibition seems to be a major factor in the development of epileptic activity (Meldrum 1975; Krnjević 1983). Sites of epileptic foci consistently show a relative loss of GABAergic inhibitory cells, perhaps owing to their particular vulnerability to damage, for example by anoxia or ischemia (Ribak *et al*. 1979). In keeping with this idea is the marked effectiveness as anticonvulsants of several drugs which potentiate GABAergic inhibition (including phenobarbitone, diazepam, etc.). On the other hand, the anticonvulsant potency of some glutamate antagonists (Meldrum *et al*. 1983) emphasizes the significance for the generation of seizures of excessive, unrestrained activity of excitatory synapses and the possibility of a new type of antiepileptic drug therapy.

Finally, the deleterious effects of prolonged over-secretion of excitatory amino acids may be responsible for the massive loss of nerve cells evident in some major neurological syndromes, including chronic recurrent epilepsy, Alzheimer type dementias, and Huntington's Chorea (e.g. Coyle *et al*. 1978).

CONCLUSION

This short survey of the history of amino acids can give only a summary indication of a most thriving field of research. Nevertheless, it may convey some idea of the rapid progress in the last few decades and the wide-ranging activity that is going on at present. So far, amino acids have been notorious for their high potency and speed of action, which justifies their characterization as fast-acting synaptic transmitters. However, it would be rash to assume that this is necessarily their only significant function. It may well be that, like other, even longer-known and more widely investigated agents—such as acetylcholine and noradrenaline—they can be involved in both fast and slow signalling: cf. the cholinergic fast nicotinic and slow muscarinic actions or the fast transmitter action of noradrenaline in the vas deferens and its postulated long-term plastic actions in the visual cortex. There is now evidence that GABA may have both a relatively fast and a much slower inhibitory action on the same hippocampal cells. Other slow actions may well be significant at many other sites.

Acknowledgements

The author's research is financially supported by the Medical Research Council of Canada.

References

Anis, N. A., Burton, N. R., and Lodge, D. (1982). Ketamine anaesthesia and the *N*-methyl-aspartate receptor: studies with the optical isomers of ketamine. *J. Physiol.* **326**, 48–9P.

Awapara, J., Landua, A. J., Fuerst, R., and Seale, B. (1950). Free γ-aminobuytric acid in brain. *J. biol. Chem.* **187**, 35–9.

Barker, J. L., McBurney, R. N., and MacDonald, J. F. (1982). Fluctuation analysis of neutral amino acid responses in cultured mouse spinal neurones. *J. Physiol.* **322**, 365–87.

Bazemore, A. W., Elliott, K. A. C., and Florey, E. (1957). Isolation of factor I. *J. Neurochem.* **1**, 334–9.

Bowery, N. G., Hill, D. R., Hudson, A. L., Doble, A., Middlemiss, D. N., Shaw, J., and Turnbull, M. (1980). (-)Baclofen decreases neurotransmitter release in the mammalian CNS by an action at a novel GABA receptor. *Nature*, **283**, 92–4.

——, Price, G. W., Hudson, A. L., Hill, D. R., Wilkin, G. P., and Turnbull, M. J. (1984). GABA receptor multiplicity. Visualization of different receptor types in the mammalian CNS. *Neuropharmacol.* **23**, 219–31.

Bradford, H. F., and Richards, C. D. (1976). Specific release of endogenous glutamate from piriform cortex stimulated *in vitro*. *Brain Res.* **105**, 168–72.

Braunstein, A. E. (1947). Transamination and the integrative functions of the dicarboxylic acids in nitrogen metabolism. *Adv. Protem. Chem.* **3**, 1–52.

Choi, D. W., Farb, D. H., and Fischbach, G. D. (1977). Chlordiazepoxide selectively augments GABA action in spinal cord cell cultures. *Nature*, **269**, 342–4.

Costa, E. & Guidotti, A. (1979). Molecular mechanisms in the receptor action of benzodiazepines. *Ann. Rev. Pharmacol. Toxicol.* **19**, 531–45.

Coyle, J. T., McGeer, E. G., McGeer, P. L., and Schwarcz, R. (1978). Neostriatal injections: a model for Huntington's Chorea. In *Kainic Acid as a Tool in Neurobiology* (eds E. G. McGeer, J. W. Olney and P. L. McGeer), pp. 139–59. Raven Press, New York.

Curtis, D. R., Duggan, A. W., Felix, D., and Johnston, G. A. R. (1970). GABA, bicuculline and central inhibition. *Nature*, **226**, 1222–4.

——, Hösli, L., and Johnston, G. A. R. (1968). A pharmacological study of the depression of spinal neurones by glycine and related amino acids. *Expl. Brain Res.* **6**, 1–18.

——, and Watkins, J. C. (1960). The excitation and depression of spinal neurones by structurally related amino acids. *J. Neurochem.* **6**, 117–41.

Dale, H. H. (1938). Acetylcholine as a chemical transmitter of the effects of nerve impulses. *J. Mt Sinai Hosp.* **4**, 401–29.

Dolphin, A. C. (1983). The excitatory amino-acid antagonist γ-D-glutamylglycine masks rather than prevents long term potentiation of the perforant path. *Neuroscience*, **10**, 377–83.

Dunlap, K., and Fischbach, G. D. (1978). Neurotransmitters decrease the calcium component of sensory nerve action potentials. *Nature*, **276**, 837–9.

Dykes, R. W., Landry, P., Metherate, R., and Hicks, T. P. (1984). Functional role of GABA in cat primary somatosensory cortex: shaping receptive fields of cortical neurons. *J. Neurophysiol.* **52**, 1066–93.

Eccles, J. C. (1969). *The Inhibitory Pathways of the Central Nervous System.* Charles C. Thomas, Springfield.

——, Ito, M., and Szentagothai, J. (1967). *The Cerebellum as a Neuronal Machine.* Springer Verlag Inc., New York.

Enna, S. J., and Snyder, S. H. (1976). Gamma-aminobutyric acid (GABA) receptor binding in mammalian retina. *Brain Res.* **115**, 174–9.

Feldberg, W. (1951). Some aspects in pharmacology of central synaptic transmission. *Arch. int. Physiol.* **59**, 544–60.

Flatman, J. A., Schwindt, P. C., Crill, W. E., and Stafstrom, C. E. (1983). Multiple actions of *N*-methyl-D-aspartate on cat neocortical neurons *in vitro*. *Brain Res.* **266**, 169–73.

Gray, J. (1982). *The Neuropsychology of Anxiety.* Oxford University Press, London and New York.

Haefely, W., Polc, P., Pieri, L., Schaffner, R., and Laurent, J-P. (1983). Neuropharmacology of benzodiazepines: synaptic mechanisms and neural basis of action. In *The Benzodiazepines: from Molecular Biology to Clinical Practice* (ed E. Costa), pp. 21–66. Raven Press, New York.

Hamill, O. P., Bormann, J., and Sakmann, B. (1983). Activation of multiple-conductance state chloride channels in spinal neurones by glycine and GABA. *Nature*, **305**, 805–8.

Hayashi, T. (1956). *Chemical Physiology of Excitation in Muscle and Nerve.* Nakayama-Shoten, Tokyo.

Hikosaka, O., and Wurtz, R. H. (1985). Modification of saccadic eye movements by GABA-related substances. I. Effect of muscimol and bicuculline in monkey superior colliculus. *J. Neurophysiol.* **53**, 266–91.

Ito, M. (1976). Roles of GABA neurons in integrated functions of the vertebrate CNS. In *GABA in Nervous System Function* (eds E. Roberts, T. N. Chase, and D. B. Tower), pp. 427–48. Raven Press, New York.

——, Sakurai, M., and Tongroach, P. (1982). Climbing fibre induced depression of both mossy fibre responsiveness and glutamate sensitivity of cerebellar Purkinje cells. *J. Physiol.* **324**, 113–34.

Iversen, L. L. (1971). Role of transmitter uptake mechanisms in synaptic neurotransmission. *Br. J. Pharmacol.* **41**, 571–91.

Karbon, E. W., Duman, R. S., and Enna, S. J. (1984). $GABA_B$ receptors and norepinephrine-stimulated cAMP production in the rat brain cortex. *Brain Res.* **306**, 327–32.

Killam, K. F., and Bain, J. A. (1957). Convulsant hydrazides. I: *in vitro* and *in vivo* inhibition of vitamin B6 enzymes by convulsant hydrazides. *J. Pharmacol. exp. Ther.* **119**, 255–62.

Kravitz, E. A., Kuffler, S. W., and Potter, D. D. (1963). Gamma-aminobutyric acid and other blocking compounds in crustacea. III. Their relative concentrations in separated motor and inhibitory axons. *J. Neurophysiol.* **26**, 739–51.

Krnjević, K. (1965). Chemical transmitters in the cerebral cortex. *Lectures in Symp. XXIII Int. physiol. Congr.*, Tokyo. pp. 435–43, Excerpta Medica Foundation, Amsterdam.

——, (1974). Chemical nature of synaptic transmission in vertebrates. *Physiol. Rev.* **54**, 418–540.

——, (1983). GABA-mediated inhibitory mechanisms in relation to epileptic discharges. In *Basic Mechanisms of Neuronal Hyperexcitability* (eds H. H.

Jasper and N. M. van Gelder), pp. 249–80. Alan R. Liss, Inc., New York.
——, and Schwartz, S. (1967). The action of γ-aminobutyric acid on cortical neurones. *Exp. Brain Res.* **3**, 320–36.
Kuffler, S. W. (1960). Excitation and inhibition in single nerve cells. *Harvey Lectures*, **54**, 176–218.
——, and Edwards, C. (1958). Mechanism of gamma aminobutyric acid (GABA) action and its relation to synaptic inhibition. *J. Neurophysiol.* **21**, 589–610.
Loewi, O. (1945). Aspects of the transmission of the nervous impulse. I. Mediation in the peripheral and central nervous system. *J. Mt Sinai Hosp.* **12**, 803–16.
MacDonald, J. F., Porietis, A. V., and Wojtowicz, J. M. (1982). L-Aspartic acid induces a region of negative slope conductance in the current-voltage relationship of cultured spinal cord neurons. *Brain Res.* **237**, 248–53.
Manske, R. H. F. (1932). The alkaloids of fumaraceous plants. II. Dicentra cucullaria (L.) bernh. *Can. J. Res.* **7**, 265–9.
Meldrum, B. S. (1975). Epilepsy and γ-aminobutyric acid-mediated inhibition. *Int. Rev. Neurobiol.* **17**, 1–36.
——, Croucher, M. J., Czuczwar, S. J., Collins, J. F., Curry, K., Joseph, M., and Stone, T. W. (1983). A comparison of the anticonvulsant potency of (±) 2-amino-5-phosphonopentanoic acid and (±) 2-amino-7-phosphonoheptanoic acid. *Neuroscience*, **9**, 925–30.
Newberry, N. R., and Nicoll, R. A. (1984). Direct hyperpolarizing action of baclofen on hippocampal pyramidal cells. *Nature*, **308**, 450–2.
Nicoll, R. A., Eccles, J. C., Oshima, T., and Rubia, F. (1975). Prolongation of hippocampal inhibitory post-synaptic potentials by barbiturates. *Nature*, **258**, 625–7.
Nowak, L., Bregestovski, P., Ascher, P., Herbet, A., and Prochiantz, A. (1984). Magnesium gates glutamate-activated channels in mouse central neurones. *Nature*, **307**, 462–5.
Olney, J. W. (1969). Brain lesions, obesity, and other disturbances in mice treated with monosodium glutamate. *Science*, **164**, 719–21.
Olsen, R. W., Wong, E. H. F., Stauber, G. B., and King, R. G. (1984). Biochemical pharmacology of the γ-aminobutyric acid receptor/ionophore protein. *Fed. Proc.* **43**, 2773–8.
Ribak, C. E., Harris, A. B., Vaughn, J. E., and Roberts, E. (1979). Inhibitory, GABAergic nerve terminals decrease at sites of focal epilepsy. *Science*, **205**, 211–4.
Roberts, E. (1976). Immunocytochemistry of the GABA system—a novel approach to an old transmitter. In *Neurotransmitters, Hormones and Receptors: Novel Approaches* (eds J. A. Ferrendelli, B. S. McEwen, and S. H. Snyder) Vol. 1, pp. 123–38. Society for Neuroscience, Bethesda, Maryland.
—— (1980). Epilepsy and antiepileptic drugs: a speculative synthesis. In *Antiepileptic Drugs: Mechanisms of Action* (eds G. H. Glaser, J. K. Penry, and D. M. Woodbury), pp. 667–713. Raven Press, New York.
——, and Frankel, S. (1950). γ-aminobutyric acid in brain: its formation from glutamic acid. *J. biol. Chem.* **187**, 55–63.
Roberts, P. J. (1981). Binding studies for the investigation of receptors for L-glutamate and other excitatory amino acids. In *Glutamate: Transmitter in the Central Nervous System* (eds P. J. Roberts, J. Storm-Mathisen, and G. A. R. Johnston), pp. 35–54. John Wiley and Sons Ltd, Chichester.

Rothman, S. (1984). Synaptic release of excitatory amino acid neurotransmitter mediates anoxic neuronal death. *J. Neurosci.* **4**, 1884–91.

Sillito, A. M. (1984). Functional considerations of the operation of GABAergic inhibitory processes in the visual cortex. In *Cerebral Cortex* (eds E. G. Jones, and A. Peters) Vol. 2, pp. 91–117. Plenum Press, New York.

Simon, R. P., Swan, J. H., Griffiths, T., and Meldrum, B. S. (1984). Blockade of N-methyl-D-aspartate receptors may protect against ischemic damage in the brain. *Science*, **226**, 850–2.

Squires, R. F. (1984). Benzodiazepine receptors. In *Handbook of Neurochemistry* (ed. A. Lajtha) Vol. 6, pp. 261–306. Plenum Press, New York and London.

Srinivasan, V., Neal, M. J., and Mitchell, J. F. (1969). The effect of electrical stimulation and high potassium concentrations on the efflux of [^{3}H]γ-aminobutyric acid from brain slices. *J. Neurochem.* **16**, 1235–44.

Sumikawa, K., Parker, I., and Miledi, R. (1984). Partial purification and functional expression of brain mRNAs coding for neurotransmitter receptors and voltage-operated channels. *Proc. Nat. Acad. Sci. USA*, **81**, 7994–8.

Takeuchi, A., and Takeuchi, N. (1964). The effect on crayfish muscle of iontophoretically applied glutamate. *J. Physiol.* **170**, 296–317.

Watkins, J. C. (1978). Excitatory amino acids. In *Kainic Acid as a Tool in Neurobiology* (eds E. G. McGeer, J. W. Olney, and P. L. McGeer), pp. 37–69. Raven Press, New York.

Weil-Malherbe, H. (1950). Significance of glutamic acid for the metabolism of nervous tissue. *Physiol. Rev.* **30**, 549–68.

Werman, R., Davidoff, R. A., and Aprison, M. H. (1968). Inhibitory action of glycine on spinal neurons in the cat. *J. Neurophysiol.* **31**, 81–95.

Willow, M., and Johnston, G. A. R. (1983). Pharmacology of barbiturates: electrophysiological and neurochemical studies. *Int. Rev. Neurobiol.* **24**, 15–49.

2

Amino acid and peptide signals in cultured CNS neurons and clonal pituitary cells

JEFFERY L. BARKER, BERNARD DUFY, AND
ROBERT N. MCBURNEY

INTRODUCTION

It has become evident in multicellular organisms that intercellular signals can be generated at a wide variety of intra- and extracellular sites, that possible sites of signal diversity are themselves increasing in number and definition as new techniques and strategies are applied in both new and old experimental systems. There is an explosion of new data on intercellular communication, including transmitter precursor-product relationships, multi-copy and co-existent transient and sustained transmitter expressions in developing and diversely differentiated cellular phenotypes, complex and intricate patterns of transmitter receptor subtypes, and fast and slow membrane and cytoplasmic signals in target cells. The clarity and speed of rapid forms of synaptic transmission at neuromuscular junctions appear relatively straightforward and understandable in the context of more recently described signals broadcast between and among functionally distinct cellular phenotypes, where physiological effects may be less easy to resolve than neuronally-directed muscular contraction. It is also evident that the same chemical substance can transmit information in quite different cellular systems and that a single substance may convey an array of biologically important signals. There is clear and convincing immunocytochemical evidence for many combinations of non-peptide and peptide transmitters in cells and circuits throughout the central and peripheral portions of the vertebrate nervous system, and not surprisingly many of these same transmitters mediate synaptic and extrasynaptic forms of communication in evolutionarily earlier forms. Finally, there is abundant electrophysiological and bio-chemical data revealing physiologically-elaborated or pharmacologically-induced interactions among various combinations of transmitters at membrane and cytoplasmic sites in target cells virtually everywhere receptor-coupled changes in cellular physiology have been studied, neuronal and extra-neuronal. Excitability and the communication of changes in excitability are obviously no longer restricted to nerve and muscle tissues. Indeed, some cells in the immune system can be stimulated to synthesize transmitters previously found to mediate neuronal and endocrine functions (Smith *et al*. 1985). In fact, some 'effector' lymphocytes contain secretory granules and effectively 'synapse' in a transient manner with specific targets, releasing toxic substances during the cell–cell conjugation process that rapidly permeabilize and kill target, but not effector cells.

A complex vocabulary and grammar of intercellular communication is emerging at an almost incomprehensibly rapid rate. In this chapter we will review evidence related to signals generated by an amino acid and by a peptide that are quite different in their cellular distributions, their time courses, their ionic mechanisms, and their functions. The first, mediated by the neutral amino acid GABA, is considered practically ubiquitous in the vertebrate nervous system and involves fast, transient generation of Cl^--dependent conductance at contiguous target cell membranes functionally post-synaptic to transmitting elements. The other, mediated by a peptide three amino acids long, is an example of a slow and sustained, extrasynaptic form of communication in the hypothalamo-adenohypophysial system.

GABA ACTIVATES Cl^- CONDUCTANCE MECHANISMS IN CULTURED CENTRAL NEURONS

The first signal to be considered—GABA-activated Cl^- conductance—has been studied extensively in many vertebrate preparations with a variety of electrophysiological assay techniques. In dissociated cell culture, for instance, virtually all neurons derived from embryonic spinal and hippocampal regions respond to GABA. GABA triggers Cl^- conductance in a molecular manner quite similar to that observed in invertebrate muscle fibres and also generally close to many other chemical signals mediated by other transmitters at other membranes (for review, see Barker 1985). At the whole-cell level of recording in cultured spinal and supraspinal neurons the response can be detected with applications of about 2–10 μM GABA (Dichter 1980; Nowak *et al*. 1982; Segal and Barker 1984a). About 10-fold lower concentrations are required if the electrophysiological assay is made as sensitive as possible using 'patch' clamp recording techniques (Jackson *et al*. 1982).

Discrete pulses of GABA along the neuronal surface reveal that there is a non-uniform topographical distribution of the Cl^- conductance response with the most intense effect typically occurring at the level of the cell body (Barker and Ransom 1978). Immunocytochemical studies of glutamic acid decarboxylase (GAD)-staining elements in spinal and hippocampal cultures show that about 25 per cent of the cells are GAD-stained and that the majority of the neurons, both GAD-positive and GAD-negative, appear to receive extensive GAD-immunoreactive inputs at both cell body and process sites (Caserta and Barker 1983).

The conductance responses activated by pulses of GABA are relatively uniform over the hyperpolarized range of membrane potential, while at depolarized, and especially at positive membrane potentials, significantly larger (10-fold) conductance changes are consistently evoked (Segal and Barker 1984a). If sufficient amino acid ($>20 \mu M$) is applied for a sustained period ($>10 s$), the current response frequently fades, sometimes with little or no decrease in the conductance response. Thus, the fading during the pharmacological response in these latter instances reflects the collapse in the existing Cl^- ion gradient rather than acute desensitization of the receptor-coupled conductance. Whenever macroscopic, whole-cell responses to GABA and the structurally related neutral amino acid glycine, which also activates Cl^- conductance in virtually every cultured spinal cord neuron, have been compared (Table 2.1), the time course of the latter has been noticeably briefer, as can be seen in Fig. 2.1. Thus, at the whole-cell level of pharmacology with similar amplitudes of current response (and with diffusion kinetics assumed to be constant for both transmitters) glycine-evoked Cl^- conductance is invariably shorter-lasting than that elicited by GABA at all sites on the cell surface where the responses can be induced.

TABLE 2.1. (A) *Comparison of electrical properties of amino acid-activated Cl^- conductances in mammalian neurons*

Electrical Properties	GABA	Glycine
1. Distribution of functional receptors		
A. CNS tissue[a]	widespread	widespread
B. SC culture[b]	virtually all cells	
C. SC neurons[c]	identical surface topography	
D. SC patch[d]	GABA + GLY:GABA:GLY:::0.45:0.30:0.25	
E. SC patch[e]	GABA + GLY:GABA:GLY:::0.83:0.11:0.06	
2. Ion selectivity	$Cl^- > Br^- > F^- > I^-$ (Identical)	

3. Elementary conductances in micron patches (pS)

A. Mouse SC (on-cell)[f]	20	not studied
B. Mouse SC[d]	30 ≫ 20	45 ≫ 30 > 20
C. Rat SC[e]	45 < 30 = 20	45 = 30 > 20
D. Rat HPC[g]	30 > 20	not studied
E. Bovine chromaffin[h]	45 > 30 > 20	no effect
F. Bovine chromaffin[i]	45 < 30 > 20	not studied

4. Estimated unitary conductances in whole cells (pS)

A. Mouse SC[j]	16–20	26–32
B. Rat HPC[k]	16–20	16
C. Rat HPC[g]	17–19	not studied
D. Rat CBLM[l]	16	not studied

5. Channel open-time (ms) at resting membrane potential and room temp

A. Mouse SC neuron[j]	20–30	5–10
B. Mouse SC patch[f,m]	5,25	not studied
C. Mouse SC patch[n]	3,29	not studied
D. Rat HPC neuron[k]	20	60
E. Rat HPC neuron, patch[g]	4,51	not studied
F. Rat cerebellum or neuron[l]	1.5,50	not studied
G. Q10 (mouse SC neuron)[o]	identical	

[a]Nistri and Constanti (1979); [b]Barker and Ransom (1978); [c]Study and Barker (1983); [d]Hamill *et al*. (1983); [e]McBurney *et al*. (1985); [f]Jackson *et al*. (1982); [g]Ozawa and Yuzaki (1984); [h]Bormann and Clapham (1984); [i]Cottrell *et al*. (1985); [j]Barker *et al*. (1982); [k]Segal and Barker (1984a); [l]Cull-Candy and Ogden (1985); [m]Redmann *et al*. (1983); [n]Sakmann *et al*. (1983); [o]Mathers and Barker (1981).

SC, spinal cord; HPC, hippocampus; CBLM, cerebellum; pS, picoSiemens.

(B) *Comparison of biochemical properties of amino acid-receptors in vertebrate tissues*

Biochemical Properties	GABA	Glycine
1. Molecular weight of receptor complex (kDaltons)		
A. Rat brain (NP40)[a]	270	
B. Bovine, rat cortex (Triton)[b]	220	
C. Rat brain (DOC, Triton)[c]	220	
D. Rat cord (Triton)[d]		246
2. Subunit molecular weight (kDaltons)		
A. Bovine cortex (DOC, Triton)[e]	53,57	
B. Rat brain (NP40)[a]	49,55	
C. Bovine cortex (CHAPS)[f]	51–53,55–57	
D. Bovine brain (DOC)[c]	50,55,62–80	
E. Rat cord (Triton)[d]		48,58,93

[a]Kuriyama and Taguchi (1984); [b]Chang and Barnard (1982); [c]Olsen *et al*. (1984); [d]Pfeiffer *et al*. (1984); [e]Sigel *et al*. (1983); [f]Sigel and Barnard (1984).

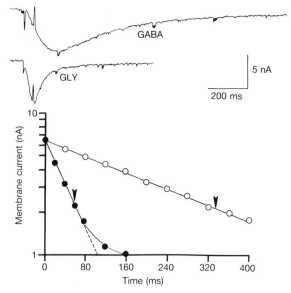

Fig. 2.1. Pharmacologically evoked amino acid conductance responses of similar magnitude last for quite different periods. The recording was made under voltage-clamp conditions with two microelectrodes filled with 3M KCl from a 3-week-old cultured mouse spinal neuron. The responses were evoked by 50 ms–20 nA pulses of iontophoretic charge applied to pipettes containing 1M GABA or 1M glycine at a distance of 2 μm from the surface of the cell body. The cell was held at − 70 mV. Both responses had identical 'reversal potentials', indicating identical driving forces acting on the membrane conductance mechanisms. Each agonist evokes a response 6.5 nA in amplitude. That triggered by glycine has a decay relatively well-fitted by a single exponential of 60 ms, while that elicited by GABA decays with a single exponential of 330 ms (J. L. Barker, unpublished observation).

GABA AND GLYCINE ACTIVATE ION CHANNELS OF DIFFERENT DURATION

The molecular electrophysiological mechanisms underlying the Cl^- conductance responses to the neutral amino acids GABA, glycine and beta-alanine have been studied at the whole-cell and, more recently, membrane-patch levels, using fluctuation analysis and patch-clamp recording techniques, respectively (for short review, see Mathers and Barker 1982; for text, see Sakmann and Neher 1983). Fluctuation analysis of current responses generated at the whole-cell level allows estimates of the average duration and conductance of the underlying molecular conductance event, while patch-clamp recordings directly access the elementary all-or-none transitions in conductance, permitting exceedingly fine sensitivity and temporal resolution of the electrical properties. Using fluctuation analysis it has been possible to estimate the average conductance and average duration of several thousand ion-channels triggered by GABA in cultured spinal cord neurons during current responses similar in amplitude to synaptically activated Cl⁻ ion current signals (Fig. 2.2). The results show that, *as a first approximation*, GABA activates elementary electrical properties that are lower in apparent conductance and longer in apparent duration that those associated with either glycine- or β-alanine-activated Cl⁻ conductance responses (Barker and McBurney 1979; Barker and Mathers 1981; Barker *et al*. 1982). Thus, each transmitter activates Cl⁻ conductance mechanisms whose elementary electrical dimensions are significantly different when the activity arising in an ion-channel population is analysed statistically. Presumably, the ion channels activated pharmacologically include those at both synaptic and extrasynaptic receptor sites. Functional Cl⁻ ion channels that are likely to be extrasynaptic can be found in the majority of patches recorded at random from neuronal cell bodies.

Patch-clamp recordings of Cl⁻ ion channels activated by GABA and glycine have provided further details of the kinetics and conductance profile of single ion-channels. At the National Institutes of Health on-cell and inside-out configurations of patch-clamp recordings from somal patches isolated on 1–3-week-old cultured mouse spinal cord neurons have consistently revealed ion-channel activities whose amplitudes are predominantly unimodal in their distribution and whose durations are consistently best-filled by the sum of two exponentials (Jackson *et al*. 1982; Redmann *et al*. 1983, 1984). The slower of the two exponentially distributed populations corresponds closely to channel duration estimates derived from fluctuation analysis of population kinetics in whole-cell current responses recorded in neurons cultured in the same manner. The elementary ion-channel behaviour described by the short time-constant has not been observed consistently or convincingly in whole-cell experiments even with extended-bandwidth-level (in the range 1–3 KHz) resolution in whole-cell recordings (but see Ozawa and Yuzaki 1984). This latter component of ion-channel activity conveys less than 10 per cent of the Cl⁻ ion charge relative to that described by the longer time-constant distribution. Inside-out and outside-out patch-clamp recordings from monolayer cultures of spinal and hippocampal neurons have revealed multiple conductance states of Cl⁻ ion channels activated by both GABA (Fig. 2.3) and glycine (Hamill *et al*. 1983; Ozawa and Yuzaki 1984; McBurney *et al*. 1985). It seems evident that while fluctuation analysis allows a first approximation of the elementary electrical properties, patch-clamp resolution of electrical details demonstrates certain aspects difficult to detect at the whole-cell level.

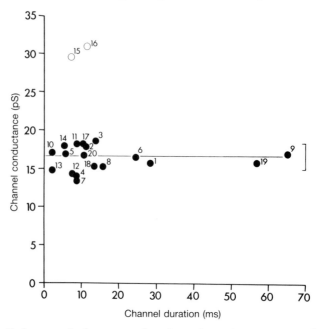

FIG. 2.2. Estimates of elementary ion-channel conductance as a function of estimates of elementary ion-channel duration. The data is derived from a series of experiments on 3-week-old cultured mouse spinal neurons at room temperature (22–24°C) using a two-(3M KCl) electrode voltage-clamp technique to hold cells at −60 mV. Amino acids (unfilled circles) and structural analogues of GABA (filled circles) were applied in sufficient concentration (10–500 μM) to evoke nA-sized Cl⁻ ion current responses for 20–30 s. The data was digitized and then analyzed statistically by computer to characterize elementary ion-current amplitude and spectral behaviour. The results show that GABA and its structural analogues all evoke current responses whose underlying elementary conductance is similar, if not identical. All of these estimates fall within one standard deviation of the mean calculated for the 18 agonists (marked by horizontal bar and vertical bracket). Glycine and β-alanine activate elementary conductance mechanisms substantially greater than that elicited by GABA. Estimates of elementary ion-channel duration show that there is about a 20-fold variation among the GABA analogues. The majority of substances trigger elementary conductances significantly different in duration from that evoked by GABA (modified from Barker and Mathers 1981; unpublished observations).

Key to numbers: (1) γ-aminobutyric acid (GABA); (2) 4,5,6,7-tetra-hydroisoxazolo[5,4-c]pyridinol-3-ol; (3) isoguvacine; (4) δ-aminovaleric acid; (5) imidazole acetic acid; (6) *trans*-4-aminocrotonic acid; (7) β-guanidinopro-pionic acid; (8) γ-amino-β-hydroxybutyric acid; (9) Muscimol; (10) Taurine; (11) 3-aminopropane sulphonic acid; (12) piperidine-4-sulphonic acid; (13) β-amino-butyric acid; (14) guanidino acetic acid; (15) β-alanine; (16) Glycine; (17) piperi-dine-4-carboxylic acid; (18) 5-methylmuscimol; (19) Dihydromuscimol; (20) *trans*-cyclopropane GABA.

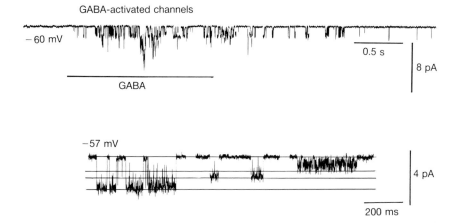

FIG. 2.3. Outside-out patch-clamp recording of microscopic transitions in Cl⁻ conductance evoked by GABA. The recording was made by excising a somal membrane patch from a rat spinal cord cell that had been maintained in culture for 3 weeks. The patch was manipulated in such a way that the external, receptor-bearing surface of the membrane was accessible to extracellularly applied GABA. The Cl⁻ ion conductance mechanism was isolated using symmetrical isomolar Tris-Cl solutions and a driving force acting on the conductance was applied using the voltage-clamp technique. Under such conditions the patch of membrane effectively becomes an ultra-sensitive, ultra-fast biological probe of receptor-coupled Cl⁻ conductance at the elementary, single receptor-channel complex. A pulse of 50 μM GABA lasting several seconds triggers a transient burst of ion-channel activity. Several discrete, quasi-unitary steps in Cl⁻ ion current movement can be detected (upper trace). Closer resolution of the activity has revealed precisely 1.5-fold steps expressed at varying levels of incidence. The most commonly expressed transition level, when extrapolated to the recording conditions utilized in previous fluctuation analysis studies, is quite close to that estimated at the whole cell level (modified from McBurney *et al.* 1985).

From these results (summarized in Table 2.1A) we can conclude that *the absolute conductance levels activated by GABA and glycine are similar, if not identical, but that the relative frequency of their activation varies both with the agonist and the cellular phenotype*. Rather interestingly, GABA activates Cl⁻ ion channels in cultured adult bovine chromaffin membranes whose primary level corresponds to that of glycine-activated channels in cultured spinal neurons, while glycine is *ineffective* in the same chromaffin membranes (Bormann and Clapham 1984). In contrast, glycine activates channels in cultured hippocampal neurons of the same apparent conductance as those activated by GABA, yet of considerably longer duration. These pharmacological actions of glycine require high agonist concentrations (>100 μM) and may represent weak-agonist properties of glycine at GABA receptors.

For comparison, we have tabulated various electrophysiological and biochemical properties of Cl$^-$ conductances activated by GABA and glycine in mammalian neurons (Table 2.1A,B). In those membranes where both transmitters are functional, the results obtained both at whole-cell and micron-patch levels suggest that *the receptors may couple to a common Cl$^-$ ion channel*. The diversity in the conductance signals generated pharmacologically by GABA and glycine at the whole-cell level (when thousands of receptor-coupled ion channels are active) lies both in channel kinetics, with GABA-activated channels conducting for substantially longer periods on average than glycine-triggered ones, as well as in channel conductance, with the principal elementary conductance level estimated varying with both transmitter and neuronal phenotype.

Biochemical analyses of the two amino acid receptor proteins isolated from the CNS with autoradiographic and immunological techniques indicate that the native receptor complexes have similar apparent molecular weights and are each comprised of several subunits whose apparent molecular weights are also close (Table 2.1B). The similarities in the biochemical properties coupled with the identical levels of elementary conductance, as well as other electrophysiological results (some of which are summarized in Table 2.1A) further support the idea that GABA and glycine share some form of common membrane specialization, a receptor-coupled Cl$^-$ ion channel complex that can be utilized by both amino acid transmitters (Barker and McBurney 1979; Barker *et al*. 1982). However, when biochemical and immunochemical techniques are used to track the movement of the subunit proteins through a gel in a charged field, preliminary results demonstrate that the various protein species do *not* precisely co-electrophorese (E. A. Barnard and H. Betz, personal communication). What accounts for the different rates of protein migration and how, if at all, this relates to the functional differences in the GABA- and glycine-activated Cl$^-$ conductance properties could be experimental lines for future investigations.

CHANNEL DURATION CORRELATES WITH MOLAR POTENCY

Fluctuation analysis of Cl$^-$ ion current responses activated by a series of structural analogues of GABA indicates that at low levels of agonist-induced response, the estimated channel conductance elicited by such analogues is indistinguishable from that activated by GABA. Estimated channel durations, however, vary over a 20-fold range, and are significantly different from those estimated to be underlying GABA-evoked

Cl$^-$ conductance responses (Fig. 2.2; Barker and Mathers 1981; Mathers and Barker 1982). For example, muscimol activates ion channels that are about two-fold longer in duration than those activated by GABA, while 3-aminopropane sulphonic acid evokes channels of considerably shorter duration. Theoretically, when the probability of channel activation is low, channel duration is *independent* of agonist concentration. Thus, for agonists activating Cl$^-$ channels to the same conductance level, channel duration reflects what might be considered as the 'molecular efficacy' of the agonist (Mathers and Barker 1982). Since the macroscopic Cl$^-$ conductance recorded is the product of the number of agonist-activated channels and their elementary average conductance and duration properties, it might be possible to account for molar potency in terms of molecular efficacy for those agonists that activate the some number of receptor-coupled channels of the same conductance. In fact, 5 μM muscimol, which activates channels about twice as long as those opened by GABA, evokes the same macroscopic Cl$^-$ conductance as 10 μM GABA (Mathers and Barker 1981b). However, a systematic examination of concentration-conductance relations for all of the agonists has yet to be carried out.

The results of dose-response studies into the relative efficacy of GABA-mimetic substances in depressing electrically stimulated excitability in the hippocampal slice correlate highly and significantly with channel duration (Table 2.2A). Although this correlation strongly suggests that the pharmacological actions in normally developed slices of hippocampal tissue involve Cl$^-$ conductance mechanisms, the complexity of this preparation (e.g. diffusion and access to a variety of receptors in neurons and glia and the presence of neuronal and glial inactivation or uptake mechanisms, etc.) makes it difficult to accept this conclusion without further experimentation. If true, however, then molecular efficacy is an elementary correlate of molar potency for these agonists and the estimated elementary conductance lifetime is inversely related to molar concentration, but independent of concentration for any one ligand. Thus, potent agonists like muscimol, which is effective at low concentrations, open Cl$^-$ channels at about the same apparent rate as weak agonists, but muscimol-activated channels remain activated and conducting for relatively longer periods. The molecular mechanisms involved in receptor-coupled activation of Cl$^-$ conductance have not yet been elucidated convincingly to model the precise reaction sequence involved.

The relationship between agonist structure and function in terms of the steric, hydrophobic, and electronic requirements for channel activation cannot yet be established owing to the many degrees of freedom in the chemical structures of the agonists tested thus far. In addition, there are several clinically important drugs with GABA-mimetic properties whose

TABLE 2.2 (A) *Channel open-time in cultured embryonic mouse SC neurons correlates with percentage block of rat hippocampal excitability for groups of GABA mimetics*

	r	b	c	n	p
Evoked population spike depression[a]	0.94	94.8	8.9	7	0.001

(B) *Channel open-time in cultured embryonic mouse SC neurons correlates with molar potencies in biochemical binding studies*

	r	b	c	n	p
Bovine cortical membranes[b]					
1. Equilibrium block of *GABA	0.93	0.4	9.0	12	0.001
a. Fast-off component	0.92	0.9	8.0	12	0.001
b. Slow-off component	0.94	0.3	7.9	12	0.001
2. Equilibrium block of *Bicuculline	0.95	5.5	11.0	9	0.001
Rat cortical membranes[c]					
1. Equilibrium block of *GABA	0.95	0.8	9.6	8	0.001
2. Equilibrium block of *Bicuculline	0.95	12.2	7.6	8	0.001
3. Equilibrium enhance of *Diazepam	0.96	28.8	2.2	8	0.001
4. Maximal diazepam K_D shift	0.83	0.3	−20.0	8	0.01

[a]Kemp *et al.* (1984); $1/EC_{50}$ $(\mu M)^{-1}$ values; KREBS saline; 23°C
[b]Olsen and Snowman (1983); $1/IC_{50}$ $(\mu M)^{-1}$ values; Tris-citrate/KSCN; 0°C.
[c]Wong and Iversen (1985); $1/IC_{50}$ $(\mu M)^{-1}$ for Expts 1 and 2, $1/EC_{50}$ $(\mu M)^{-1}$ for Expt. 3 and % change in K_D for Expt. 4; Krebs-Bicarb; 23°C.
  r, correlation coefficient; b, slope; c, intercept; n, number of agonists; p, significant level (Student's *t*-test).

molecular mechanism can also be described in the terms of Cl⁻ ion-channel mechanisms (Mathers and Barker 1980; Study and Barker 1983). In the case of (-)pentobarbital, however, 20–50 times higher concentration than GABA is required to activate Cl⁻ conductance at the whole-cell and micron-patch levels of recording yet channel duration is typically three to five times longer (Barker and Mathers 1981; Jackson *et al.* 1982). Thus, molar potency does not correlate with molecular efficacy for all GABA-mimetic agonists.

CHANNEL DURATION CORRELATES WITH BINDING CONSTANTS

For a structurally-related series of GABA-mimetic analogues, Cl⁻ ion channel duration (estimated at room temperature and resting membrane potential) is highly and significantly correlated with various molecular binding parameters measured in membranes isolated from rodent or bovine cortex, either at $0°C$ or at room temperature, and under non-physiological or physiological ionic conditions. The regression equations are summarized in Table 2.2B. The concentration of agonist displacing 50 per cent of isotopically labelled GABA binding to the membranes under equilibrium conditions is linearly and inversely related to estimated channel duration. This correlation holds true for two independent studies of 8 and 12 GABA-mimetics assayed biochemically by investigators using two quite different experimental conditions. Despite these differences the regression equations have relatively similar 'slope' and 'y-intercept' values. Fast and slow unbinding rates from an equilibrated state of binding (corresponding to low and high affinity binding, respectively) have recently been detected. The relative displacements by GABA-mimetics of unbinding at both fast and slow rates also correlate highly and significantly with channel duration.

Competitive block, by the same series of GABA-mimetics, of the binding reaction involving labelled bicuculline, a competitive antagonist of GABA's activation of Cl^- conductance, is likewise highly and significantly correlated with channel duration. Again, the terms in the regression equations relating these competitive reactions against bicuculline binding with channel duration are relatively similar for the two independent biochemical binding studies despite the different experimental conditions employed. The y-intercepts are similar to those obtained in correlations of channel duration with competitive block of GABA binding, while the slopes are an order of magnitude greater. Finally, channel duration also correlates highly and significantly with agonist-induced shift in the equilibrium binding constant for labelled diazepam, as well as the maximal extent of the shift.

All of these correlations in GABA-mimetic structure and activity strongly suggest that some aspect of ligand binding important in activation of receptor-coupled Cl^- channels is being measured in the biochemical experiments and further, that functional GABA-receptor coupled ion channels *in vitro* resemble GABA receptors normally developed *in vivo*. The more potent an agonist is in competitively displacing GABA or bicuculline, the more sustained is its activation of Cl^- ion channel mechanisms.

CHANNEL DURATION CORRELATES WITH IPSC DECAY

Neurons cultured from the embryonic rat hippocampus express functional GABA receptors and Cl^--dependent inhibitory postsynaptic conductance changes (Segal and Barker 1984a,b). The inhibitory post-synaptic currents (IPSCs) are modulated by drugs that alter GABA evoked Cl⁻ conductances in the same cells. The close correspondence between the ionic mechanisms of the physiological and pharmacological conductances, and their parallel sensitivities to the same drugs suggest that the IPSCs are GABA-mediated. Further support for this notion comes from comparison of the estimates of GABA-activated channel duration (τ_{GABA}) and the time constant of IPSC decay (τ_{IPSC}) under control conditions at different potentials and in the presence of different drugs (Table 2.3). The results show that the rates of channel closing in the pharmacological and physiological experiments are both exponential and comparable under different conditions. If the IPSCs are indeed GABA-mediated, then we can tentatively conclude that the *drug actions on the IPSCs are primarily postsynaptic* and *likely involve changes in GABA-activated Cl⁻ ion channel kinetics* rather than conductance. Using the single-channel conductance estimates for GABA-activated responses, we can determine that about several thousand channels conduct at the peak of the evoked IPSC, remaining open for variable periods. τ_{IPSC} corresponds closely both to

TABLE 2.3. *Comparison of* τ_{IPSC} *and* τ_{GABA} *in rat hippocampal neurons*

	τ_{IPSC} (ms)	τ_{GABA} (ms)	Ref.
Culture			
−60 mV	20	20	A
−40 mV (WCR)	25	—	B
+10 mV	30	30	A
Picrotoxin	18	21	A
Pentobarbital	70	50	A
Diazepam	30	26	A
Diazepam (WCR)	30	—	B
Alphaxalone (WCR)	135	—	B
Slice			
−60 mV	12	—	C
+10 mV	20	—	C
Pentobarbital	43	—	C

Data obtained in culture with either two-electrode or whole-cell patch recording (WCR) techniques and in slice with single-electrode voltage-clamp methods using either KCl- or Kgluconate-filled microelectrodes at 23°C in physiological salines.

A. Segal and Barker (1984b); B. Harrison *et al.* (1985); C. Collingridge *et al.* (1984).

τ_{GABA}, a value derived from the activities of several thousand GABA-activated ion channels at equilibrium in whole-cell recordings, and to the long time-constant kinetics recorded at the single channel level in micron patches. These latter results imply that the time course of the synaptic conductance decay in these cultured neurons is set primarily, if not entirely, by the physiological characteristics of the receptor-coupled Cl^- ion-channel complex in the post-synaptic membrane, particularly by the exponential kinetics in channel closing. Most of the thousand or so channels remain open for relatively brief periods, with fewer and fewer remaining open for longer and longer durations. Since the closing rate for the synaptically triggered population of channels is similar to that measured at the level of the pharmacologically activated single channel, which probably involves extra-synaptic GABA receptors, there may be a single, relatively uniform class of such GABA-receptor/Cl^- conductance mechanisms commonly expressed in neurons cultured from spinal and supraspinal regions. Values for τ_{IPSC} are quite comparable when recorded at the same potential and temperature, and drug condition in embryonic rat neurons cultured from the hippocampus and in fully differentiated hippocampal neurons studied in tissue slices prepared from the adult rat (Table 2.3). This suggests that the synaptic signals recorded in culture are a relatively faithful reconstitution of the conductance generated *in vivo*. A close correspondence between $\tau_{\text{transmitter}}$ and τ_{IPSC} has been reported for a variety of synaptically activated conductances in vertebrate and invertebrate tissues. Thus, the pharmacologically activated kinetics may be a useful reference for identifying other synaptic signals in the CNS.

τ_{GABA} is similar, if not identical whether it is recorded using two-electrode or single-electrode voltage clamp techniques or whole-cell patch-clamp assay methods. This latter result indicates that the GABA-receptor coupled Cl^- conductance mechanism is relatively insensitive to the loss of soluble factors reported to occur during patch-clamp-type recordings and to account for the 'rundown' of several types of membrane excitability. From this we may conclude that generation of this relatively rapid Cl^- conductance signal at synapses in the CNS is not immediately dependent on soluble second-messenger substances in contrast to the slower peptide-mediated hormonal communications at extrasynaptic sites on pituitary tumour cells described below.

TRH ALTERS THE EXCITABILITY OF CLONAL PITUITARY CELLS

Radiochemical and spectrophotometric assays of TRH's stimulation of clonal GH3 cells to release prolactin have shown that nanomolar levels of the peptide induce about a 10-fold rise in intracellular $[Ca^{2+}]$ ($[Ca^{2+}]_i$)

and coincident *efflux* of Ca^{2+} lasting less than a minute, which is then followed by a longer period of two-fold elevation in $[Ca^{2+}]_i$ and influx of Ca^{2+} ions (for recent text, see MacLeod *et al*. 1985). The transient, marked elevation in $[Ca^{2+}]_i$ is now considered to arise primarily from cytoplasmic stores (endoplasmic reticulum and mitochondrion) while the sustained low-level increase apparently reflects influx of Ca^{2+} through voltage-gated Ca conductance mechanisms. It has also recently become clear from other studies on TRH activation of GH3 cells that changes in inositol lipid metabolism occur whose time course closely parallels that reported for $[Ca^{2+}]_i$ fluctuations. From these observations has emerged the hypothesis that TRH-induced prolactin release is mediated in some intricate, yet ill-defined manner by receptor-coupled alterations in 'second messenger' inositol lipid substances and fluctuations in $[Ca^{2+}]_i$.

Electrophysiological recordings of GH3 cells have shown that the peptide has complex effects on cell excitability that typically last for several minutes following a brief exposure. The time course of these actions generally corresponds to that established for $[Ca^{2+}]_i$ and inositol lipid changes. There is, on average, a delay of more than a second after TRH has reached the cell before changes in membrane excitability can be detected (Ozawa 1985). Furthermore, successive applications of the peptide lead to longer delays and less well-expressed changes in cell excitability, indicating some significant desensitization to the peptide-mediated signal (Ozawa 1985).

The most commonly reported sequence of changes in excitable membrane properties involves an initial, transient increase in membrane potential associated with an increase in membrane conductance (for review, see Ozawa 1985; Fig. 2.4A1). Under voltage-clamp there is an outward current response and associated increase in conductance (Fig. 2.4A2) that can be evoked over a wide range of membrane potential and then reversed in polarity in the hyperpolarized range. This conductance response, which is thought to involve K^+ ions primarily, if not exclusively (Barker *et al*. 1983; Ozawa 1985; Dubinsky and Oxford 1985), typically lasts for 10–30 s and is often, but not always followed by a sustained period (1–2 min) of spontaneous action potential activity, with (Ozawa 1985) or without concomitant depolarization of the cell (Fig. 2.4A1). Using experimental conditions where voltage-activated Ca^{2+} conductance mechanisms can be studied in relative isolation, it is evident that TRH also induces a rapid, slowly reversible depression in the amplitude of inwardly directed Ca current (Dufy *et al*. 1985; Fig. 2.4B). Finally, the peptide rapidly and reversibly depresses a transient-type outward (K^+) current response (Barker *et al*. 1983; Dubinsky and Oxford 1985; Fig. 2.5A), primarily by shifting its activation curve in a depolarizing direction (Fig. 2.5B).

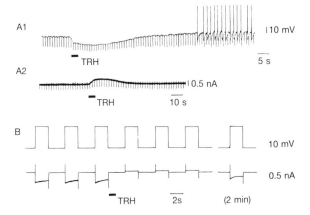

FIG. 2.4. TRH alters the excitability of GH3/6 cells. The cell was recorded with a conventional high-resistance microelectrode filled with 3M KCl at room temperature. (A1) Brief application of 50 nM TRH induces a transient hyperpolarization of the cell and increase in membrane conductance (reflected in the decrease in amplitude of the downward-going voltage responses to constant-current hyperpolarizing stimuli). This transient phase is followed by the generation of spontaneous (Ca^{2+}-dependent) action potential activity. (A2) In the (switching) voltage-clamp configuration of the single microelectrode recording, TRH evokes a transient outward current and associated conductance increase (reflected in the increase in applitude of downwardly-directed current responses to constant hyperpolarizing voltage commands). (B) With the cell bathed in 10 mM Ba^{2+} rather than 10 mM Ca^{2+}, inwardly-directed voltage-gated Ba^{2+} currents can be isolated. Ba^{2+} currents were activated by 50 mV, 2-s commands from a holding potential of -70 mV. 50 nM TRH completely blocks their activation in a reversible manner (from Dufy *et al.* 1985).

From these results it is obvious that TRH has complex, long-lasting actions on the excitability of clonal prolactin-secreting pituitary cells involving (Ca^{2+} and/or voltage-gated) cationic conductance mechanisms. A role for Ca^{2+} ions in the activation of K^+ conductance mechanisms is strongly suggested by activation of the conductance mechanisms with extracellular application of the Ca^{2+} ionophore A23187 and by elimination of both peptide- and ionophore-activated conductance responses when recording with microelectrodes containing elevated concentrations of a Ca^{2+} chelator (Ozawa 1985). Thus, the K^+ conductance phase of TRH stimulation likely corresponds to the rapid rise in $[Ca^{2+}]_i$ detected by spectrophotometry. During this phase there is also a virtual *inactivation* of voltage-gated Ca^{2+} conductance mechanisms under experimental conditions that isolate such mechanisms for electrophysiological assay. These findings are complementary to, and consistent with the relative insensitivity of the TRH-induced increase in $[Ca^{2+}]_i$ to pharmacological

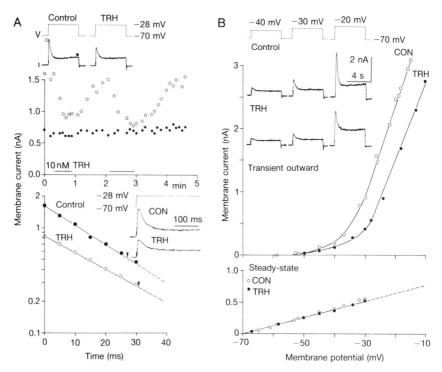

FIG. 2.5. TRH attenuates transient outward current (TOC) responses in a GH3/6 cell. The voltage-clamp recording was made with two microelectrodes filled with KCl. TRH was applied by pressure from a closely positioned pipette containing 10 nM peptide. (A) TRH reduces the amplitude of the TOC evoked upon stepping to -28 mV with little, if any, effect on the steady-state current flowing at the end of the 5 s command. (Upper panel) Plot of TOC amplitude and steady-state current before, during, and after successive applications of TRH. (Lower panel) Plot of the TOC decay evoked at -28 mV (see inset) before and during TRH application. The time constant (arrowhead) is not appreciably changed. (B) TOCs (upper panel) and steady-state currents (lower panel) are plotted for a series of voltage commands under control and in the presence of TRH. (Inset) Specimen records. TRH shifts the activation curve for the TOC in a depolarizing direction while having little effect on the steady-state current (From Barker *et al.* 1983).

antagonists of voltage-gated Ca conductance mechanisms. Since prolactin release can be detected during this early phase, TRH-triggered release of Ca^{2+} from organelle stores must play a significant role in TRH-induced prolactin release. The second phase of TRH-evoked stimulation of GH3 cell excitability, which involves generation of spontaneous Ca-dependent action potential activity, may account for the later, sustained elevation in $[Ca^{2+}]_i$ detected spectrophotometrically. Action potential activity and this phase of elevated $[Ca^{2+}]_i$ are both eliminated by Ca conductance antagonists.

Precise roles for 'second messenger' mechanisms underlying the TRH stimulation of $[Ca^{2+}]_i$ changes and subsequent prolactin release have yet to be elucidated. Resolution of the details involved in the relationship between such cytoplasmic events and membrane excitability may be possible with the advent of 'whole-cell-patch' recording techniques, which provide both an avenue for experiments in intracellular biochemical pharmacology and a means for recording ion conductances at high levels of temporal resolution.

CONCLUSION

In this chapter we have reviewed recent electrophysiological observations on amino acid- and peptide-mediated signals studied in monolayer culture. It is evident from the results discussed here that some receptor-coupled signals have evolved that are predominantly, if not exclusively membrane in origin while others appear to involve both membrane and cytoplasmic components. Such diversity in signalling mechanisms at synaptic and extrasynaptic sites, which is representative of chemical signalling mechanisms discovered throughout evolution, allows for and underlies fast and slow modes of information transfer within the 'hard-wired' web of the CNS, and between the CNS and one of its important target systems, the endocrine pituitary.

References

Barker, J. L. (1985). GABA and glycine: ion channel mechanisms. In *Neurotransmitter Actions in the Vertebrate Nervous System* (eds M. A. Rogawski and J. L. Barker). Plenum, New York, pp. 71–100.

——, Dufy, B., Owen, D., and Segal, M. (1983). Excitable membrane properties of cultured CNS neurons and clonal pituitary cells. In *48th Cold Spring Harbor Symposium on Quantitative Biology: Molecular Neurobiology*, pp. 259–68.

——, and Mathers, D. A. (1981). GABA analogues activate channels of different duration on cultured mouse spinal neurons. *Science*, **212**, 358–61.

——, and McBurney, R. N. (1979). GABA and glycine may share the same conductance channel on cultured mammalian neurones. *Nature (Lond.)* **277**, 234–6.

——, ——, and MacDonald, J. F. (1982). Fluctuation analysis of neutral amino acid responses in cultured mouse spinal neurons. *J. Physiol. (Lond.)* **322**, 365–87.

——, and Ransom, B. R. (1978). Amino acid pharmacology of mammalian central neurones grown in tissue culture. *J. Physiol.* **208**, 331–54.

Bormann, J., and Clapham, D. (1984). Gamma-aminobutyric acid receptor channels in adrenal chromaffin cells: a patch-clamp study. *Proc. Nat. Acad. Sci.* **82**, 2168–72.

Caserta, M. T., and Barker, J. L. (1983). Development of glutamic acid decarboxylase immunoreactivity in mouse spinal cord cultures. *Soc. Neurosci. Abs.* **9**, 7.

Chang, L-R., and Barnard, E. A. (1982). The benzodiazepine/GABA receptor complex: molecular size in brain synaptic membranes and in solution. *J. Neurochem.* **39**, 1507–18.

Collingridge, G. L., Gage, P. W., and Robertson, B. (1984). Inhibitory post-synaptic currents in rat hippocampal CA1 neurones. *J. Physiol.* **356**, 551–64.

Cottrell, E. A., Lambert, J. J., and Peters, J. A. (1985). Chloride currents activated by GABA in cultured bovine chromaffin cells. *J. Physiol.* **365**, 90P.

Cull-Candy, S. G., and Ogden, D. C. (1985). Ion channels activated by L-glutamate and GABA in cultured cerebellar neurons of the rat. *Proc. Roy. Soc. Lond. B* **224**, 367–73.

Dichter, M. A. (1980). Physiological identification of GABA as the inhibitory transmitter for mammalian cortical neurons in cell culture. *Brain Res.* **190**, 111–21.

Dubinsky, J. M., and Oxford, G. S. (1985). Dual modulation of K channels by thyrotropin releasing hormone in clonal pituitary cells. *Proc. Nat. Acad. Sci.* **82**, 4282–6.

Dufy, B., Dupuy, B., Georgescauld, D., and Barker, J. L. (1985). Calcium-mediated inactivation of calcium conductance in a prolactin-secreting cell line. In *Prolactin: Basic and Clinical Correlates* (eds R. M. MacLeod, M. D. Thorner, and U. Scapagnini), Fidia Research Series, Vol. 1, pp. 177–83. Liviani Press, Padova, Italy.

Hamill, O. P., Bormann, J., and Sakmann, B. (1983). Activation of multiple-conductance state chloride channels in spinal neurons by glycine and GABA. *Nature*, **305**, 805–8.

Harrison, N. L., Vicini, S., Owen, D. G., and Barker, J. L. (1985). Steroid modulation and mimickry of GABA-activated chloride conductance in cultured mammalian central neurons. *Soc. Neurosci. Abs.* **11**, 281.

Jackson, M. B., Lecar, H., Mathers, D. A., and Barker, J. L. (1982). Single channel currents activated by GABA, muscimol, and (-)pentobarbital in cultured mouse spinal neurons. *J. Neurosci.* **2**, 889–94.

Kemp, J. A., Marshall, G. R., and Woodruff, G. N. (1984). Quantitative analysis of GABA agonists and antagonists using the rat hippocampal slice preparation. *Br. J. Pharmacol.* **82**, 199P.

Kuriyama, K., and Taguchi, J. (1984). Properties of purified gamma-aminobutyric (GABA) receptor. In *Neurotransmitter Receptors: Mechanisms of Action and Regulation* (eds S. Kito, T. Segawa, K. Kuriyama, H. I. Yamamura, and R. W. Olsen), pp. 221–32. Plenum Press, New York.

MacLeod, R. M., Thorner, M. A., and Scapagnini, U. (1985). *Prolactin: Basic and Clinical Correlates*, Fidia Research Series, Vol. 1. Liviana Press, Padova, Italy.

Mathers, D. A., and Barker, J. L. (1980). (-)Pentobarbital opens channels of long duration on cultured mouse spinal neurons. *Science*, **209**, 507–9.

——, and —— (1981a). GABA- and glycine-induced Cl⁻ channels in cultured mouse spinal neurons require the same energy to close. *Brain Res.* **224**, 441–5.

——, and —— (1981b). GABA and muscimol open channels of different lifetimes on cultured mouse spinal neurons. *Brain Res.* **204**, 242–7.

——, and —— (1982). Chemically-induced ion channels in nerve cell membranes. *Int. Rev. Neurobiol.* **23**, 1–34.

McBurney, R. N., Smith, S. M., and Zorec, R. (1985). Conductance states of gamma-aminobutyric acid (GABA)- and glycine-activated chloride (Cl⁻) channels in rat spinal neurones in cell culture. *J. Physiol.* **365**, 87P.

Nistri, A., and Constanti, A. (1979). Pharmacological characterization of different types of GABA and glutamate receptors in vertebrates and invertebrates. *Proc. Neurobiol.* **13**, 117–235.

Nowak, L., Young, A. B., and Macdonald, R. L. (1982). GABA and bicuculline actions on mouse spinal cord and cortical neurons in cell culture. *Brain Res.* **244**, 155–64.

Olsen, R. W., and Snowman, A. (1983). [³H] Bicuculline methochloride binding to low affinity GABA receptor sites. *J. Neurochem.* **41**, 1653–63.

Olsen, R. W., Wong, E. H. F., Stauber, G. H., Murakami, D., King, R. G., and Fischer, J. B. (1984). Biochemical properties of the GABA/barbiturate/benzo-diazepine receptor chloride ion channel complex. In *Neurotransmitter receptors: mechanism of action and regulation* (eds S. Kito, T. Seqaura, K. Kuriyama, H. I. Yamamura, and R. W. Olsen), pp. 205–19. Plenum Press, New York.

Ozawa, S. (1985). Electrophysiology of clonal anterior pituitary cells. In *The Electrophysiology of the Secretory Cell* (eds A. M. Poisner and A. Trifaro), pp. 221–39. Elsevier, Holland.

——, and Yuzaki, M. (1984). Patch-clamp studies of chloride channels activated by gamma-aminobutyric acid in cultured hippocampal neurones of the rat. *Neurosci. Res.* **1**, 275–93.

Pfeiffer, F., Simler, R., Grenningloh, G., and Betz, H. (1984). Monoclonal antibodies and peptide mapping reveal structural similarities between the subunits of the glycine receptor of rat spinal cord. *Proc. Nat. Acad. Sci.* **81**, 7224–7.

Redmann, G. A., Lecar, H., and Barker, J. L. (1983). Single muscimol-activated ion channels show voltage-sensitive kinetics in cultured mouse spinal neurons. *Soc. Neurosci. Abs.* **9**, 507.

——, ——, and —— (1984). Diazepam increases GABA-activated single channel burst duration in cultured mouse spinal neurons. *Biophys. J.* **45**, 386a.

Sakmann, B., Hamill, O. P., and Bormann, J. (1983). Patch-clamp measurements of elementary chloride currents activated by the putative inhibitory transmitters GABA and glycine in mammalian spinal neurons. *J. Neural Transmiss.* Suppl. 1, 83–95.

——, and Neher, E. (1983). *Single-Channel Recording.* Plenum Press, New York.

Segal, M., and Barker, J. L. (1984a). Rat hippocampal neurons in culture: properties of GABA-activated Cl⁻ ion conductance. *J. Neurophysiol.* **52**, 469–87.

——, and —— (1984b). Rat hippocampal neurons in culture: voltage clamp analysis of inhibitory synaptic connections. *J. Neurophysiol.* **52**, 469–87.

Sigel, E., and Barnard, E. A. (1984). A gamma-aminobutyric acid/benzodiazepine receptor complex from bovine cerebral cortex. Improved purification with preservation of regulatory sites and their interactions. *J. Biol. Chem.* **259**, 7219–32.

——, Stephenson, F. A., Mamalaki, C., and Barnard, E. A. (1983). A gamma-aminobutyric acid/benzodiazepine receptor complex of bovine cerebral cortex. *J. Biol. Chem.* **258**, 6955–71.

Smith, E. M., Harbour-McMenamin, D. and Blalock, J. E. (1985). Lymphocyte production of endorphins and endophrin-mediated immunoregulatory activity. *J. Immunol.* **135**, 7795–825.

Study, R. E., and Barker, J. L. (1983). Neurotransmitter-activated chloride channels in cultured mouse spinal neurons are also voltage regulated. *Soc. Neurosci. Abs.* **9**, 410.

Wong, E. H. F., and Iversen, L. L. (1985). Modulation of [^{3}H]diazepam binding in rat cortical membranes by GABA$_A$ agonists. *J. Neurochem.* **44**, 1162–7.

3

Fast synaptic responses in the hippocampus and dorsal lateral geniculate nucleus of the rat

JOHN S. KELLY, VINCENZO CRUNELLI, SUSAN FORDA, NATHALIE LERESCHE, AND MARIO PIRCHIO

INTRODUCTION

By and large most of what is known of fast synaptic transmission in the central nervous system of mammals has been inferred from the study of just three preparations: the amphibian and arthropod neuromuscular junction and the squid giant synapse (Katz 1966; Martin 1977; Takeuchi 1977) and it is only very recently that intracellular recording from *in vivo* and *in vitro* preparations of the mammalian central nervous system has reached a level of sufficient sophistication and precision to allow some of the following basic premises about synaptic transmission in the central nervous system to be tested. In this chapter we review in some detail how our results (Crunelli, *et al.*, 1982, 1983, 1984, 1985a,b; Godfraind and Kelly 1981) from two *in vitro* brain slice preparations have been used to study synaptic physiology in two regions where both the afferent and efferent pathways are known to be myelinated, fast conducting, and likely to be connected by fast and efficient synaptic mechanisms. In both the dentate gyrus of the hippocampus and the dorsal lateral geniculate nucleus our results show the synaptic events evoked *in vivo* by stimulation of the perforant path and the optic nerve to be similar if not identical to the peripheral synaptic events mentioned above. However, the post-synaptic membrane of both the granule cell of the dentate gyrus and the principal neurones of the dorsal lateral geniculate nucleus are much more specialized and, in all probability contribute additional speed and efficiency to fast synaptic events. Whether or not these specializations are features of, or additional to, other roles of dendrites and dendritic spines remains to be seen (Crick 1982; Perkel & Perkel 1985; Miller *et al.* 1985).

The studies at the neuromuscular junction mentioned above have led to the view that the postsynaptic action of the transmitter at fast synapses involves a specific interaction between the transmitter and a receptor, followed directly by the opening of an ion channel or channels, known as ionophores, linked to the receptor. This movement of ions through the ionophore brings about electrical changes recorded from the post-synaptic neurone. This simple proposition has led to two relatively straightforward tests of transmitter identity at single synapses known as *identity of action* and *pharmacological antagonism*. The first states quite simply that the effect of putative exogenous substance applied to the receptor ionophore complex must be identical in every respect to that of the transmitter release from the presynaptic nerve terminals by electrical stimulation and the second that during pharmacological antagonism the action of the putative transmitter and the synaptic event should remain identical, but reduced in amplitude by the presence of the pharmacological antagonist. In other words, the antagonists of the synaptic events must be shown, preferably on the same preparation, *not* to act on the mechanisms controlling the synthesis, storage, or release of the transmitter or indeed the integrity of the ionophore.

Although in this chapter we do not wish to dwell on the technical details of our work it is of interest to draw attention to a number of features of the two preparations which make them almost ideal for this type of work. In both preparations the key feature is the ease with which in each experiment the same class of neuron and the same type of synapse can be located. In addition the morphology of the post-synaptic neuron can be verified by filling the neurone with an intracellular marker at the end of the recording session and then compared with descriptions of Golgi material. The careful placement of the stimulating electrodes under visual control on either a discrete point in the medial perforant pathway or rather more grossly on the optic nerve is also necessary for synapse identification. As yet we have not attempted to fill the presynaptic nerve terminal with marker dyes and thus locate much more precisely the number and position of the synapses under study as has been done so elegantly in the spinal cord *in vivo* by Redman and Walmsley (1983a,b). Secondly, the extracellular fluid bathing the preparation can be changed several times while maintaining long term stable intracellular recordings. However, only in the dorsal lateral geniculate preparation have we used a bath similar to that developed by Haas *et al.* (1979) which allows rapid and efficient changes of the extracellular media. Thirdly, both tissues allow the synapse under study to be stimulated monosynaptically. The dentate preparation is particularly suitable since discrete stimulation of the medial perforant pathway leads to a pure excitatory post-synaptic potential (e.p.s.p.) and no additional steps are required to eliminate the inhibitory post-synaptic potential (i.p.s.p.) which invariably follows a monosynaptic e.p.s.p. in other preparations. In the dorsal lateral geniculate the potential and conductance changes associated with the i.p.s.p. which follows the e.p.s.p. were eliminated by perfusing the preparation with the GABA antagonist bicuculline. The fourth criteria is much more difficult to satisfy and concerns the quantitative analysis of the synaptic events which are best conducted in spherical neurones free from dendrites and therefore known to be isopotential. Fortunately, in the case of the granule cell a number of workers have shown these neurones to be essentially isopotential (Brown *et al*. 1981) and the topography of the tissue allows the iontophoretic pipette used to apply putative transmitters to be moved out along the dendrites in an attempt to apply the exogenous material to the same region of the dendrite served by the synapses of the medial perforant path. In the dorsal lateral geniculate, however, both the orientation of the dendrites and the position of the synapses under study, and indeed the degree of isopotentiality is unknown. It is not possible to discount these factors when comparing disparities between the reversal level of the e.p.s.p. evoked by optic nerve stimulation and the potential evoked by the application of exogenous substances. The fifth desirable feature of the synapse under study is the ability to resolve clearly single quantal events above the background noise of the recording electrode (Johnston and Brown 1984). Although in the dentate gyrus our present technique for discrete stimulation of the medial perforant pathway allowed us to show indirectly the quantal nature of this synaptic event, a similar analysis in the dorsal lateral geniculate nucleus will require the development of more sophisticated techniques for stimulating the optic nerve.

HIPPOCAMPUS

Fast synaptic events in granule cells of the dentate gyrus

In the first brain slice preparation we have studied the granule cells of the dentate gyrus and the e.p.s.p. evoked by stimulation of the perforant pathway which is the major excitatory input to the granule cells from the entorrhinal cortex. As predicted by the simple model of fast synaptic transmission the peak amplitude of the e.p.s.p. was accompanied by a marked decrease in membrane resistance and varied in amplitude with the membrane potential of the cell in a predictable manner, i.e. the amplitude of the e.p.s.p. increased as the cell was hyperpolarized and decreased as the cell was depolarized. Voltage current plots of the peak of the e.p.s.p. and the membrane potential allowed the potential at which the amplitude of the e.p.s.p. would be zero, i.e. the reversal potential, to be determined by extrapolation (Crunelli *et al*. 1982). Although these plots showed that there was no significant deviation from linearity over a relatively large range of membrane potentials in the range (-70 to -40 mV) measurements of the true reversal of the e.p.s.p. proved impossible using conventional techniques.

Responses evoked by excitatory amino acids

The response to short iontophoretic applications of glutamate consisted of rapid and dose-dependent depolarizations that were only rarely accompanied by an appropriate change in membrane resistance and were as often as not accompanied by either no change in membrane resistance or an increase. In addition, these changes in membrane resistance appeared to be dependent on the rate and magnitude of the depolarization which in turn was critically dependent upon the position of the iontophoretic pipette. As the pipette approached the area of maximum glutamate sensitivity the rate of rise and the amplitude of the depolarization evoked by the same iontophoretic pulse of current reached a maximum, and this depolarization was invariably accompanied by a decrease in membrane resistance. As shown in Fig. 3.1 at the point of maximum glutamate sensitivity the latency to the onset of the depolarization could be as short as 10 ms and the maximum rate of depolarization approached between 0.05–0.07 mV/s.

On the same cells the response to N-methyl-D-aspartate (NMDA) was distinctly different from that to glutamate. For instance, at the point of maximum sensitivity to glutamate even low doses of NMDA tended to produce a burst pattern of firing and high doses resulted in a huge depolarization of up to 45 mV from which the cell could take several minutes to recover. However, when comparing approximately equipotent doses the maximum rate of depolarization was somewhat slower (0.01 mV/ms). Although the depolarization evoked by NMDA could be accompanied by both small decreases and no change in membrane resistance, the apparent increases in resistance were more pronounced than those seen with glutamate.

In experiments in which the depolarization evoked by the amino acids was negated by using manual voltage clamp, the responses evoked by glutamate, NMDA, and quisqualate were still accompanied by an apparent increase in membrane resistance in the absence of a change in voltage and thus the apparent increase in resistance must be additional to the increases seen during passive depolarizations of the post-synaptic cells.

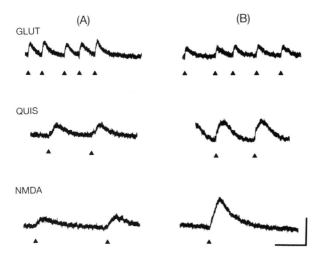

FIG. 3.1. Differential sensitivity of the same granule cell in rat hippocampal slice to glutamate, quisqualate, and NMDA. In (A) the voltage records show the response to all three amino acids when the position of the pipette was at an optimum for glutamate sensitivity and in (B) when the pipette was at an optimum for NMDA sensitivity. In (A) the maximum mean rate of depolarization for glutamate was 0.07 mV/ms compared with 0.01 and 0.006 mV/ms for quisqualate and NMDA, respectively. In (B) the maximum rates of depolarization were: glutamate 0.04 mV/ms, quisqualate 0.04 mV/ms and NMDA 0.01 mV/ms. Triangles indicate 50, and 70 ms ionophoretic applications of glutamate, quisqualate, and NMDA, respectively. The ionophoretic currents were: glutamate 103 nA, quisqualate 130 nA, and NMDA 100 nA. Calibration bars equal 10 mV and 4 s. Reproduced with permission from Crunelli *et al.* (1984).

Effect of TTX, Co^{2+}, and Mg^{2+} on the depolarizations evoked by the amino acids

In the presence of tetrodotoxin (TTX) all four amino acids continued to be equally effective in depolarizing of granule cells, and indeed, on some occasions, TTX appeared to enhance both the depolarization and the increase in membrane resistance. Presumably in the absence of cell firing there is less calcium entry and less activation of restorative potasium currents. In the presence of Co^{2+} the depolarizations produced by glutamate, aspartate, and quisqualate persisted, but were no longer associated with increases in membrane resistance and the response to NMDA was abolished. One interpretation of these results might be that the increases in resistance seen with glutamate and aspartate are due to an action of these amino acids on NMDA receptors and that their action on the NMDA receptor is abolished by Co^{2+} (Mayer & Westbrook 1984).

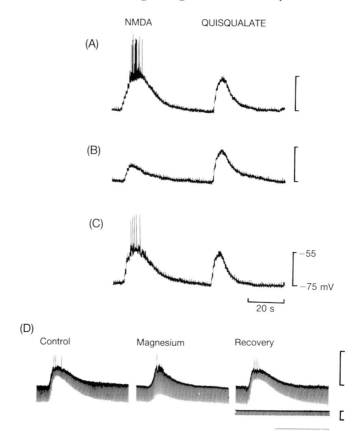

FIG. 3.2. The selective and channel blocking action of Mg^{2+} on NMDA evoked depolarization of a CA1 hippocampal pyramidal cell. (A) Intracellular voltage records show the depolarizing response of a CA1 pyramidal neurone to pulsed iontophoretic application of NMDA and quisqualate obtained in a solution containing 5 mM Ca^{2+} and 0 mM Mg^{2+}. (B) In the same cell, after a drop of medium containing 2 mM Mg^{2+} was put on the surface of the slice close to the recording electrode the depolarization produced by NMDA is very much reduced (30 per cent of control), while the response to quisqualate remains unaffected. (C) In the presence of Mg^{2+}, the ejection time of NMDA has been doubled in order to obtain a depolarization of similar amplitude to that shown in (A). The initial resting membrane potential was -62 mV and the input resistance, 39 MΩ. Iontophoretic currents were quisqualate 70 nA and NMDA 38 nA.

In (D, control), the depolarization produced by NMDA in another cell in the absence of Mg^{2+} is accompanied by a decrease in membrane resistance. Following the application of a drop of medium containing 2 mM Mg^{2+} to the surface of the slice (magnesium) the depolarization produced by NMDA is accompanied by an apparent increase in membrane resistance. Recovery was recorded 20 min after the drop application. The records are from one granule cell that was impaled 5 h after

Earlier Ault *et al.* (1980) showed Co^{2+}, Mg^{2+}, and other divalent cations to be competitive antagonists of the depolarizing action of NMDA on the neonatal spinal cord and this was confirmed on hippocampal pyramidal neurones (Dingledine 1983). Recently, the increase in resistance which accompanies the action of NMDA receptors on a variety of neuronal membranes was explored in great detail by a number of authors using both voltage clamp and single channel techniques, and shown to require the presence of magnesium ions in the extracellular medium and to be due to the entry of magnesium into the channels opened by the NMDA receptor followed by block (Mayer *et al.* 1984; Nowak *et al.* 1984). As shown in Fig. 3.2, the application of Mg^{2+} on granule cells selectively antagonizes the depolarization evoked by NMDA and only in the presence of Mg^{2+} can the depolarization be shown to be accompanied by an increase in resistance. Thus, in the absence of Mg^{2+}, NMDA appears to behave as a classical excitatory transmitter. Although the precise significance of the Mg^{2+} blockade of NMDA receptors is not clear it has recently been suggested by Thomson *et al.* (1985) that certain synaptic events in the cerebral cortex are mediated by NMDA receptors since they could be blocked by a variety of agents thought to be specific NMDA antagonists.

The reversal potential of the amino acid and e.p.s.p. evoked depolarization

In the same study we were also able to differentiate the depolarizing response evoked by NMDA from that evoked by glutamate, quisqualate, kainate, and indeed the e.p.s.p. evoked by stimulation of the perforant path by showing the reversal level of the NMDA response to be different from that of the others. This study was only possible when the intracellular electrodes were filled with Cs^{2+} chloride and the injection of Cs^{+} used to eliminate the voltage-sensitive rectifying properties of the cells (Hablitz and Langmoen 1982). As shown in Figs 3.3 and 3.4 the injection of Cs^{2+} showed the responses to the amino acids and the e.p.s.p. to be relatively pure events, and the depolarizing and hyperpolarizing envelopes on either side of the reversal potential are very nearly mirror images of each other. Using linear regression the reversal level for NMDA was shown to be significantly different from that of the other amino acids and the e.p.s.p.

---

the slice was transferred to a medium containing 0 mM Mg^{2+} and 5 mM Ca^{2+}. The resting membrane potential was (-68 mV). Downward deflections in the voltage records are due to electrotonic potentials used to measure input resistance. After Mg^{2+} was applied to the slice, the response to NMDA was strongly reduced, so that the amount of charge used to eject NMDA was doubled from 160 to 320 nC in order to obtain a depolarization of similar amplitude to that recorded in control. Calibration bars equal 30 mV, 1 nA, and 30 s. (Reproduced with permission from Crunelli & Mayer, 1984.)

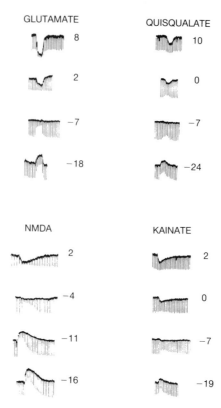

FIG. 3.3. Reversal of the response to glutamate, quisqualate. NMDA, and kainate in a Cs$^+$-loaded granule cell. Typical records of each amino-evoked response are shown at four different membrane potentials. The reversal levels calculated by regression analysis were E$_{glutamate}$: -5.4 mV; E$_{quisqualate}$: -7.3 mV; E$_{NMDA}$: -4.4 mV and E$_{kainate}$: -8.6 mV. Downard deflexions of the trace are the voltage response of the cell to pulses of hyperpolarizing current that were kept constant during each individual drug application. The mean input resistance of the cell was 62 MΩ. Ionophoretic applications were: glutamate 405 nA, 4 s; quisqualate: 130 nA, 3 s; NMDA: 393 nA, 1 s; kaintate 98 nA, 600 ms. Calibration bars equal 10 mV and 10 s. (Reproduced with permission from Crunelli *et al.* 1984.)

The use of pharmacological antagonists to identify the receptor

The possibility that the e.p.s.p. evoked by stimulation of the perforant path might be mediated by NMDA receptors was also excluded by experiments which showed the e.p.s.p. to be blocked by γ-D glutamylglycine (DGG), but not (±)-2-amino-5-phosphonovalerate (APV). As shown in Fig. 3.5 the iontophoretic application of DGG markedly and reversibly reduced the amplitude of the e.p.s.p. evoked by stimulation of the perforant path

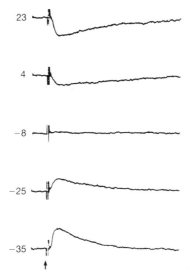

FIG. 3.4. Reversal of the e.p.s.p. evoked by stimulation of the medial perforant path in a granule cell loaded with Cs^+. Typical voltage records show the e.p.s.p. at five different membrane potentials. The reversal level calculated by regression analysis was -9.8 mV. The arrow marks stimulation of the medial perforant path and calibration bars equal 10 mV and 20 ms. (Reproduced with permission from Crunelli *et al*. 1984.)

without causing any change in the resting membrane potential. Additional experiments showed the blockade to occur in the absence of a change in membrane resistance or excitability of the cell under study. The post-synaptic action of the blockade was confirmed by showing the iontophoretic application of DGG to have no effect on the amplitude or duration of the mass presynaptic fibre volley that can be recorded from within the cell by the intracellular electrodes (Crunelli *et al*. 1982). By extrapolating voltage-current plots derived on and off the peak of the e.p.s.p. in the presence and absence of DGG the antagonism was shown not to evoke a significant change in the e.p.s.p. reversal level. Thus, the predictable decline in the amplitude of the e.p.s.p. and the associated decrease in membrane resistance is in keeping with the idea that the amplitude of the e.p.s.p. is determined by the summation of the events generated by the opening of a relatively constant number of transmitter specific channels and that specific antagonists simply reduce the number of available channels.

As shown in Fig. 3.6 it was also possible to show using trains of 150 or more e.p.s.p.'s recorded in the presence and absence of DGG that the decline in amplitude of the e.p.s.p. in the presence of the antagonist can be

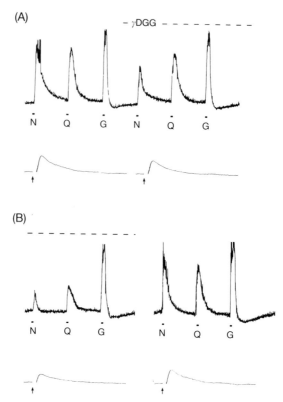

Fɪɢ. 3.5. Intracellular voltage recordings from a granule cell showing the effect of DGG on the medial perforant path-evoked e.p.s.p. and the depolarization induced by pulsed iontophoretic applications of NMDA (N), quisqualate (Q), and glutamate (G). The upper trace is the intracellular voltage response to drug application, and the lower trace the computer average of 10 e.p.s.ps. In (A) a 13-s application of DGG was sufficient to reduce the NMDA-induced depolarization to 59 per cent of control, while the e.p.s.p. was unaltered. In (B) the continuing application of DGG reduced the response to quisqualate and the e.p.s.p. amplitude decreased from 11 to 5 mV. (A) and (B) are consecutive records and the gap in (B) is 1.5 min. The action potentials evoked by the drug-induced depolarization have been attenuated by the computer plotting subroutine. The durations of the amino acid applications are indicated by the black bars below the voltage records. Drug ejection currents where (nA): NMDA 100, quisqualate 150, glutamate 200, and DGG 105. Calibration bars represent 20 mV and 30 s for the upper trace, 20 mV and 32 ms for the lower one. Solid arrows mark stimulation of the perforant path. (Reproduced with permission from Crunelli *et al*. 1982.)

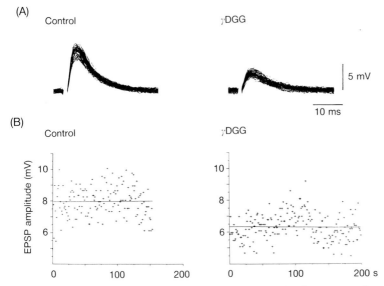

FIG. 3.6. Quantal analysis of the medial perforant path evoked e.p.s.p. to show the post-synaptic nature of DGG antagonism. (A) Superimposed (*n*-150) computer drawn e.p.s.ps. recorded before (control) and during (DGG) the ionophoretic application of 200 nA of DGG. A clear reduction in fluctuation accompanies the decline of the e.p.s.p. amplitude produced by DGG. This is reflected in a decrease in quantal size (q) with no change in quantal content (m). (B) Plots of e.p.s.p. amplitude versus time to show that no clear sign of synaptic habituation occurs both in control condition and during the ionophoretic application of DGG, when reasonable steady-state conditions have been reached. Note that the decrease in the e.p.s.p. amplitude is associated with a reduction in the fluctuation (decrease in the variance of the mean e.p.s.p. amplitude) indicating a post-synaptic action of DGG (data uncorrected for non-linear summation). (Reproduced with permission from Crunelli *et al*. 1983.)

accounted for almost entirely by a reduction in quantal size rather than quantal content. Thus, according to the quantal model of synaptic transmission (Katz 1966) the action of DGG is due to a reduction in post-synaptic sensitivity to the endogenous transmitter and does not involve a change in transmitter release.

On many cells a further attempt was made to identify the transmitter by comparing the action of DGG on the e.p.s.p. evoked by perforant path stimulation with its action on amino acids induced depolarization of the same cell. As shown in Fig. 5 the reduction in the e.p.s.p. amplitude only occurred when the application of DGG was of sufficient magnitude to reduce the depolarization evoked by quisqualate and/or kainate. Even smaller applications of DGG blocked the response evoked by NMDA. The failure of an effective dose of DGG to block the response to glutamate

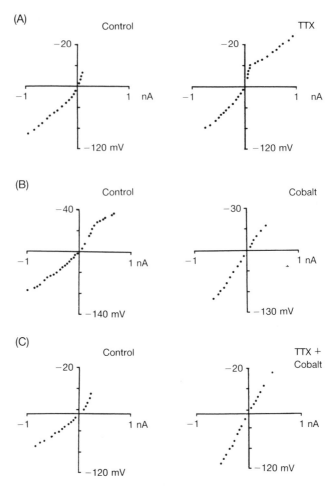

FIG. 3.7. Voltage-current plots from three different granule cells in normal medium and in the presence of TTX and Co^{2+}-containing medium. In (A) the apparent increase in input resistance at membrane potentials between -60 and -40 mV were unaffected by TTX. The resting membrane potential was -60 mV and input resistance (42 MΩ) were unaffected by TTX. In (B) in control solutions, a region of apparent increase in input resistance can be clearly observed between -70 and -50 mV. In the presence of Co^{2+} (10 mM), the cell depolarized by 10 mV, the resting membrane input resistance increased from 52 to 80 MΩ, and the anomalous rectification disappeared. In (C) the anomalous rectification present between -60 and 45 mV was abolished following the application of TTX and Co^{2+}. The membrane input resistance increase from 49 to 82 MΩ at the membrane potential of -63 mV was unchanged. (Reproduced with permission from Crunelli *et al.* 1984.)

shown in Fig. 3.5 was not unusual. Similar experiments confirmed APV's greater selectivity for NMDA receptors and its inability to block responses evoked by quisqualate, kainate, and glutamate.

Membrane properties of granule cell which may facilitate synaptic transmission

In the majority of granule cells, a steady depolarization of the cell by the injection of current caused a 25 per cent increase in the voltage deflection evoked by a constant pulse of current. As shown in Fig. 3.7 this result shows up as an apparent increase in the membrane input resistance and is visible as a hump on the voltage-current plots which persisted in the presence of TTX (1×10^{-5}M). Indeed, in the presence of TTX the hump becomes more pronounced since in the absence of cell firing a greater range of depolarizing current can be used to explore the region of interest which lies just subthreshold to the level required for the generation of action potentials. In addition, the presence of action potentials leads to the initiation of restorative potassium currents which are associated with a decrease in membrane resistance.

In the presence and absence of TTX the addition of Co^{2+} (5–15 mM) to the surface of the 'slice' abolishes the apparent increase in membrane resistance.

Clearly the summation of increases in resistance evoked in part by the passive depolarization of the cell and in part by the action of the amino acids will potentiate the depolarizing action of synaptic currents mediated by the opening of ion channels. The apparent increase in resistance will not only lead to greater voltage deflection in response to the same current injection, but it will effectively shorten the length of the dendrites, and thus reduce the distance between the active synapses and the region of the cell where fast sodium spikes are initiated. However, even if the apparent increase in membrane resistance is an artefact in that it is due not so much to the closure of potassium channels, but to the opening of additional Ca^{2+} channels, the additional voltage will again make firing more certain and the slow onset and longer duration of the voltage change produced by the entry of Ca^{2+} ions when compared with the peak of the e.p.s.p., will effectively shorten the distance between the synaptic event and the site at which fast sodium spikes are initiated. Recently, it has been suggested that the enhancement of synaptic events by the entry of calcium may also involve more dramatic changes in the postsynaptic membrane such as structural alterations to the shape of dendritic spines (Crick 1982).

LATERAL GENICULATE NUCLEUS

Optic nerve evoked e.p.s.p. in the dorsal lateral geniculate nucleus

The e.p.s.p. recorded from neurones of the dorsal lateral geniculate nucleus showed all the features of fast excitatory synaptic events described earlier, and in particular increased in amplitude during hyperpolarization of the cell under study and decreased in amplitude during depolarization and as shown in Fig. 3.8 reversed in polarity as the membrane potential approached 0 mV. As mentioned earlier the e.p.s.p. was followed by a long lasting hyperpolarization that showed all the electrophysiological and pharmacological features of an inhibitory postsynaptic potential mediated by γ-aminobutyric acid. In particular it could be antagonized reversibly by bicuculline (10^{-3}–10^{-4}M) and on the dorsal lateral geniculate nucleus preparation, all the experiments described in this chapter were performed in the presence of this antagonist.

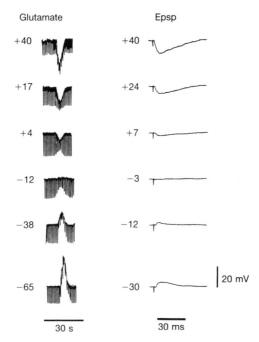

FIG. 3.8. Reversals of the e.p.s.p. evoked by optic nerve stimulation and the glutamate evoked depolarization of a principal neurone of the dorsal lateral geniculate nucleus. Records show the e.p.s.p. and glutamate evoked response recorded from the same cell at six different membrane potentials following the intracellular injection of Cs^- ions. Downward deflections on the glutamate response are electronic potentials evoked by a constant hyperpolarizing pulse of current.

A comparison of the reversal potential of the e.p.s.p. evoked by optic nerve stimulation with that of glutamate

Using caesium filled electrodes for the reasons mentioned earlier it proved possible to examine the changes in the e.p.s.p. and the response to the iontophoretic application of glutamate over the membrane potential range $+40$ to -70 mV. in both instances, a clear cut reversal was obtained and the depolarizing and hyperpolarizing potential recorded on each side of this potential were not only symmetrical, but showed no signs of heterogeneity. Although the mean reversal potentials for both events in 11 experiments were not statistically different, in six experiments the reversal level of glutamate was, as shown in Fig 3.8, 10–17 mV more hyperpolarizing than that of the e.p.s.p.

Pharmacological antagonists of the optic nerve evoked e.p.s.p.

As in the earlier experiments on the perforant path evoked e.p.s.p., DGG added to the perfusion media reversibly blocked the optic nerve evoked e.p.s.p. in 98 per cent of the cells tested. This action of DGG was not associated with any change in resting membrane potential, input resistance or excitability of the cell. Results from those cells in which two or more concentrations of DGG were tested, were used to construct dose-response curves and the IC_{50} for DGG antagonism of the e.p.s.p. calculated to be in the 5 mM range. In other neurones stimulus response curves in the absence and presence of DGG applied at the IC_{50} concentration of 4.7 mM, showed DGG to produce a parallel shift of the stimulus-response curve to the right. This result was seen as suggestive evidence for the view that DGG blocked the endogenous transmitter in a competitive manner.

The potency of DGG against the excitatory amino acids was of the same order of magnitude as that against the e.p.s.p., i.e. DGG at a concentration of 4.7 mM was able to reversibly reduce the amplitude of the depolarizations evoked by glutamate and quisqualate by about 50 per cent.

Although D-2-amino-5-phosphonvalerate (APV) applied at a concentration of 10^{-5}M completely blocked the response to NMDA, it had no effect on the optic nerve evoked e.p.s.p.

Even when the D-isomer of 2-amino-4-phosphonobutyrate (APB), another excitatory amino acid antagonist was used at a concentration of 10^{-2}M it was completely inactive on the e.p.s.p. evoked by optic nerve stimulation. However, at a concentration of 5×10^{-3}M in about 20 cells the L-isomer caused a decrease in the membrane resistance of 30–60 per cent without altering the resting membrane potential and this non-specific change was accompanied by a similar reduction in the e.p.s.p.

It is perhaps worthwhile mentioning an attempt on this preparation to examine the possibility that the failure of APV to antagonize the e.p.s.p. is due to the presence of magnesium ions. Indeed, in the majority of experiments the removal of magnesium had no effect on the e.p.s.p. and the insensitivity of the e.p.s.p. to high concentrations of APV was retained. Although in a few experiments removal of the magnesium and enhancement of the calcium concentration to 4 M increased the amplitude of the e.p.s.p. as described by others working on CA1 pyramidal cells of the hippocampus (Coan *et al.* 1985), APV was without effect on the optic nerve e.p.s.p.

An additional Ca^{2+} dependent potential seen only in dorsal lateral geniculate neurones

In the dorsal lateral geniculate nucleus and not in the ventral, hyperpolarization of the membrane potential led, as shown in Fig. 3.9, to an unexpected enhancement of the peak and following phase of the e.p.s.p. due to the development of an additional potential with comparatively slow rising and falling phases. The additional potential was clearly voltage dependent and further hyperpolarization of the cell led to further enhancements, which eventually triggered fast TTX sensitive action potentials. Similarly shaped slow potentials were also evoked by the injection of depolarizing current into the cell under study and by the transient depolarization evoked by an antidromic action potential (Fig. 3.9). The slowly developing potentials were insensitive to TTX (10^{-6}M), and blocked in the presence of 2 mM Co^{2+}, 0.2 mM Cd, by

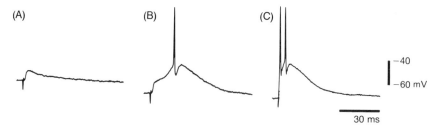

FIG. 3.9. The effect of hyperpolarization on the e.p.s.p. evoked by optic nerve stimulation. (A) and (B) show the e.p.s.p. evoked by constant electrical stimulation of the optic nerve in the dorsal lateral geniculate nucleus at two different membrane potentials −60 and −75 mV. At membrane potentials more hyperpolarizing than −60 mV a slow rising and falling potential is evoked on the peak of the e.p.s.p. which readily reaches the threshold for fast action potentials. (C) At a membrane potential of −75 mV the antidromic spike was also followed by a Ca^{2+} mediated slow potential.

Ca^{2+}-free perfusing media and the intracellular injection of EGTA, and were thus likely to be mediated by a calcium current. Thus, in these cells there appears to be a calcium current which is inactive at the normal resting potential and only becomes activated and ready for triggering during hyperpolarization of the cell under study. Recently, detailed accounts have been published describing low threshold, normally inactivated calcium currents which appeared to be characteristic of thalamic neurones *in vitro* (Llinas & Jahnsen 1982; Jahnsen & Llinas, 1984a,b) and *in vivo* (Steriade & Deschenes 1984).

DISCUSSION

In both preparations the most conclusive evidence in support of an excitatory amino acid as the transmitter of fast synaptic events is the antagonism of the e.p.s.p. by DGG. The present classification of excitatory amino acid receptors (Watkins & Evans 1981) favours the existence of three receptors: the NMDA, the quisqualate and the kainate receptors. NMDA receptors, that are activated by NMDA, are potently and selectively blocked by Mg^{2+} (and other divalent cations) and by APV; quisqualate and kainate receptors are preferentially activated by quisqualate and kainate respectively, and blocked by DGG, *cis*-2,3-piperidine dicarboxylic acid and other antagonists with different potencies and much less selectivity. Thus, the lack of action of APV on the e.p.s.p. indicates that NMDA receptors do not mediate these particular e.p.s.p. and that no NMDA-sensitive component of the optic nerve e.p.s.p. is blocked at physiological Mg^{2+} concentration. Moreover, in the presence of Mg^{2+} the lack of any voltage-sensitive reduction of the e.p.s.p. amplitude in the range -30 to -80 mV would also strongly argue against synaptic NMDA receptors.

The apparently competitive nature of the antagonism evoked by DGG also suggests that the synaptic receptors activated by nerve stimulation in both preparations are of the quisqualate/kainate type. This conclusion is supported in the dorsal lateral geniculate nucleus by the similarity in the IC_{50} of DGG against the e.p.s.p. and the glutamate (and quisqualate) evoked depolarizations, and the lack of action of this antagonist against other non-amino acid excitants. Unfortunately, differentiation between quisqualate and kainate receptors is not possible at the moment because of the lack of a suitable antagonist. However, an IC_{50} in the millimolar range must raise doubts about the value of this approach. Unfortunately, no data for comparison are available from other central neurones, but even in the frog and the immature rat spinal cord, where the IC_{50} is about 5–10 times less, this situation is far from satisfactory.

References

Ault, R., Evans, R. H., Francis, A. A., Oakes, D. J., and Watkins, J. S. (1980). Selective depression of excitatory amino acid induced depolarizations by magnesium ions in isolated spinal cord preparations. *J. Physiol.* **307**, 413–28.

Brown, T. H., Fricke, R. A., and Perkel, D. A. (1981). Passive electrical constants in three classes of hippocampal neurons. *J. Neurophysiol.* **46**, 812–27.

Coan, E. J., Collingridge, G. L., Herron, C. E., and Lester, R. A. J. (1985). Demonstration of a phencyclidine-sensitive N-methyl-D-aspartate-receptor-mediated component of an e.p.s.p. in rat hippocampal slices. *J. Physiol.* **365**, 45P.

Crick, F. (1982). Do dendritic spines twitch? *Trends Neurosci.* **5**, 44–6.

Crunelli, V., Forda., S., Collingridge, G. S., and Kelly, J. S. (1982). Intracellular recorded synaptic antagonism in the rat dentate gyrus. *Nature*, **300**, 450–2.

——, ——, and Kelly, J. S. (1983). Blockade of amino acid-induced depolarizations and inhibition of excitatory post-synaptic potentials in rat dentate gyrus. *J. Physiol.* **341**, 627–40.

——, ——, and —— (1984). The reversal potential of excitatory amino acid action on granule cells of the rat dentate gyrus. *J. Physiol.* **351**, 327–42.

——, Leresche, N., and Pirchio, N. (1985a). Lack of calcium potentials in principal neurones of the rat ventral LGN *in vitro*. *J. Physiol.* **365**, 39P.

——, ——, and —— (1985b). Non-NMDA receptors mediate the optic nerve input to the rat LGN *in vitro*. *J. Physiol.* **365**, 40P.

—— and Mayer, M. L. (1984). Mg^{2+} dependence of membrane resistance increases evoked by NMDA in hippocampal neurones. *Brain Res.* **311**, 392–6.

Dingledine, R. (1983). N-methyl aspartate activates voltage-dependent calcium conductance in rat hippocampal pyramidal cells. *J. Physiol.* **343**, 385–406.

Godfraind, J. M. and Kelly, J. S. (1981) Intracellular recording from thin slices the lateral geniculate nucleus of rats and cats. In *Electrophysiology of Isolated Mammalian CNS Preparations* G. A. Kerkut and H. V. Wheel, pp. 257–84. Academic Press, London.

Haas, H. L., Schaerer, B., and Vosmansky, M. (1979). A simple perfusion chamber for the study of nervous tissue slices *in vitro*. *J. Neurosci. Meth.* **1**, 232–5.

Habblitz, J. J. and Langmoen, I. A. (1982). Excitation of hippocampal pyramidal cells by glutamate in the guinea-pig and rat. *J. Physiol.* **325**, 317–31.

Jahsen, H. and Llinas, R. (1984a). Electrophysiological properties of mammalian thalamic neurones. An *in vitro* study. *J. Physiol.* **349**, 205–26.

—— and —— (1984b). Ionic basis for the electrical activation and the oscillatory properties of thalamic neurones *in vitro*. *J. Physiol.* **349**, 227–47.

Johnston, D. and Brown, T. H. (1984). Biophysics and microphysiology of synaptic transmission in hippocampus. In *Brain Slices* (ed. Dingledine), pp. 51–86. Plenum Press, New York.

Katz, B. (1966). *Nerve, Muscle, and Synapse*. McGraw Hill, New York.

Llinas, R. and Jahnsen, H. (1982). Electrophysiology of mammalian thalamic neurones *in vitro*. *Nature*, **297**, 406–8.

Martin, A. R. (1977). Junctional Transmissional.II. Presynaptic mechanisms. In *Handbook of Physiology*, (eds J. M. Brookhart, and V. B. Mountcastle) Vol. 1, pp. 329–55. Waverley Press, Baltimore.

Mayer, M. L. and Westbrook, G. L. (1984). Mixed-agonist action of excitatory amino acids on mouse spinal cord neurones under voltage clamp. *J. Physiol.* **354**, 29–53.

——, —— and Guthrie, P. B. (1984). Voltage-dependent block by Mg^{2+} of NMDA responses in spinal cord neurones. *Nature*, **309**, 261–3.

Miller, V. P., Rall, W., and Rinzel, V. (1985). Synaptic amplification by active membrane in dendritic spines. *Brain Res.* **325**, 325–30.

Nowak, L., Bregestovski, P., Ascher, P., Herbet, A., and Prochiantz, A. (1984). Magnesium gates glutamate-activated channels in mouse central neurones. *Nature*, **307**, 462–5.

Perkel, D. H. and Perkel, D. J. (1985). Dendritic spines: role of active membrane in modulating synaptic efficiency. *Brian Res.* **325**, 331–5.

Redman, S. and Walmsley, B. (1983a). Amplitude fluctuations in synaptic potentials recorded in cat spinal motoneurones at identified group la synapses. *J. Physiol.* **343**, 135–45.

—— and —— (1983b). The time course of synaptic potentials evoked in cat spinal motoneurones at identified group la synapses. *J. Physiol.* **343**, 117–33.

Steriade, M. and Deschenes, M. (1984). The thalamus as a neuronal oscillator. *Brain Res. Rev.* **8**, 1–63.

Takeuchi, A. (1977). Junctional transmission. I. Postsynaptic mechanisms. In *Handbook of Physiology*, (eds J. M. Brookhart and V. B. Mountcastle) Vol 1, pp. 295–327. Baltimore: Waverley Press.

Thomson, A. M., West, D.C., and Lodge, D. (1985). An N-methylaspartate receptor-mediated synapse in rat cerebral cortex: a site of action of ketamine? *Nature*, **313**, 479–81.

Watkins, J. C. and Evans, R. H. (1981). Excitatory amino acid transmitters. *Ann. Rev. Pharmacol. Toxicol.* **21**, 165–204.

4

Fast and slow chemical signalling in the visual cortex: an evaluation of GABA- and neuropeptide-mediated influences

A. M. SILLITO

GENERAL INTRODUCTION

The mammalian visual cortex is involved in the fast processing of those aspects of the retinal input utilized to generate the internal model of the visual world which constitutes the basis to visual perception. It is virtually self-evident that this requires rapid transitions in cell activity patterns in order to reflect the dynamics of the image flux on the retina. There is evidence to show that many of the stimulus selective response properties of visual cortical cells, relevant to visual perception, are secondary to synaptic inputs mediated by amino acids. However, there is also evidence to show that peptide-containing neurones form a significant component of the circuitry in the visual cortex. Judged from the viewpoint of visual processing, the function of the potentially slow peptide mediated events in circuits generating response selectivity is less clear. On the other hand, some of the peptide containing cells may have a special role to play in the context of the control of cortical blood flow and metabolism, and this is clearly commensurate with slow signals. In the following account I shall firstly illustrate the role of a fast amino acid-mediated process by reference to the influence of GABAergic mechanisms, and then consider the available evidence for the role of the various peptides.

Implicit in the experimental approach outlined here, is the assumption that the specific stimulus response properties of visual cortical cells reflect the operation of the synaptic circuitry in which they are embedded. Thus, a perturbation in a component of the circuitry influencing a particular cell will be reflected by a corresponding perturbation in some aspect of its visual response properties. The unique thing about the visual system in this respect, is that it is possible to document these response properties, using computer controlled stimuli, with a speed and precision that is virtually unobtainable for any other higher brain system. This applies to even quite complex and subtle attributes of the 'visual response properties'. It is important to appreciate how distinct the properties of visual cortical cells are from those of the input cells in the dorsolateral geniculate nucleus (dLGN). Whilst dLGN cells have concentric receptive fields, are well activated by flashing spots of light and respond to an elongated contour (such as a light dark edge or a bar of light) moving over their receptive field in any orientation or direction, cortical cells are much more selective (Hubel and Wiesel 1962). They are, for example, generally best activated by a moving contour passing over their receptive field, respond only to a limited range of orientations of the contour (orientation selectivity) and often show preference to one of the two directions of motion at a particular orientation (directional selectivity). Other facets of the response selectivity of visual cortical cells that are notably different from dLGN cells include length preference, disparity tuning, velocity preference, and spatial and temporal frequency tuning (Hubel and Wiesel 1962, 1968; Movshon 1975; Poggio and Fischer 1977; De Valois and Tootell 1983). The issue here is that each of these components of visual cortical cell selectivity derive from specific synaptic interactions pertaining in the cortex. There is already good evidence to support the view that for example orientation, direction and length preference involve different mechanisms (Hubel and Wiesel 1968; Sillito 1979, 1985; Hammond 1981; Ganz and Felder 1984). Insight into the nature of the transmitter involved in a component of the cortical circuitry introduces the possibility of manipulating it pharmacologically and assessing the impact of this manipulation on the receptive field organization. The evidence detailed in this chapter involves the use of micro-iontophoretic techniques to produce a localized manipulation of components of the circuitry in the visual cortex and utilizes the evidence from this to evaluate the functional role of the circuitry in question. Whilst the conclusions are phrased in terms of the functions of the visual cortex, the similarities between the synaptic organization of the visual cortex and other neocortical areas, supports the view that they can be generalized to enhance our understanding of chemical signalling throughout the neocortex.

GABAERGIC PROCESSES

The evidence for the role of GABA in the visual cortex is now very strong. It is released in a calcium dependent fashion (Iversen *et al.* 1971), exerts a powerful inhibitory effect on all types of visual cortical cell, and iontophoretic application of the GABA antagonist bicuculline blocks visually elicited inhibitory effects (Sillito 1975a,b, 1984). Recent evidence has shown that neurones considered to be inhibitory on the basis of

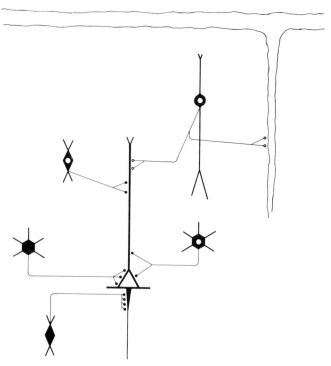

FIG. 4.1 Diagram summarizing the main types of GABAergic and peptidergic cells in the visual cortex. All the connections are shown with reference to synaptic contacts made on a pyramidal cell (cell body shown as open pyramid). This receives inhibitory input from two types of exclusively GABAergic cell (solid cell body), one a chandelier cell to the initial segment, the other a basket cell to the cell body (but this may also contact proximal and distal dendrites). Inputs to the cell body and dendrites also derive from cells where GABA is coexistent with a peptide, these include fusiform cells (open spindle for cell body) and short axon multipolar cells (open hexagon for cell body). These latter cells may contain GABA in association with either CCK or SSt and NPY (see text for further detail). A final group of cells, the bipolar cells (cell body open circle), contain the peptide VIP or possibly in some circumstances CCK, but not GABA; they make synaptic contacts with dendrites and probably also blood vessels. All the GABA containing cells appear to make the type II synaptic contacts associated with inhibitory function, these are shown as closed terminals, whilst the non-GABAergic bipolars can apparently make either type I or type II synapses and as a group these are shown as open terminals on the diagram.

morphological criteria (Gray 1963; Le Vay 1973) show GABA uptake and GAD immunoreactivity (Ribak 1978; Somogyi *et al*. 1981, 1984). The distribution and source of inhibitory synapses on visual cortical cells is interesting because it shows evidence for a range of distinct types of GABAergic influence. Inhibitory synapses are most densely aggregated in the vicinity of the cell body and proximal dendrites, although there are significant inputs to the more remote dendritic processes (Le Vay 1973; Peters and Regidor 1981; Martin *et al*. 1983; Somogyi *et al*. 1983). The diagram in Fig. 4.1 presents a summary of some of the types of inhibitory input converging on a pyramidal cell in the visual cortex. Inputs reach the cell body and proximal dendrites from basket cells and short axon multipolar cells (Peters and Regidor 1981; Martin *et al*. 1983; Somogyi *et al*. 1983), the initial segment from chandelier cells (Somogyi 1977) and the distal parts of apical and basal dendrites from bitufted cells and basket cells (Peters and Regidor 1981; Somogyi and Cowey 1981; Martin *et al*. 1983). The variation in the pattern of axonal and dendritic arborization of these inhibitory cells means they will both integrate differing patterns of input and distribute the resulting output over differing spatial domains. This in turn suggests that the complex visual response properties of the cells will include components generated by differing types of inhibitory mechanisms, a view which is supported by a wide range of data extending from receptive field analysis (e.g. De Valois and Tootell 1983; Ganz and Felder 1984), to intracellular studies (Creutzfeldt *et al*. 1974; Innocenti and Fiore 1974) and neuropharmacological evidence (Sillito 1975a,b 1977, 1979, 1984; Sillito and Versiani 1977; Tsumoto *et al*. 1979). I shall here provide

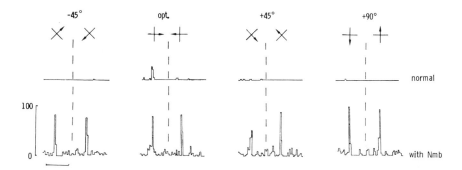

FIG. 4.2. The responses of a simple cell to a bar of light moving backwards and forwards over the receptive field at a range of different orientations, before and during the iontophoretic application of the GABA antagonist N-methyl bicuculline (Nmb) to the cell. Each record is a peristimulus time histogram averaged from 25 cycles of the stimulus motion. Orientation and direction of motion of the bars is shown by the crossed arrow above each record. The actual orientation is shown in degrees clockwise (+) or anticlockwise (−) to the optimum for the cell (opt). Prior to drug application (upper records) the cell responds to only the optimum orientation and to only direction of motion at that orientation. During drug application (lower records), the cell responds to all testing orientations and both directions of motion at each orientation. Bin size 50 ms, vertical calibration indicates range corresponding to 0-100 counts/bin, horizontal calibration 1 s.

examples of the neuropharmacological evidence concerning the role of inhibitory mechanisms in directional selectivity and orientation selectivity.

The records in Fig. 4.2 show the averaged responses of a simple cell in layer IV of the visual cortex to the motion of a bar of light over the receptive field at a range of different orientations. It is apparent from the records taken in the control situation, that the cell only responds to one of the testing orientations (it is orientation selective), and to only one of the two directions of motion at the effective orientation (it is directionally selective). During iontophoretic application of the GABA antagonist bicuculline, three changes occur in the cell's responsiveness: the response magnitude increases, the orientation selectivity is lost, and the directional selectivity is lost. This type of observation is typical for simple cells. The effects of the bicuculline are reversible and correlate with a blockade of the action of iontophoretically applied GABA (Sillito *et al*. 1980; Sillito 1984). The loss of the receptive field selectivity occurs at response magnitudes below that involving a response saturation; moreover, it is not secondary to a simple increase in excitability. Increasing cell excitability and background discharge to the levels seen with bicuculline, by application of an excitatory amino acid, does not cause a loss of response selectivity. Paradoxically, such an increase in background discharge level without concomitant GABAergic blockade, can actually enhance visually elicited inhibitory effects, possibly because of recurrent collateral feedback to the inhibitory interneurones (Sillito 1979).

The implication of the bicuculline induced changes in response selectivity seen in Fig. 4.2 is that GABAergic mechanisms are essential to the generation of that selectivity. This structuring of the response selectivity via the inhibitory mechanism requires that the spatial distribution of the zones in visual space driving the inhibitory input to the cell are such that stimuli moving in the preferred direction, or at non-optimal orientations, activate these zones and elicit a 'wave' of inhibition that pre-empts the capability of the cell to respond to its excitatory input. An optimal stimulus, on the other hand, although eliciting some component of drive from these zones, does not maximally activate them, and hence the cell is not prevented from responding to its excitatory input. The fact that optimal stimuli do often invoke inhibitory inputs (Creutzfeldt *et al*. 1974) is important because it accounts for the increase in response magnitude to optimal stimuli during inhibitory blockade (Sillito 1975a,b). The point at issue is that non-optimal stimuli evoke inhibitory inputs which greatly attenuate or block the capability of the excitatory input to drive the cell, whilst optimal stimuli evoke a weaker input. Inspection of orientation tuning curves shows that the attenuation of the response from the optimum declines gradually from the optimum, suggesting a progressive attenuation of responsiveness (e.g., Sillito 1979,

1984). Much of what is required here can be achieved by a simple partially shifted overlap of circular inhibitory and excitatory fields (Heggelund 1981a,b; Sillito 1984), basically a variant of 'classical' lateral inhibition; although this arrangement may only be a first level of interaction in a process which progresses through several levels, with feedback between each (Sillito 1984, 1985).

The range of changes in response selectivity evoked by bicuculline application is not uniform across the population of visual cortical cells. This is documented in Fig. 4.3, which compares the action of the drug on the directional selectivity of a simple cell, a complex cell, and a superficial layer cell with strong length preference ('hypercomplex'). The drug abolishes or greatly reduces the selectivity of the simple cell and the complex cell, but has no effect on that of the 'hypercomplex cell', despite causing a very significant increase in response magnitude (see Sillito 1977

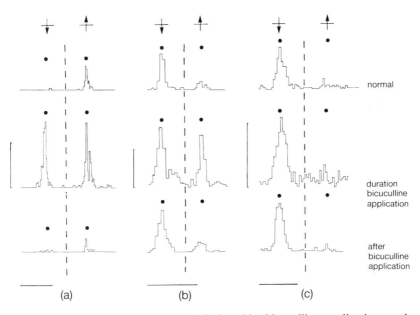

FIG. 4.3. Action of inhibitory blockade induced by bicuculline application on the directional selectivity of three different types of visual cortical cell. Peristimulus time histograms show the response to an optimally oriented bar in each case, which is arbitrarily represented as horizontal by the crossed arrows above the records. Dots above the records show points at which the bar crosses the receptive field of the cell. (a) Simple cell, 20-ms bins. (b) Complex cell, 50-ms bins. (c) Hypercomplex cell, 50-ms bins. Vertical calibration range corresponding to 0–100 counts/bin. Horizontal calibration 1 s. Bicuculline application modifies the directional selectivity of the simple and complex cell, but not the hypercomplex cell.

for further discussion). This variation in action on directional selectivity across the cortical cell population also applies to orientation selectivity. Only approximately half of the complex cells sampled showed a loss of orientation selectivity during bicuculline application, the remainder exhibiting a reduction in selectivity but retaining a bias to the original optimum (Sillito 1979). This data is consistent with the view that some complex cells receive an intracortically-mediated excitatory drive only, which may be presumed to be orientation biased, whilst others receive direct geniculate input (e.g. Martin and Whitteridge 1984). One alternative possibility that needs to be considered, is that the variation in effect on selectivity reflects a variation in the effective drug distribution at the relevant inhibitory synapses. This cannot be excluded, but needs to be considered in the context of several additional comments. In all experiments the data were only considered valid when we observed that the bicuculline application had produced a block of the effect of application of GABA, with an ejection current of sufficient magnitude to stop the visually driven response of the cell to an optimal stimulus. Hence, the blockade achieved was from the pharmacological viewpoint, uniform across the sample. Further to this, it was often possible to demonstrate a loss of one aspect of cell selectivity despite the fact that another was apparently resistant or only partially affected. Thus, some complex cells showed a complete loss of directional selectivity, but only a partial reduction in orientation selectivity (Sillito 1979). This suggests that in these instances the drug was effectively blocking at least some of the inhibitory synapses acting on the cell. In fact, for a multibarrelled micropipette positioned to effectively record cell activity, it seems likely that the drug would be adequately distributed to those synapses in the vicinity of the cell body and proximal dendrites, but unlikely that it would be adequately distributed to inhibitory synapses on the more remote apical dendrites of pyramidal cells. Consequently, although the data is ambiguous, it may be taken to reflect either that the contribution of the excitatory mechanism to the facets of cell selectivity in question varies, or that the inhibitory synapses mediating different aspects of this response selectivity have a differential distribution over the cell body and dendrites. Clearly from what is known, both these possibilities are likely to occur. The data may in this sense then accurately reflect a variation in the mechanisms processing particular receptive field characteristics, but not delineate the cause of the variation. However, what is clear is that GABAergic mechanisms play a major role in establishing the response selectivity of visual cortical cells and that virtually all aspects of response selectivity can be shown to be dependent on or strongly enhanced by such processes at some level in the visual cortex (see Sillito 1984 for further discussion).

PEPTIDES THAT ARE COEXISTENT WITH GABA

The peptide containing cells in the visual cortex can be broadly subdivided into two groups, those where the peptide appears to be coexistent with GABA and those where it is not. The presently available evidence suggests that somatostatin (SSt), cholecystokinin (CCK) and neuropeptide Y (NPY) are found in cells that show immunoreactivity to glutamic acid decarboxylase (GAD) (MacDonald *et al.* 1982a, 1982c; Morrison *et al.* 1983; Hendry *et al.* 1984a, b). Whilst the cells that contain CCK are distinct from those that contain SSt, it seems that NPY may occur in the SSt cells (Hendry *et al.* 1984a,b). Not all the GABAergic cells appear to contain peptides, two groups in particular, the basket cells and chandelier cells do not show immunoreactivity to any of the currently identified peptides (Hendry *et al.* 1983). The peptide containing GABAergic cells fall into the multipolar and bitufted cell categories (see Fig. 4.1). This is of interest because the two groups of GABAergic cells likely to exert a particularly potent control over cell activity as a consequence of their heavy input to the cell body and initial segment, seem to lack the 'co-transmitter' capable of mediating 'slow synaptic events'. Thus, following from the suggestion that the different types of inhibitory interneurone mediate different facets of receptive field organization, it seems likely that we might anticipate some aspects of this organization to be subject to a 'slow even modulation' whilst others are not. In attempting to provide a preliminary overview of the possible action of SSt and CCK in terms of cortical processing, we have examined the effect of the iontophoretic application (and in some cases pressure ejection) of these drugs in relation to the following questions.

(1) Is response magnitude to an optimal visual stimulus increased or decreased?

(2) Where a change in response magnitude occurs does it result in an increase or a decrease in the signal to noise level?

(3) Utilizing directional selectivity as an example, can we ascertain any evidence for a modulation of a natural, stimulus specific GABAergic input?

(4) Does iontophoretically applied SSt modulate the effect of iontophoretically applied GABA?

(5) Where effects are seen do they correlate with cell type as judged from receptive field classification or cortical lamination?

The logic underlying these questions is interlinked and draws on a number of simple hypotheses. Firstly, we have assumed that the peptide may modulate the effect of GABA either by enhancing its effectiveness following from the model established in the periphery for vasoactive intestinal polypeptide (VIP) and acetylcholine (ACh) (Lundberg *et al.* 1982) or possibly by decreasing its effectiveness. If the effect of SSt is secondary to a modulation of a GABAergic process, it might be presumed to either increase or decrease the overall responsiveness of a cell to optimal stimuli as a result of a modulation of any GABAergic influences elicited by these stimuli (see above). Further to this, where a cell is directionally selective and that selectivity follows from a GABAergic input initiated by stimulus motion in the non-preferred direction, then an SSt induced modulation of the synaptically-released GABA might be expected to either increase or decrease the directional selectivity. The argument here is that the effect of the modulation will be on the component of the visual response most affected by a GABAergic input, viz the response in the non-preferred direction. This assumes that any potentiation of the GABAergic effect is not such as to increase the duration of the effect of the input initiated in the non-preferred direction to a point where it extends as an equal magnitude influence into the time domain occupied by the stimulus motion in the preferred direction. An assumption of this type is virtually obligatory if the modulation following from the 'natural' release of the peptide is to be seen in terms of an effect on the stimulus specific responses of a cell. Thus, the peptide would potentiate the action of GABA in the context of the time period of the fast stimulus dependant release, without extending it beyond this. This does not exclude the possibility that the duration of the peptide effect in terms of its capability to potentiate GABA action should not be extended for a period considerably beyond that of the fast responses, but simply that the potentiation of the GABA effect would not increase the duration of the GABA action. Whilst it is not self evident (see below) that the effects of the peptides associated with the GABAergic cells must be reflected in terms of an action on stimulus specific responses, it is clear that this is an issue that should be tested.

The data we have obtained for the action of SSt on the responses of visual cortical cells is consistent with the view that it is potentially capable of modulating their responses, but does not lead to any insight into the way this is achieved or its functional significance (Salt and Sillito 1984). In our sample of 40 cells, 15 showed a facilitation of visual responses during SSt application, 14 a depression of visual responses, and 11 no change. In some cases these effects were marked, in others they were small and variable. An example is shown in Fig. 4.4 where iontophoretic application of SSt produces a marked enhancement of the visually elicited responses of a simple cell. In the control and recovery situation this cell exhibited a weak, but directionally selective response to an optimally oriented stimulus. During drug application, the response to the preferred direction was

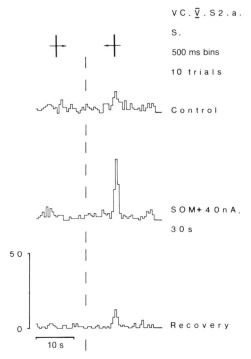

FIG. 4.4. Action of somatostatin (SOM) on the responses of a simple type cell to an optimally oriented bar moving over the receptive field. In the control situation the cell was poorly responsive to even its optimal stimulus, during drug application there is a marked increase in the response magnitude in the preferred direction of motion. The effect is reversible. The drug record was taken 30 seconds after the onset of SOM application with a 30 nA ejecting current. Vertical calibration indicates range corresponding to 0–50 counts/bin. Bin size 500 ms. Horizontal calibration 10 s. Peristimulus time histograms averaged for 10 trials.

greatly increased but there was no significant response in the non-preferred direction. Several factors are of note here; firstly that the SSt potentiates cell responsiveness, which seen in terms of an interaction with GABAergic processes would suggest that the effect of these was diminished. Secondly, that the directional selectivity of the simple cell was arguably enhanced, certainly not diminished by the drug application. As preceding evidence has shown that antagonism of GABAergic mechanisms invariably eliminates directional selectivity in simple cells (Sillito 1975a, 1977), an action of SSt that served to diminish the effects of GABAergic mechanisms would be expected to reduce directional selectivity. Thus, the effects of SSt on this cell are not consistent with any simple logic following from a modulation of GABAergic mechanisms. To date, we have been unable to isolate any clear association between the overall action of SSt on cell responsiveness (i.e. whether visual responses are enhanced or reduced) and an appropriate change in response selectivity reflecting the operation of a visually evoked GABAergic input (directional selectivity being the example chosen here). Similarly, we have been unable to detect any obvious modulation of the effects of iontophoretically applied GABA, other than that which would follow from a simple additive or subtractive interaction with the facilitatory or depressive effects of SSt. Although we have studied a much smaller population of cells (11) the pattern of effect with CCK seems to be very similar to that obtained with SSt.

Previous evidence has shown SSt to have both facilitatory and inhibitory effects (Renaud *et al*. 1975; Ioffe *et al*. 1978; Dodd and Kelly 1981; Pittman and Siggins 1981) and in this sense the present data are entirely consistent with what would be expected. The difficulty lies in relating this to any facet of present knowledge regarding the function of synaptic processes in the visual cortex. A not insignificant component of the problem is that of deciding precisely what facet of the visual response profile of cortical cells might be expected to be subject to a functionally relevant 'slow' modulation. This is not immediately clear and the present data seem to underline the fact that the answer is unlikely to lie in a simple global modulation of 'fast' GABAergic processes relevant to the more immediately accessible components of receptive field organization. However, several issues merit consideration in relation to the further analysis of the situation. Firstly, it is notable that the effects of GABA on cell bodies and dendrites of both neocortical and hippocampal cells are qualitatively different, involving a hyperpolarization on the one hand and a depolarization on the other (e.g. Anderson *et al*. 1980; Kemp 1984); with the caveat that the dendritic effect might mediate a localized discriminative inhibition whilst at the same time even enhancing the cell's responsiveness to other excitatory inputs. Clearly, an SSt induced enhancement of the dendritic effect of GABA might both enhance cell responsiveness and potentiate the 'discriminative inhibition'. This is not incompatible with some of the data reported here and could account for the type of observation shown in Fig. 4.4. Following from this, the depressive effects of SSt might relate to actions that are dominated by the cell body as opposed to the dendritic component of GABAergic effects. Thus, the apparent disassociation of the logic of the SSt effects in terms of GABAergic mechanisms might follow from interactions with two distinct components of the GABAergic control of cortical cells, with possibly equally balanced, and hence mixed effects on some cells. Certainly, this could account for some of the more confusing aspects of our data. There is evidence to support the view that there may be two distinct populations of SSt receptors (Delfs and Dichter 1983; Reubi 1984) and it would be very interesting to learn more about their distribution and functional effects on cells. The question of the potential SSt interaction with GABA needs to also take note of the possibility that it be phrased in terms of the GABA-B receptor, rather than (or as well as), the GABA-A receptor. At present we have no real insight into the likely functional role of effects mediated by the GABA-B receptor, but it would be very relevant to check for a modulation of the effects of baclofen by SSt. Further to this, in consideration of the recent evidence suggesting that the SSt cells may also contain neuropeptide Y (Hendry *et al.* 1984a,b), the whole question of the action of SSt has to be approached from the viewpoint that functionally relevant effects could require the combined presence of SSt, NPY and GABA.

PEPTIDES THAT ARE NOT COEXISTENT WITH GABA: VIP AND THE CHOLINERGIC MODULATION OF CORTICAL FUNCTION

The VIP containing bipolar cells (MacDonald *et al.* 1982b; Morrison *et al.* 1984) constitute the main group of peptide cells lacking GAD immunoreactivity, although there may be a small subgroup of CCK bipolars that also fall into this category (Hendry *et al.* 1983). Both these groups of bipolars are distinguished by the fact that they appear to make synaptic contact with blood vessels as well as other cortical cells. In the rat it seems that the VIP cells may also be cholinergic (Eckenstein and Baughman 1984), although to date it appears that the cat visual cortex does not contain cholinergic cells (Stichel and Singer 1985). Present evidence supports a role for the VIP bipolars in the control of neocortical blood flow and metabolism (Magistretti *et al.* 1981; Lee *et al.* 1984; Morrison *et al.* 1984). Taking this evidence, together with the fact that the VIP bipolars show a relatively sparse, albeit uniform, distribution through the neocortex (Morrison *et al.* 1984), it seems unlikely that their synaptic interactions with other cortical cells would be central to the mechanisms elaborating receptive field selectivity. Thus, there are grounds for considering that they may exert a modulatory, rather than a stimulus specific influence on other cortical cells. We have examined this possibility by checking the influence of iontophoretically applied VIP on the visually driven responses of visual cortical cells. In addition, following the evidence that VIP is known to modulate the effects of ACh in the periphery, we have checked for a possible interaction between these two substances in the visual cortex, utilizing either ACh or the muscarinic agonist methacholine.

Iontophoretic application of VIP to cortical cells was observed to produce changes in the visual responsiveness of the majority of cells tested (15/17). The effects involved either a facilitation of responses or a

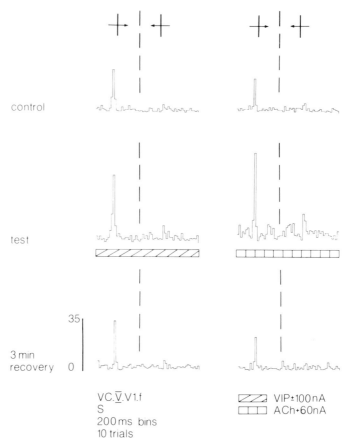

FIG. 4.5. A comparison of the effects of iontophoretic application of vasoactive intestinal polypeptide (VIP) and ACh on the responses of a simple cell to the motion of an optimally oriented bar of light over its receptive field. In the test situation both ACh and VIP cause an increase in the cells directionally selective response. Drug type indicated by patterned bar under the records. Vertical calibration indicates range corresponding to 0–35 counts/bin. Bin size 200 ms. Peristimulus histograms averaged for 10 trials. VIP ejection current from one barrel of 100 nA, balanced by opposite polarity current of the same magnitude applied through a second VIP barrel.

depression and were repeatable for any given cell (Grieve *et al.* 1985). In terms of both the time course and the nature of the action on stimulus specific responses they were directly comparable to the previously reported observations on the effects of ACh on visual cortical cells (Sillito and Kemp 1983). Of particular interest is that within our present sample ACh and VIP exerted the same polarity of effect on the majority of cells tested

(11/15). Thus, where ACh or methacholine facilitated visual responses, VIP also exerted a facilitatory effect, and an example of this is shown in Fig. 4.5. A summary of the presently available data regarding the correlation of effect is given in Fig. 4.6. When both substances were applied together the effects were additive in most cases. This raises the possibility that there may be an interaction between the extrinsic cholinergic modulatory input to the visual cortex and the VIP containing bipolars. At present, we cannot exclude the possibility that the action of VIP is, itself, secondary to a modulation of the effect of the endogenous ACh released by the cholinergic input and experiments are in progress to resolve this issue.

These observations should be judged in the context of the pattern of effect previously reported for ACh, where cells facilitated by ACh were observed through all cortical layers, whilst those inhibited by ACh were restricted to the lamina III/IV region (Sillito and Kemp 1983), suggesting a very specific modulation of the neuronal organization of the visual cortex by the cholinergic input. The facilitatory effects of ACh, where the high

Correlation of cholinergic effects with those of VIP

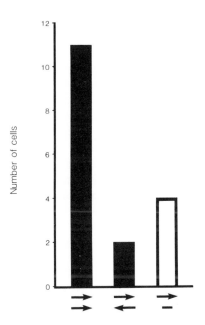

FIG. 4.6. Histogram showing the number of cells where iontophoretic application of ACh (or methacholine) had the same effect as VIP (arrows in the same direction), opposite effects (arrows in opposite directions) or where only one of the drugs was effective (one arrow only). See text for further discussion.

frequency stimulus specific responses are strongly enhanced, are consistent with present evidence suggesting that ACh modulates the voltage and calcium dependent potassium channels (Krnjevic *et al.* 1971; Halliwell and Adams 1982). The facilitatory effects of VIP on visual responses are very similar and suggest that VIP may either act directly on these mechanisms or as suggested above, modulate the action of ACh. The possibility of an interaction between ACh and VIP in the cortex has a number of potentially important functional implications. The cholinergic input to the cortex, originating in the vicinity of the nucleus basalis of Meynert, appears to be topologically organized (Lamour *et al.* 1982; Price and Stern 1983; Saper 1984). This suggests that there is a potential for activation of discrete cortical regions in a way in which, as far as the facilitatory effects of ACh are concerned, it would increase the signal to noise ratio, and could be an important part of the neocortical component of selective attention. Certainly, changes in visual cortical cell responsiveness in the arousal from sleep can mirror those produced by ACh (Livingstone and Hubel 1981; Sillito and Kemp 1983). If the cholinergic input produces a global change in responsiveness in a given neocortical region, then the sparse, but relatively uniform mosaic of VIP bipolars (Morrison *et al.* 1984) would be in a unique position to modulate this in terms of the specific distribution of activity over the cortical region generated by the 'fast processing' of the particular event engaging that region. The point being that the bipolar cell dendrites sample the particular region of the cortical mosaic in which they lie and thus it can be assumed that the output of the cell is likely to reflect the activity of that region. Moreover, the bipolars appear to have the capability to translate this into actions on blood flow and metabolism (Magistretti *et al.* 1981; Lee *et al.* 1984). In essence I am suggesting a local modulation of an extrinsic modulatory influence, that serves to enhance the match between the cortical processes modulated and the specific patterns of activity involved in those processes. The interactions between these different levels of modulatory influence may have a bearing on our understanding of the deficit that develops in Alzheimer's disease, a disease which involves amongst other things damage to the extrinsic cholinergic input to the neocortex.

Acknowledgements

I am grateful for the help of Dr K. L. Grieve, Dr J. A. Kemp, Dr P. C. Murphy, and Dr T. E. Salt who have all contributed to parts of the data and ideas discussed above. The work would not have been possible without the ongoing support of the Medical Research Council and Wellcome Trust.

References

Anderson, P., Dingledene, R., Gjerstad, L., Langmoen, I. A., and Laursen, A. M. (1980). Two different responses of hippocampal pyramidal cells to application of Gamma-aminobutyric acid. *J. Physiol.* **305**, 279–96.

Creutzfeldt, U., Kuhnt, O. D., and Benevento, L. A. (1974). An intracellular analysis of visual cortical neurones to moving stimuli: responses in a co-operative neuronal network. *Exp. Brain Res.* **21**, 251–74.

Delfs, J. R. and Dichter, M. A. (1983). Effects on somatostatin on mammalian cortical neurons in culture; physiological actions and unusual dose response characteristics. *J. Neurosci.* **3**, 1176–88.

De Valois, K. K. and Tootell, R. B. H. (1983). Spatial frequency specific inhibition in cat striate cortex cells. *J. Physiol.* **336**, 359–76.

Dodd, J. and Kelly, J. S. (1981). The actions of cholecystokinin and related peptides on pyramidal neurones of the mammalian hippocampus. *Brain Res.* **205**, 337–50.

Eckenstein, F. and Baughman, R. W. (1984). Two types of cholinergic innervation in the cortex, one co-localised with vasoactive intestinal polypeptide. *Nature* **309**, 153–5.

Ganz, L. and Felder, R. (1984). Mechanism of directional selectivity in simple neurons of the cat's visual cortex analysed with stationary flash sequences. *J. Neurophysiol.* **51**, 294–324.

Gray, E. G. (1963). Electron microscopy of presynaptic organelles of the spinal cord. *J. Anat.* **97**, 101–6.

Grieve, K. L., Murphy, P. C., and Sillito, A. M. (1985). An evaluation of the role of CCK and VIP in the cat visual cortex. *J. Physiol.* **365**, 42P.

Halliwell, J. V. and Adams, P. R. (1982). Voltage-clamp analysis of muscarinic excitation in hippocampal neurons. *Brain Res.* **250**, 71–92.

Hammond, P. (1981). Simultaneous determination of directional tuning of complex cells in cat striate cortex for bar and texture motion. *Exp. Brain Res.* **41**, 364–9.

Heggelund, P. (1981a). Receptive field organization of simple cells in cat striate cortex. *Exp. Brain Res.* **42**, 89–98.

—— (1981b). Receptive field organization of complex cells in cat striate cortex. *Exp. Brain Res.* **42**, 99–107.

Hendry, S. H. C., Jones, E. G., and Beinfeld, M. C. (1983). CCK immunoreactive neurons in rat and monkey cerebral cortex make symmetric synapses and have intimate associations with blood vessels. *Proc. Nat. Acad. Sci.* **80**, 2400–3.

——, ——, De Felipe, J., Schmecheal, D., Brandon, C., and Emson, P. C. (1984a). Neuropeptide containing neurons of the cerebral cortex are also GABAergic. *Proc. Nat. Acad. Sci.* **81**, 6520–30.

——, ——, and Emson, P. C. (1984b). Morphology, distribution, and synaptic relations of somatostatin and neuropeptide Y immunoreactive neurons in rat and monkey neocortex. *J. Neurosci.* **4**, 2497–517.

Hubel, D. H. and Wiesel, T. N. (1962). Receptive fields, binocular interaction and functional architecture in the cat's visual cortex. *J. Physiol.* **160**, 106–54.

—— and —— (1968). Receptive fields and functional architecture of monkey striate cortex. *J. Physiol.* **195**, 215–43.

Innocenti, G. M. and Fiore, I. (1974). Postsynaptic inhibitory components of the response to moving stimuli in area 17. *Brain Res.* **80**, 122–6.

Ioffe, S., Haulicek, V., Friesen, H., and Chernick, V. (1978). Effect of somatostatin (SRIF) and L-glutamate on neurons of the sensorimotor cortex in the awake halo-tuated rabbits. *Brain Res.* **153**, 414–8.

Iversen, L. L., Mitchell, J. F., and Srinivasan, V. (1971). The release of gamma-amino butyric acid during inhibition in the cat visual cortex. *J. Physiol.* **212**, 519–34.

Kemp, J. A. (1984). Intracellular recordings from rat visual cortical cells *in vitro* and the action of GABA. *J. Physiol.* **349**, 13P.

Krnjevic, K., Pumain, R., and Renaud, L. (1971). The mechanism of excitation by acetylcholine in the cerebral cortex. *J. Physiol.* **215**, 247–268.

Lamour, Y., Dutar, P., and Jobert, A. (1982). Spread of acetylcholine sensitivity in the neocortex following lesions of the nucleus basalis. *Brain Res.* **252**, 377–81.

Lee, T. J-F., Saito, A., And Berezin, I. (1984). Vasoactive intestinal polypeptide-like substance: the potential transmitter for cerebral vasodilation. *Science*, **224**, 898–901.

Le Vay, S. (1973). Synaptic patterns in the visual cortex of the cat and monkey, Electron microscopy of Golgi preparation. *J. Comp. Neurol.* **150**, 53–86.

Livingstone, M. S. and Hubel, D. H. (1981). Effects of sleep and arousal on the processing of visual information in the cat. *Nature*, **291**, 554–61.

Lundberg, J. M., Anggard, A., and Fahrenkrug, J. (1982). Complementary role of vasoactive intestinal polypeptide (VIP) and acetylcholine for cat submandibular gland blood flow and secretion III. Effects of local infusions. *Acta Physiol. Scand.* **114**, 329–37.

MacDonald, J. K., Parnavelas, J. G., Karamanlidis, A. N., Brecha, N., and Koenig, J. I. (1982a). The morphology and distribution of peptide-containing neurons in the adult and developing visual cortex of the rat, I. Somatostatin. *J. Neurocytol.* **11**, 809–24.

——, ——, ——, and —— (1982b). The morphology and distribution of peptide containing neurones in the adult and developing visual cortex of the rat, II. Vasoactive intestinal polypeptide. *J. Neurocytol.* **11**, 825–37.

——, ——, ——, Rosenquist, G. and Brecha, N. (1982c). The morphology and distribution of peptide containing neurons in the adult and developing visual cortex of the rat. III. Cholecystokinin. *J. Neurocytol.* **11**, 881–95.

Magistretti, P. J., Morrison, J. H., Shoemaker, W. J., Sapin, V., and Bloom, F. E. (1981). Vasoactive intestinal polypeptide induces glycogenolysis in mouse cortical slices: a possible regulatory mechanisms for the local control of energy metabolism. *Proc. Nat. Acad. Sci.* **78**, 6535–9.

Martin, K. A. C. and Whitteridge, D. (1984). Form, function and intracortical projections of spiny neurones in the striate visual cortex of the cat. *J. Physiol.* **353**, 463–504.

——, Somogyi, P., and Whitteridge, D. (1983). Physiological and morphological properties of identified basket cells in the cat's visual cortex. *Exp. Brain Res.* **50**, 193–200.

Morrison, J. H., Benoit, R., Magistretti, P. J., and Bloom, F. E. (1983). Immunohistochemical distribution of pro-somatostatin related peptides in cerebral cortex. *Brain Res.* 262, 344–351.

——, Magistretti, P. J., Benoit, R., and Bloom, F. E. (1984). The distribution and morphological characteristics of the intracortical VIP positive cell: An immunohistochemical analysis. *Brain Res*. **292**, 269–82.

Movshon, J. A. (1975). The velocity tuning of single units in cat striate cortex. *J. Physiol*. **249**, 445–68.

Peters, A. and Regidor, J. (1981). A reassessment of the forms of non-pyramidal neurons in area 17 of cat visual cortex. *J. Comp. Neurol*. **203**, 685–716.

Pittman, Q. J. and Siggins, G. R. (1981). Somatostatin hyperpolarises hippocampal pyramidal cells *in vitro*. *Brain Res*. **221**, 402–8.

Poggio, G. F. and Fischer, B. (1977). Binocular interaction and depth sensitivity in striate and prestriate areas of behaving rhesus monkey. *J. Neurophysiol*. **40**, 1392–405.

Price, J. L. and Stern, R. (1983). Individual cells in the nucleus basalis-diagonal band complex have restricted axonal projections to the cerebral cortex in the rat. *Brain Res*. **269**, 352–4.

Renaud, L. P., Martin, J. B., and Brazeau, P. (1975). Depressant action of TRH, LH-RH and somatostatin on activity of central neurones. *Nature*, **255**, 233–5.

Reubi, J. C. (1984). Evidence for two somatostatin-14 receptor types in rat brain cortex. *Neurosci. Lett*. **49**, 259–63.

Ribak, C. E. (1978). Aspinous and sparsely-spinous stellate neurons in the visual cortex of rats contain glutamic acid decarboxylase. *J. Neurocytol*. **7**, 461–478.

Salt, T. E. and Sillito, A. M. (1984). The action of somatostatin (SSt) on the response properties of cells in the cat's visual cortex. *J. Physiol*. **350**, 28P.

Saper, C. B. (1984). Organisation of cerebral cortical afferent systems in the rat. I Magnocellular basal nucleus. *J. Comp. Neurol*. **222**, 313–42.

Sillito, A. M. (1975a). The effectiveness of bicuculline as an antagonist of GABA and visually evoked inhibition in the cat's striate cortex. *J. Physiol*. **250**, 287–304.

—— (1975b). The contribution of inhibitory mechanisms to the receptive field properties of neurones in the striate cortex of the cat. *J. Physiol*. **250**, 305–29.

—— (1977). Inhibitory processes underlying the directional specificity of simple complex and hypercomplex cells in the cat's visual cortex. *J. Physiol*. **271**, 699–720.

—— (1979). Inhibitory mechanisms influencing complex cell orientation selectivity and their modification at high resting discharge levels. *J. Physiol*. **289**, 33–53.

—— (1984). Functional considerations of the operation of GABAergic inhibitory processes in the visual cortex. In *The Cerebral Cortex*, (eds A. Peters and E. G. Jones) Vol. 2A, 91–117. Plenum Press, New York.

—— (1985). Inhibitory circuits and orientation selectivity in the visual cortex. In *Models of the Visual Cortex* (eds D. Rose and V. Dobson). John Wiley, Chichester, 396–407.

—— and Kemp, J. A. (1983). Cholinergic modulation of the functional organisation of the cat visual cortex. *Brain Res*. **289**, 143–55.

——, ——, Milson, J. A. and Berardi, N. (1980) A reevaluation of the mechanisms underlying simple cell orientation selectivity. *Brain Res*. **194**, 517–20.

—— and Versiani, V. (1977). The contribution of excitatory and inhibitory inputs to the length preference of hypercomplex cells in layers II and II of the cat's striate cortex. *J. Physiol*. **273**, 775–90.

Somogyi, P. (1977). A specific 'axo-axonal' interneuron in the visual cortex of the rat. *Brain Res.* **136**, 345–50.

—— and Cowey, A. (1981). Combined Golgi and electron microscopic study on the synapses formed by double bouquet cells in the visual cortex of the cat and monkey. *J. Comp. Neurol.* **195**, 547–66.

——, ——, Halasz, N., and Freund, T. F. (1981). Vertical organization of neurons accumulating ^{3}H-GABA in the visual cortex of the rhesus monkey. *Nature*, **294**, 761–3.

——, Freund, T. F., and Kisvarday, Z. F. (1984). Different types of ^{3}H-GABA accumulating neurons in the visual cortex of the rat. Characterization by combined autoradiography and golgi impregnation. *Exp. Brain Res.* **54**, 45–56.

——, Kisvarday, Z. F., Martin, K. A. C., and Whitteridge, D. (1983). Synaptic connections of morphologically identified and physiologically characterised large basket cells in the striate cortex of the cat. *Neurosci.* **10**, 261–94.

Stichel, C. C. and Singer, W. (1985). Organisation and morphological characteristics of cholineacetyltransferase containing fibres in the visual thalmus and striate cortex of the cat. *Neurosci. Lett.* **53**, 155–60.

Tsumoto, T., Eckhart, W., and Creutzfeldt, O. D. (1979). Modification of orientation sensitivity of cat visual cortex neurones by removal of GABA mediated inhibition. *Exp. Brain Res.* **34**, 351–63.

5

GABA and glutamic acid agonists of pharmacological interest

POVL KROGSGAARD-LARSEN, LONE NIELSEN, ULF MADSEN, AND ELSEBET Ø. NIELSEN

INTRODUCTION

The amino acid 4-aminobutyric acid (GABA) is an inhibitory neurotransmitter concerned with the control of neuronal activity in the mammalian central nervous system (CNS). Impaired transmission of GABA-operated synapses appears to be implicated in certain neurological disorders, and GABA is known to be involved in the central regulation of a variety of physiological functions including cardiovascular mechanisms, endocrine functions, and the sensation of pain and anxiety (Johnston 1978; Krogsgaard-Larsen *et al.* 1979; Di Chiara and Gessa 1981; Okada and Roberts 1982; Enna 1983).

These findings have brought GABAergic compounds into focus as potential therapeutic agents. Although the relative susceptibility of the GABA synaptic mechanisms to pharmacological manipulation is far from being elucidated, the most direct approach to stimulation of the GABA-mediated neurotransmission is activation of the postsynaptic GABA receptors. Consequently, design and development of specific GABA agonists suitable for pharmacological and clinical studies has been, and continues to be, an active research field.

(S)-Glutamic acid (GLU) amd (S)-aspartic acid (ASP) are putative central excitatory neurotransmitters. The possible involvement of these transmitters in the mechanisms underlying the neuronal degenerations and alterations of neuronal pathways observed in some diseases such as Huntington's chorea and epilepsy (McGeer *et al.* 1978; Coyle *et al.*.1981) has focused much interest on central excitant amino acid receptors as sites for pharmacological and therapeutic attack (Watkins 1984). In principle, compounds with antagonistic effects at these receptors have therapeutic interest. The availability of specific agonists is, however, a necessary condition for satisfactory receptor characterization and for the development of specific antagonists at these receptors.

GABA AGONISTS

Molecular pharmacology

By systematic alterations of the structural parameters characterizing the molecule of GABA, a wide variety of GABA analogues has been developed (for reviews see Allan and Johnston 1983; Krogsgaard-Larsen *et al.* 1983, 1985b).

A necessary, but not sufficient, condition for GABA agonist activity is the presence of a negatively and a positively charged group. In Fig. 5.1 a

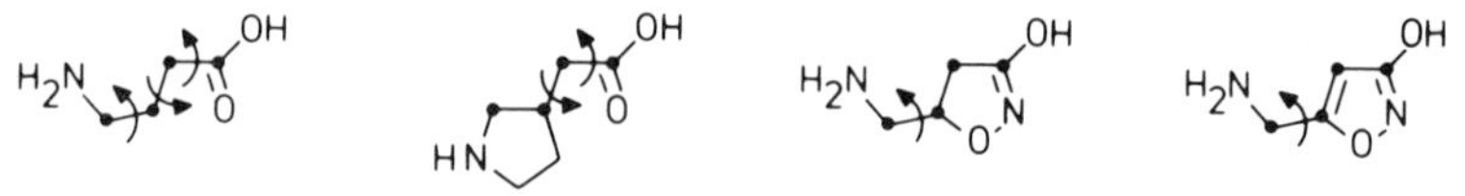

	GABA	Homo-β-proline	DHM	Muscimol
A	– – –	– – –	– – – –	– – – –
B	0.03	0.3	0.008	0.006
U·N	15	75	1500	2500
U·G	35	20	4000	2000

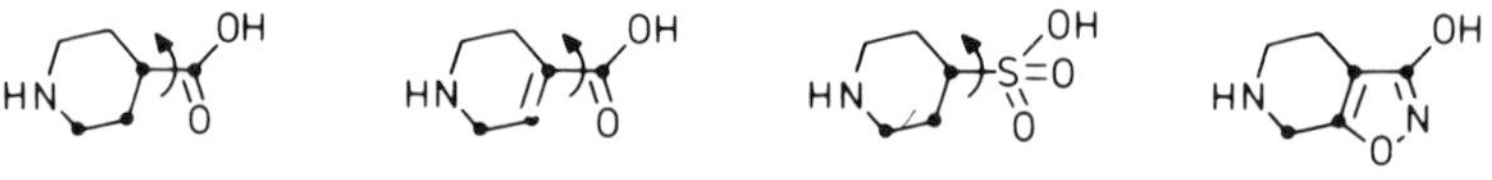

	Isonipecotic acid	Isoguvacine	P4S	THIP
A	– – –	– – – –	– – – –	– – – (–)
B	0.3	0.04	0.03	0.1
U·N	>5000	>5000	>5000	>5000
U·G	>5000	>5000	>5000	>5000

Fig. 5.1. Structure, conformational flexibility, and biological activity of GABA and a number of GABA analogues. (A) BMC-sensitive GABA agonist (GABA-A agonist) activity relative to that of GABA (– – –) as determined microelectrophoretically (Curtis *et al.* 1971; Falch *et al.* 1985). (B) Inhibition of GABA-A receptor binding (IC$_{50}$, μM) using rat brain membranes (Krogsgaard-Larsen *et al.* 1985c); U-N and U-G: inhibition of neuronal and glial GABA uptake using brain slices and cultured astrocytes, respectively, at GABA concentrations of 10^{-6} M (IC$_{50}$, μM) (Schousboe *et al.* 1979).

number of zwitterionic GABA agonists, sensitive to the antagonist bicuculline methochloride (BMC) (GABA-A agonists), are illustrated in the unionized form. A considerable degree of conformational flexibility is a trait of the molecule of GABA (Fig. 5.1). This molecular characteristic apparently is essential for the synaptic activity of GABA, but it inherently makes GABA itself inapplicable for studies of its 'active conformations'. An approach to these problems involves design of GABA agonists of more rigid structure. In Fig. 5.1 the structures of some of these compounds are depicted in which various parts of the molecule of GABA are immobilized. While homo-β-proline interacts effectively with the GABA-A receptors as well as the GABA uptake systems, dihydromuscimol (DHM) and muscimol, interact very potently and quite selectively with these receptors. The potent and specific GABA-A agonists isonipecotic acid, isoguvacine, piperidine-4-sulphonic acid (P4S), and 4,5,6,7-tetrahydroisoxazolo-[5,4-c]pyridin-3-ol (THIP) apparently reflect the 'receptor-active conformation' of GABA. Since the GABA-A receptors do seem to prefer agonists with flattened anionic structures, this effect of P4S carrying a bulky tetrahedrally arranged sulphonate group is remarkable.

While conformational flexibility of GABA agonists does not seem to be a prerequisite for activation of GABA-A receptors (Fig. 5.1), this structural parameter apparently is of greater importance for the ability of such compounds to stimulate benzodiazepine (BZD) binding to the post-synaptic GABA-BZD receptor complex (Falch *et al*. 1985). The lower 'efficacy' of conformationally restricted GABA agonists in this *in vitro* test system probably reflects certain aspects of the molecular mechanisms of this receptor complex.

The approximate charged structures of GABA-A agonists must be considered in structure-activity studies (Krogsgaard-Larsen *et al*. 1983). The pronounced potency of DHM (Fig. 5.2) as a GABA-A agonist is consistent with its structural similarity with GABA, and the much lower activity of isomuscimol is readily understandable in light of its very different charged structure. As illustrated in Fig. 5.2, the degree of delocalization of the positive charges of the imidazole group of imidazole-4-acetic acid (IAA), the isothiouronium groups of 2-aminothiazoline-4-acetic acid (ATAA) (Bristow *et al*. 1985) and *cis*-3-[(aminoiminomethyl)thio]propenoic acid (I) (Johnston and Allan 1984), and the guanidino group of 3-guanidinopropionic acid (II) is different from that of the amino group of GABA. A high degree of delocalization of the positive charges does seem to be tantamount to reduced GABA-A agonist activity.

Stereostructure-activity studies have disclosed different effects of the optical isomers of chiral GABA analogues on the GABA synaptic mechanisms as exemplified in Fig. 5.3 (Krogsgaard-Larsen *et al*. 1984;

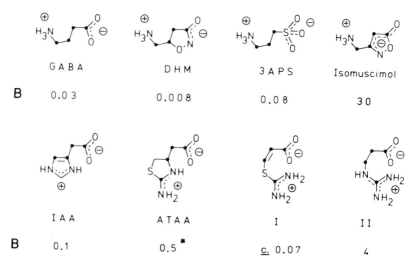

FIG. 5.2. Structure, approximate charge delocalization, and *in vitro* activity of GABA and a number of GABA analogues. (B) Inhibition of GABA-A receptor binding (IC$_{50}$, μM) (see legend for Fig. 5.1). *This IC$_{50}$ value has been determined under conditions, where the IC$_{50}$ values for GABA and isoguvacine were 0.1 and 0.49 μM, respectively (Bristow *et al*. 1985).

Schousbo 1979). The degree of stereoselectivity does, however, depend on the structure of the asymmetric GABA analogues, being a function of the conformational flexibility of the compounds. Thus, the (*S*)- and (*R*)-forms of the flexible GABA analogue 4-methyl-GABA (4-Me-GABA) are equally effective at the GABA-A receptors, and both of these isomers interact with the neuronal as well as the glial GABA uptake systems. Introduction of double bonds into the molecules of these compounds to give the (*S*)- and (*R*)-forms of *trans*-4-amino-pent-2-enoic acid (4-Me-*trans*-ACA) has quite dramatic effects. While the (*S*)-form of this conformationally restricted GABA analogue specifically and quite effectively interacts with the GABA-A receptors, (*R*)-(+)-4-Me-*trans*-ACA interacts with the GABA uptake systems without affecting significantly the receptors (Fig. 5.3). Similarly, the (*S*)- and (*R*)-forms of the conformationally flexible GABA analogue 3-hydroxy-4-aminobutyric acid (3-OH-GABA) show a low degree of stereoselectivity, whereas the less flexible bioisosteres of these optical isomers, (*S*)-(+)- and (*R*)-(−)-DHM, respectively, have very different pharmacological profiles. While (*S*)-(+)-DHM is the most potent GABA-A agonist so far described, showing no affinity for GABA uptake, (*R*)-(−)-DHM binds much less tightly to the GABA-A receptors, and this antipode interacts with the neuronal as well as the glial GABA uptake

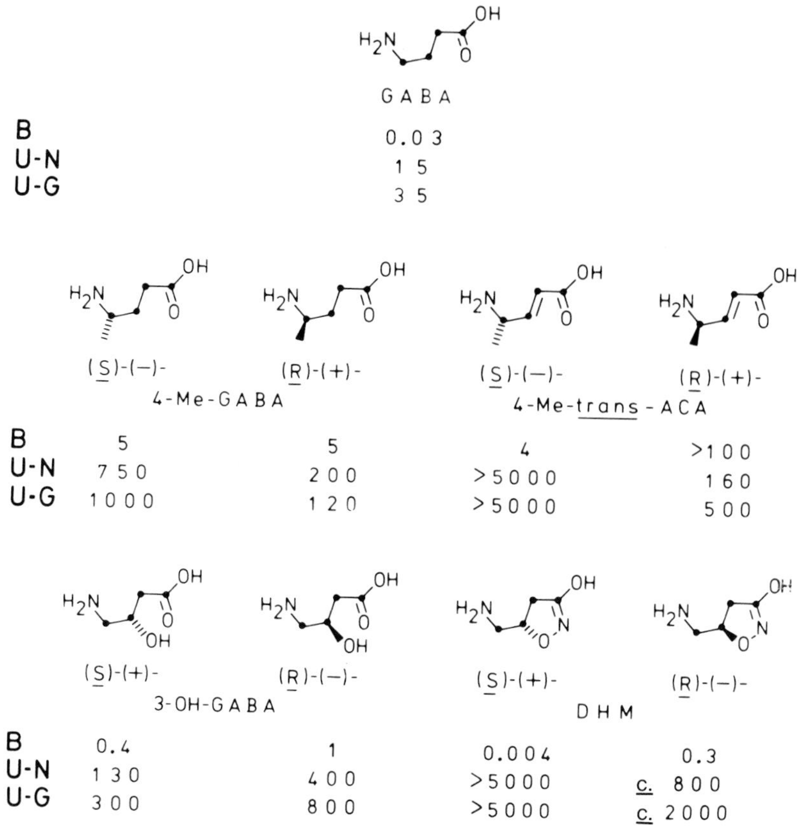

GABA

B	0.03
U-N	1 5
U-G	3 5

(S)-(−)-   (R)-(+)-   (S)-(−)-   (R)-(+)-
4-Me-GABA   4-Me-trans-ACA

B	5	5	4	>100
U-N	7 5 0	2 0 0	>5 0 0 0	1 6 0
U-G	1 0 0 0	1 2 0	>5 0 0 0	5 0 0

(S)-(+)-   (R)-(−)-   (S)-(+)-   (R)-(−)-
3-OH-GABA   DHM

B	0.4	1	0.004	0.3
U-N	1 3 0	4 0 0	>5 0 0 0	c. 8 0 0
U-G	3 0 0	8 0 0	>5 0 0 0	c. 2 0 0 0

FIG. 5.3. Structure and *in vitro* activity of GABA and a number of chiral GABA analogues with known absolute stereochemistry. (B) U-N and U-G: inhibition of GABA-A receptor binding, neuronal GABA uptake, and glial GABA uptake, respectively (IC_{50}, μM) (see legend for Fig. 5.1).

system (Fig. 5.3; Krogsgaard-Larsen *et al.* 1985c). These stereostructure-activity relationships, which support earlier findings of different 'active conformations' of GABA at the receptor and uptake sites (Krogsgaard-Larsen *et al.* 1983, 1984b), might be of importance for the design of new drugs with specific actions at synapses at which GABA is the transmitter.

Pharmacokinetic aspects

As mentioned earlier, all compounds so far known with specific actions at GABA-A receptors have zwitterionic structures as exemplified in Fig. 5.2, and the early stage of the pharmacology of GABA-A agonists reflects the

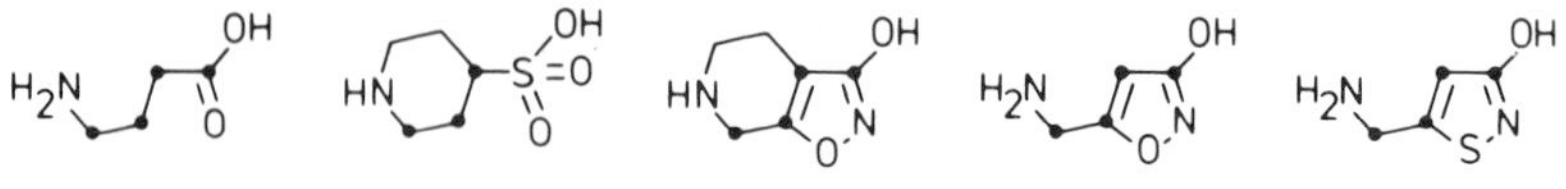

	GABA	P4S	THIP	Muscimol	Thiomuscimol
A	– – –	– – – –	– – –(–)	– – – –	– – – –
B	0.03	0.03	0.1	0.006	0.02
G-T, Km	1.92	No effect	No effect	1.27	c. 1.0
pK	4.0;10.7	<1;10.3	4.4 ; 8.5	4.8 ; 8.4	6.1 ; 8.9
I/U	800 000	>1 000 000	1 500	900	13
BBB	No	No	Yes	Yes	Yes

FIG. 5.4. Structure, biological activity, and pharmacokinetic properties of GABA and a number of GABA analogues. (A) and (B) relative GABA-A agonist activity and inhibition of GABA-A receptor binding, respectively (IC_{50}, μM) (see legend for Fig. 5.1). G-T, Km: effect on GABA aminotransferase (Km, mM); BBB: ability to penetrate the blood-brain barrier.

difficulties in designing such compounds with satisfactory pharmacokinetic properties.

The ability of neutral amino acids, for which active transport processes do not exist, to penetrate the blood-brain barrier (BBB) appears to be a function of the ratio between the concentrations of ionized and un-ionized molecules in solution (I/U ratio) (Krogsgaard-Larsen et al. 1982, 1983). A low value of this ratio, which is a function of the difference between the pK values of the amino acids, is tantamount to an increased ability of the compounds to penetrate the BBB (Fig. 5.4). Thus, in contrast to for example GABA and P4S, THIP and muscimol penetrate the BBB very easily (Moroni et al. 1982). The I/U ratio for thiomuscimol is exceptionally low. The very short-lived effect of this compound after systemic administration to animals probably reflects that thiomuscimol (L. J. Fowler and P. Krogsgaard-Larsen 1985, unpublished), like muscimol (Fowler et al. 1983), is a substrate for GABA aminotransferase (G-T) (Fig. 5.4).

In contrast to muscimol, THIP is well tolerated by various animal species and man (Christensen et al. 1982), it is active after oral administration, and it is excreted unchanged and, to some extent, in a conjugated form in the urine from animals and humans (Schultz et al. 1981).

As illustrated in Fig. 5.5, even minor alterations of the structure of THIP result in pronounced or complete loss of affinity for the GABA-A receptors (Krogsgaard-Larsen et al. 1982, 1983; Nordmann et al. 1985). On the basis of these observations, the potency and specificity of THIP as a GABA-A agonist, and its favourable pharmacokinetic and toxicological

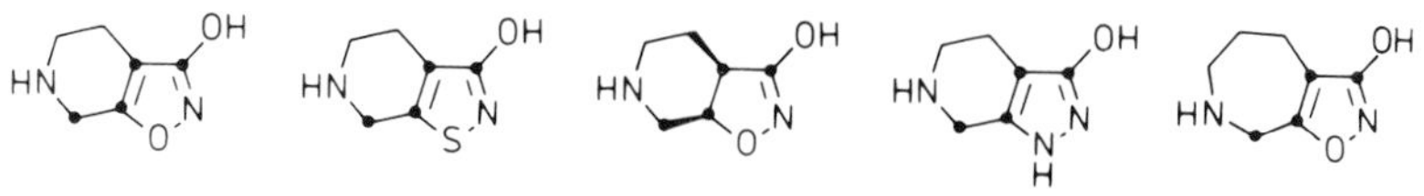

	THIP	Thio-THIP	Cis-DH-THIP	Aza-THIP	THIA
A	— — —(—)	—	N.t.	N.t.	N.t.
B	0.1	40	>100	>100	>100

FIG. 5.5. Structure and biological activity of THIP and a number of THIP analogues. (A) and (B) Relative GABA-A agonist activity and inhibition of GABA-A receptor binding, respectively (IC_{50}, μM) (see legend for Fig. 5.1).

properties, THIP has been the subject of comprehensive pharmacological and clinical studies.

THIP: PHARMACOLOGICAL AND CLINICAL STUDIES

The pharmacological and clinical studies on THIP have been reviewed (Christensen *et al.* 1982; Korgsgaard-Larsen *et al.* 1984a, 1985b).

The anticonvulsant effects of THIP and muscimol have been compared in a variety of animal models. THIP typically is two to five times weaker than muscimol in suppressing seizure activities. In mice and gerbils with genetically determined epilepsy systemically administered THIP has proved very effective in suppressing seizure activity, and THIP is capable of reducing audiogenic seizures in DBA/2 mice. However, THIP failed to protect baboons with photosensitive epilepsy against photically-induced myoclonic responses. In acute as well as chronic studies THIP is capable of reducing seizures in mice induced by various agents. An analysis of these effects and the effects of THIP on pento-geniculo-occipital (PGO) activity in cats seems to indicate an as yet unclarified involvement of the serotonin system. THIP has been subjected to a single-blind controlled trial in patients with epilepsy, in which THIP was added to the concomitant antiepileptic treatment. Under these conditions no significant effects of THIP were detected, although a trend was observed for lower seizure frequency during a period of submaximal doses of THIP.

The effects of THIP on ethanol withdrawal symptoms in rats have recently been studied. Whilst intracisternally administered THIP proved effective in reducing audiogenic clonic-tonic seizures, no effects on forelimb tremor were observed. These selective effects may have clinical interest.

THIP has been found to induce anaesthesia in rodents. It was shown to be equipotent with thiopental, ketamine, and midazolam, but its anaesthetic action lasted longer than those of these generally used anaesthetics, and was less toxic.

The demonstration of potent analgesic effects of THIP in different animal models has made studies of the clinical prospects of GABA-mediated analgesia possible. THIP-induced analgesia is insensitive to naloxone, indicating that the effect is not mediated by the opiate receptors. Quite surprisingly, THIP analgesia can not be reversed by bicuculline, which may indicate the involvement of a distinct class of GABA receptor. On the other hand, THIP-induced analgesia can be reduced by atropine and potentiated by physostigmine, reflecting as yet unclarified functional interactions between GABA and acetylcholine neurones and the central opiate systems rather than a direct action of THIP on muscarinic receptors.

THIP and morphine are approximately equipotent as analgesics, although their relative potencies are dependent on the animal species and experimental models used. Acute injection of THIP potentiates morphine-induced analgesia, and chronic administration of THIP produces a certain degree of functional tolerance to its analgesic effects. In contrast to earlier findings the results of recent studies have been interpreted in terms of some cross tolerance between THIP and morphine. In contrast to morphine, THIP does not cause respiratory depression. Clinical studies on post-operation patients and patients with chronic pain of malignant origin have disclosed potent analgesic effects of THIP, in the latter group of patients at doses of 5–30 mg (IM) of THIP. In these cancer patients and also in patients with chronic anxiety the desired effects of THIP were accompanied by side effects, notably sedation, nausea, and in a few cases euphoria. The side effects of THIP have been described as mild and similar in quality to those of other GABA-mimetics. These undesirable effects of THIP may to a certain extent be ascribed to the non-optimal pharmacokinetics of THIP, emphasizing the need for a sustained release preparation of THIP.

It is assumed that the post-synaptic GABA receptor complex is mediating the anxiolytic effects of the BZD, and, consequently, it is of interest to see whether GABA agonists have anxiolytic effects. Muscimol has proved effective in conflict tests, though with a pharmacological profile different from that of diazepam, and in humans muscimol in low doses was found to sedate and calm some schizophrenic patients. In a recent study on a number of patients with chronic anxiety the anxiolytic effects of THIP were assessed on several measures of anxiety. Although these effects were accompanied by side effects, the combination of analgesic and anxiolytic effects of THIP would seem to have therapeutic prospects.

There is very strong evidence that GABA is involved in the regulation of cardiovascular mechanisms. Whilst ICV-administered THIP reduced blood pressure as well as heart rate, systemically administered THIP did not affect these functions significantly. On the other hand, systemically administered GABA and isoguvacine had significant cardiovascular effects, which could be blocked by BMC. Since THIP, but not GABA and isoguvacine, are capable of penetrating the BBB these results seem to indicate that peripheral GABA receptors are involved in the regulation of cardiovascular functions. It is interesting to note that GABA and a number of GABA-A agonists produce a dose-dependent dilation of isolated cat and dog cerebral artery segments apparently by activation of GABA receptors with pharmacological characteristics similar to those of central postsynaptic GABA receptors.

The results of pharmacological studies on the spastic mouse are consistent with a role of GABA in spasticity. Systemic administration of the GABA agonists muscimol, THIP, and isoguvacine to cat affected spinal cord activities. Since isoguvacine does not readily penetrate the BBB, its pharmacological effects in this animal model may suggest that some parts of the spinal cord are not effectively protected by a BBB. THIP has recently been studied in spastic patients. At oral doses of 15–25 mg, THIP clearly reduced the monosynaptic T-reflexes without affecting the flexor threshold significantly.

Studies in recent years have disclosed very complex interactions between different neurotransmitter systems in the basal ganglia. These interactions have been extensively studied with the intension of getting better insight into the mechanisms underlying schizophrenia, Parkinson's disease, and different dyskinetic syndromes and, furthermore, of developing new strategies for the treatment of these severe diseases. Much interest has been focused on the nigrostriatal dopamine (DA) neurones, which form part of the nigrostriatal 'feedback' pathway, of which the striatonigral GABA neurones terminate within the substantia nigra (SN) pars reticulata, possibly on cholinergic neurones. The DA neurones of this system originating in SN pars compacta as well as the mesolimbic DA neurones involved in an analogous 'feedback' loop are assumed to be under inhibitory GABA control. These aspects have opened up the prospects of using GABA-stimulating therapies in the treatment of schizophrenic patients, and, consequently, THIP has been quite extensively studied in different animal models.

Whilst activation of GABA receptors in SN pars reticulata of rats has dramatic behavioural consequences the DA neurones in SN pars compacta and the mesolimbic DA neurones are much less sensitive. Direct application of THIP in the respective brain areas actually has weak inhibitory effects on both types of DA neurones, whereas systemically administered THIP weakly stimulates these DA neurones. No simple explanation of these apparently self-contradictory observations has, so far, been forwarded. The behavioural effects of acute and chronic administration of THIP have been studied. DA agonist-induced locomotor activity and stereotypies are altered by simultaneous treatment with GABA agonists, the former activity being depressed and the latter intensified. From a clinical point of view the interactions between THIP and neuroleptics may be particularly interesting. Most neuroleptic drugs inhibit DA-induced stereotypy and induce catalepsy in animals, the former effect being related to clinical antipsychotic effects and the latter to extrapyramidal side-effects of neuroleptics. Since THIP, and also scopolamine, antagonize the antistereotypic effects of some neuroleptics it has been tentatively concluded that GABA agonists such as THIP would probably not potentiate the antipsychotic effect of neuroleptics, but rather antagonize it.

The interactions between THIP and the central DA systems have also been studied in monkeys. Analyses of the complex pharmacological profile in this animal of THIP, which to some extent was similar to that of diazepam, led to the conclusion that THIP would probably have limited therapeutic effect in different kinds of dyskinesia, and THIP actually has proved ineffective in reducing the symptoms of dyskinetic patients.

GLUTAMIC ACID AGONISTS: MOLECULAR PHARMACOLOGY

Pharmacological investigations *in vivo* and receptor binding studies *in vitro* have disclosed heterogeneity of central excitatory amino acid receptors, which, at present, are most conveniently subdivided into three classes (Watkins 1981, 1984; Foster and Fagg 1984): (1) Quisqualic acid (QUIS) receptors, at which GLU diethyl ester (GDEE) is a weak but relatively selective antagonist; (2) *N*-methyl-(*D*)-aspartic acid (NMDA) receptors, at which AP5, α-aminoadipic acid (α-AA) (Fig. 5.7), and various other compounds are selective antagonists; (3) kainic acid receptors, which are relatively insensitive to GDEE or AP5.

Although the physiological relevance of this receptor classification is unclear, QUIS receptors are assumed to represent the postsynaptic receptors for GLU, and NMDA receptors those for ASP. Consequently, there is a particular interest in selective agonists and antagonists for these classes of receptors. The naturally occurring amino acid QUIS (Fig. 5.6) actually does not interact specifically with the QUIS receptors. It also binds very tightly to kainic acid receptors. Ibotenic acid (IBO), also a naturally occurring amino acid, interacts primarily with the NMDA receptors, showing only little effect on QUIS receptors. AMPA, which is an analogue

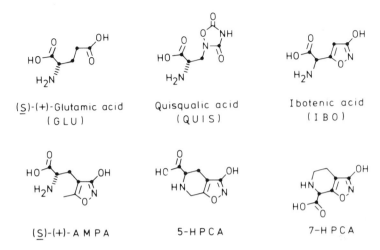

FIG. 5.6. Structure of GLU, QUIS, IBO, and the specific QUIS agonists (S)-(+)-AMPA, 5-HPCA, and 7-HPCA.

of IBO, does, on the other hand, interact potently and with a high degree of selectivity with QUIS receptors (Krogsgaard-Larsen *et al.* 1980), (S)-(+)-AMPA being the more active enantiomer. AMPA has proved to be a useful tool for studies of QUIS receptors (Honoré *et al.* 1982). 5-HPCA, which is a conformationally restricted analogue of AMPA, shows a pharmacological profile almost identical with that of AMPA, suggesting that 5-HPCA represents the 'active conformation' of AMPA at the QUIS receptors (Krogsgaard-Larsen 1985a). Similarly, the biological effects of 7-HPCA are almost indistinguishable from those of AMPA and 5-HPCA (Krogsgaard-Larsen *et al.* 1984c). Detailed structure-activity studies have shown that the fully ionized molecules of 5- and 7-HPCA are virtually superimposable. Attempts to convert these specific QUIS receptor agonists into antagonists at these receptors are in progress.

Besides NMDA itself, relatively few selective NMDA agonists are known such as *cis*-1,3-ADCP and AMAA (Fig. 5.7). Although the conformational flexibility of these cyclic NMDA agonists is less pronounced than that of the parent compound, *cis*-1,3-ADCP as well as AMAA can still adopt a variety of conformations. In an attempt to shed light on the 'active conformation(s)' of these compounds, the bicyclic AMAA analogue, 4-HPCA, was synthesized and tested biologically (Madsen *et al.* 1985). This compound, in which only the carboxyl group can rotate relatively freely was, however, shown to be completely inactive, indicating that 4-HPCA does not reflect the 'active conformation(s)' of NMDA, *cis*-1,3-ADCP, or AMAA.

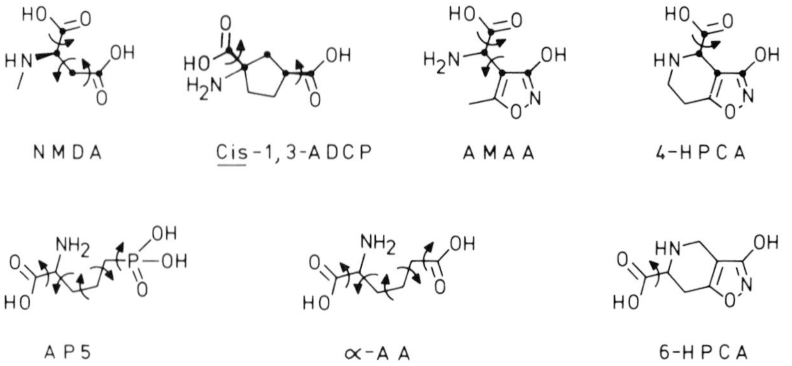

FIG. 5.7. Structure and conformational flexibility of NMDA, the NMDA agonists *cis*-1,3-ADCP and AMAA, the NMDA antagonists AP5 and α-AA, and the bicyclic analogues 4-HPCA and 6-HPCA.

Similarly, attempts to elucidate the conformations, in which the NMDA antagonists AP5 and α-AA bind to the receptors, via synthesis and structure-activity studies of 6-HPCA were only to a limited extent a success. Although 6-HPCA is an analogue of AP5 and α-AA, it did not significantly reduce neuronal excitations induced by NMDA. Thus, among the numerous conformations accessible to AP5 and α-AA (Fig. 5.7) the partially folded conformation represented by 6-HPCA does not seem to reflect the folding of these antagonists during their blockade of the NMDA receptors.

Acknowledgements

This work was supported by grants from the Danish Medical Research Council. The collaboration with Professor D. R. Curtis, Canberra, Drs A. Schousboe, L. Brehm, E. Falch, and K. Schaumburg, Copenhagen, and the secretarial and technical assistance of Mrs B. Hare and Mr S. Stilling are gratefully acknowledged.

References

Allan, R. D. and Johnston, G. A. R. (1983). Synthetic analogs for the study of GABA as a neurotransmitter. *Medicinal Res. Rev.* **3**, 91–118.
Bristow, D. R., Campbell, M. M., Iversen, L. L., Kemp, J. A., Marshall, G., Watling, K. J., and Wong, E. H. F. (1985) (submitted).
Christensen, A. V., Svendsen, O., and Krogsgaard-Larsen, P. (1982). Pharmacodynamic effects and possible therapeutic uses of THIP, a specific GABA-agonist. *Pharm. Weekbl. Sci. Ed.* **4**, 145–53.

Coyle, J. T., Bird, S. J., Evans, R. H., Gulley, R. L., Nadler, J. V., Nicklas, W. J., and Olney, J. W. (1981). Excitatory amino acid neurotoxins: selectivity, specificity, and mechanisms of action. *Neurosci. Res. Prog. Bull.* **19**, 333–427.

Curtis, D. R., Duggan, A. W., Felix, D., and Johnston, G. A. R. (1971). Bicuculline, an antagonist of GABA and synaptic inhibition in the spinal cord of the cat. *Brain Res.* **32**, 69–96.

Di Chiara, G. and Gessa, G. L. (eds) (1981). *GABA and the Basal Ganglia.* Raven Press, New York.

Enna, S. J. (ed.) (1983). *The GABA Receptors.* The Humana Press, Clifton, New Jersey.

Falch, E., Jacobsen, P., Krogsgaard-Larsen, P., and Curtis, D. R. (1985). GABA-mimetic activity and effects on diazepam binding of aminosulphonic acids structurally related to piperidine-4-sulphonic acid. *J. Neurochem.* **44**, 68–75.

Foster, A. C. and Fagg, G. E. (1984). Acidic amino acid binding sites in mammalian neuronal membranes: their characteristics and relationship to synaptic receptors. *Brain Res. Rev.* **7**, 103–64.

Fowler, L. J., Lovell, D. H., and John, R. A. (1983). Reaction of muscimol with 4-aminobutyrate aminotransferase. *J. Neurochem.* **41**, 1751–4.

Honoré, T., Lauridsen, J., and Krogsgaard-Larsen, P. (1982). The binding of [^{3}H]AMPA, a structural analogue of glutamic acid, to rat brain membranes. *J. Neurochem.* **38**, 173–8.

Johnston, G. A. R. (1978). Neuropharmacology of amino acid inhibitory transmitters. *Ann. Rev. Pharmacol. Toxicol.* **18**, 269–89.

—— and Allan, R. D. (1984). GABA agonists. *Neuropharmacol.* **23**, 831–2.

Krogsgaard-Larsen, P., Brehm, L., Johansen, J. S., Vinzents, P., Lauridsen, J., and Curtis, D. R. (1985a). Synthesis and structure-activity studies on excitatory amino acids structurally related to ibotenic acid. *J. Med. Chem.* **28**, 673–9.

——, Falch, E., and Christensen, A. V. (1984a). Chemistry and pharmacology of the GABA agonists THIP (Gaboxadol) and isoguvacine. *Drugs of the Future*, **9**, 597–618.

——, —— and Hjeds, H. (1985b). Heterocyclic analogues of GABA: chemistry, molecular pharmacology and therapeutic aspects. *Prog. Med. Chem.* **22**, 67–120.

——, ——, Mikkelsen, H., and Jacobsen, P. (1982). Development of structural analogs and pro-drugs of GABA agonists with desirable pharmacokinetic properties. In *Optimization of drug delivery* (eds. H. Bundgaard, A. B. Hansen, and H. Kofod), pp. 225–35. Munksgaard, Copenhagen.

——, Honoré, T., Hansen, J. J., Curtis, D. R., and Lodge, D. (1980). New class of glutamate agonist structurally related to ibotenic acid. *Nature, Lond.* **284**, 64–6.

——, Jacobsen, P., and Falch, E. (1983). Structure-activity requirements of the GABA receptor. In *The GABA receptors* (ed. S. J. Enna), pp. 149–76. The Humana Press, Clifton, New Jersey.

——, Lenicque, P., and Jacobsen, P. (1984b). GABA-ergic drugs: configurational and conformational aspects. In *Handbook of stereoisomers: drugs in psychopharmacology* (ed. D. F. Smith), pp. 369–99. CRC Press, Boca Raton, Florida.

——, Nielsen, E. Ø., and Curtis, D. R. (1984c). Ibotenic acid analogues. Synthesis and biological and *in vitro* activity of conformationally restricted agonists at central excitatory amino acid receptors. *J. Med. Chem.* **27**, 585–91.

——, Nielson, L., Falch, E., and Curtis, D. R. (1985c). GABA agonists. Resolution absolute stereochemistry and enantioselectivity of (*S*)-(+)- and (*R*)-(−)-dihydromuscimol. *J. Med. Chem.* **28**, 1612–7.

——, Scheel-Krüger, J., and Kofod, H. (eds) (1979). *GABA-neurotransmitters. Pharmacochemical, Biochemical and Pharmacological Aspects.* Munksgaard, Copenhagen.

Madsen, U., Schaumburg, K., Brehm, L., and Krogsgaard-Larsen, P. (1985). Ibotenic acid analogues. Synthesis and biological testing of two bicyclic 3-isoxazolol amino acids. *Acta Chem. Scand.* **B39**, (in press).

McGeer, E. G., Olney, J. W., and McGeer, P. L. (eds) (1978). *Kainic Acid as a Tool in Neurobiology.* Raven Press, New York.

Moroni, F., Forchetti, M. C., Krogsgaard-Larsen, P., and Guidotti, A. (1982). Relative disposition of the GABA agonists THIP and muscimol in the brain of the rat. *J. Pharm. Pharmacol.* **34**, 676–8.

Nordmann, R., Graff, P., Maurer, R., and Gähwiler, B. H. (1985). A dihydroanalogue of THIP: *cis* DH-THIP. *J. Med. Chem.* **28**, 1109–11.

Okada, Y. and Roberts, E. (eds) (1982). *Problems in GABA Research from Brain to Bacteria.* Excerpta Medica, Amsterdam.

Schousboe, A., Thorbek, P., Hertz, L., and Krogsgaard-Larsen, P. (1979). Effects of GABA analogues of restricted conformation on GABA transport in astrocytes and brain cortex slices and on GABA receptor binding. *J. Neurochem.* **33**, 181–9.

Schultz, B., Aaes-Jørgensen, T., Bøgesø, K. P., and Jørgensen, A. (1981). Preliminary studies on the absorption, distribution, metabolism, and excretion of THIP in animal and man using ^{14}C-labelled compound. *Acta Pharmacol. Toxicol.* **49**, 116–24.

Watkins, J. C. (1981). Pharmacology of excitatory amino acid receptors. In *Glutamate: transmitter in the central nervous system* (eds. P. J. Roberts, J. Storm-Mathisen, and G. A. R. Johnston), pp. 1–24. John Wiley, Chichester.

Watkins, J. C. (1984). Excitatory amino acids and central synaptic transmission. *Trends Pharmacol. Sci.* **5**, 373–6.

6

Selective antagonists define sub-classes of excitatory amino acid receptors

J. C. WATKINS

INTRODUCTION

Twenty-five years have passed since detailed structure-activity relationships for responses produced by excitatory amino acids led to the conclusion that the receptor (considered at that time to be a single entity) probably had a negative site for interaction with the protonated amino group of the amino acid, and either one or two positive sites, for interaction with one or both of the anionic groups of the excitant (Curtis and Watkins, 1960; Curtis *et al*. 1961). These alternatives gave rise to the 'two-point' or 'three-point' receptor concepts shown in Fig. 6.1. The two-point concept had the advantage of leaving one anionic group of the excitant free. If this free anionic group projected into a membrane pore (pre-existing or created by the complex formation) it could mediate the trans-membrane transport of monovalent cations, particularly Na^+, and thus readily explain the depolarization produced by the amino acids. The three-point receptor, which was considered to function by molecular perturbation leading to the creation of new membrane pores during its interaction with the amino acids, readily accounted for the stereochemical requirements of the receptor that were later demonstrated, and has been adopted by many workers in the field as the more representative model. However, it would be difficult from purely pharmacological considerations to choose between the two-point and three-point models, since even the two-point model does not preclude stereoselectivity and in any case still required the presence of the other anionic group, even though the latter group does not itself interact with a corresponding group in that particular model.

EXCITATORY AMINO ACID RECEPTORS

The concept of a single type of receptor, 'three-point' or otherwise, is no longer tenable. Over 100 excitatory amino acids are now known, including

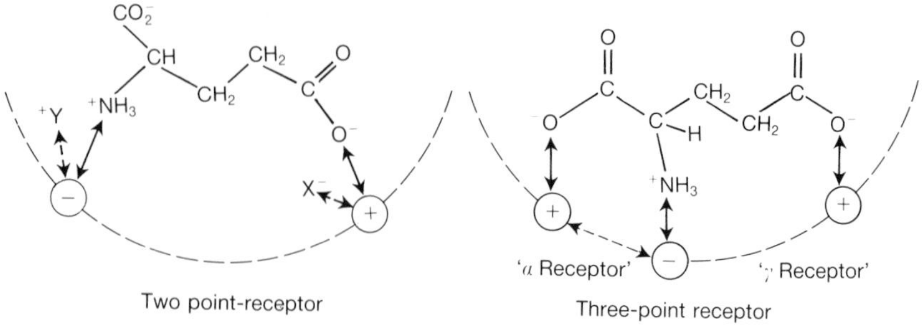

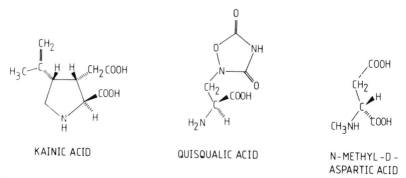

FIG. 6.1. Early 'two-point' and 'three-point' receptor models for excitatory amino acids (Curtis & Watkins 1960).

the potent excitants, N-methyl-D-aspartate (NMDA), kainic acid and quisqualic acid (Fig. 6.2) which are currently considered to act at different sub-types of excitatory amino acid receptors. That the receptor-ionophore complexes activated by some excitants are different from those activated by others became clear when, firstly, low concentrations of Mg^{2+} (Evans *et al*. 1977; Davies and Watkins 1977; Ault *et al*. 1980) and, later, a range of mono- and diamino dicarboxylic acids (Biscoe *et al*. 1977; Evans *et al*. 1978, 1979; Davies and Watkins 1979), were shown to produce a particular pattern of depression with respect to different agonists. Thus, as shown in Table 6.1 and Figs 6.3 and 6.4, both Mg^{2+} and $(\pm)$-α, ε-diaminopimelic acid (DAP) depressed responses produced by excitants in the following order of agonist susceptibility: N-methyl-D-aspartate (NMDA) > L-homocysteate > L-aspartate > L-glutamate > kainate > quisqualate (unaffected). Clearly, this indicated the existence of at least two different classes of excitatory amino acids receptor, one class activated by NMDA and subject to depression by both Mg^{2+} and DAP, and one or more other types of receptor that are activated by such agonists as

KAINIC ACID   QUISQUALIC ACID   N-METHYL-D-ASPARTIC ACID

FIG. 6.2. Structure of some potent amino acid excitants now considered to act at different types of receptors.

TABLE 6.1. *Depressant effects of Mg^{2+} and α, ε-diaminopimelic acid (DAP) on depolarizing responses of frog motoneurones to a range of excitatory amino acids*[1]

Excitant	Percentage control responses in presence of	
	Mg^{2+} (1 mM)	α, ε-DAP (0.5 mM)
Quisqualate	101 ± 3 (4)	101 ± 4 (4)
Kainate	96 ± 3 (4)	98 ± 2 (4)
L-glutamate	83 ± 1 (5)	83 ± 3 (4)
L-aspartate	69 ± 4 (4)	71 ± 3 (4)
L-homocysteate	22 ± 1 (5)	34 ± 3 (4)
NMDA	12 ± 3 (3)	14 ± 1 (4)

[1]Evans *et al.* 1978, and unpublished observations.

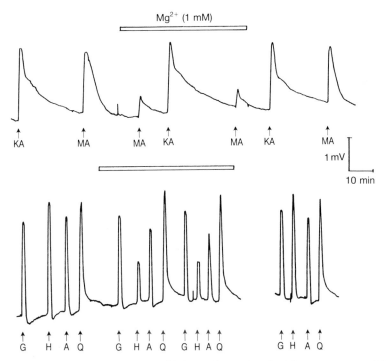

FIG. 6.3. Depressant effects of Mg^{2+} (1 mM) on depolarizing responses of frog motoneurones to excitatory amino acids. Responses were recorded from ventral roots and the medium contained tetrodotoxin (Ault *et al.* 1980). KA, kainate; MA, N-methyl-D-aspartate; G, L-glutamate; H, L-homocysteate; A, L-aspartate; Q, quisqualate.

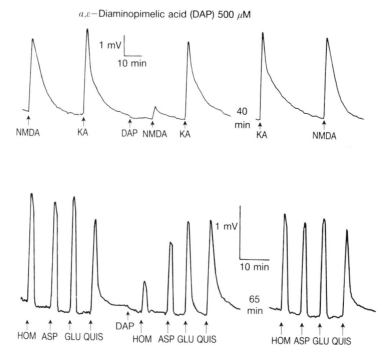

FIG. 6.4. Depressant effect of $(\pm)$-α, ε-diaminopimelic acid (DAP, 500 μM) on depolarizing responses of frog motoneurones to excitatory amino acids. Responses were recorded from ventral roots and the medium contained tetrodotoxin (Evans *et al.* 1978 and unpublished records). Note similarity to effects of Mg^{2+} shown in Fig. 6.3. NMDA, N-methyl-D-aspartate; KA, kainate; HOM, L-homocysteate; ASP, L-aspartate; GLU, L-glutamate; QUIS, quisqualate.

quisqualate or kainate, but not subject to depression by either the organic or inorganic antagonists. These receptors have become known simply as the NMDA and non-NMDA types. The maximum extent to which responses produced by a particular excitatory amino acid can be depressed by specific NMDA antagonists indicates the relative extent to which the NMDA type of receptor (or, more precisely, the NMDA receptor-ionophore mechanism activated) contributes to the responses produced by that agonist. This is exemplified for Mg^{2+} in Fig. 6.5 while Table 6.2 indicates the relative extents to which responses to a larger range of agonists likewise appear to be mediated by NMDA receptors in the frog and cat spinal cord. It should be noted, however, that, for mixed agonists, the extent to which a particular type of receptor is activated will depend on the particular test conditions used (e.g. tissue, medium composition, and agonist concentration).

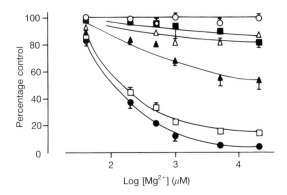

FIG. 6.5. Antagonism by increasing concentration of Mg^{2+} of frog ventral root depolarization induced by a range of amino acid excitants O, quisqualate; ■, kainate; △, L-glutamate; ▲, L-aspartate; □, L-homocysteate; ●, NMDA.

NMDA-receptor antagonists

A range of structurally-related antagonists, including the D forms of α-aminoadipate and longer chain α, ω-dicarboxylic α-amino acids (Fig. 6.6) acted in a similar way to Mg^{2+} and DAP (Davies *et al*. 1979a). That D-α-aminoadipate (DαAA) and DAP acted in a competitive way with NMDA at its receptor sites was suggested by the parallel shift in the NMDA dose-response relationship obtained in the presence of either antagonist, (Evans *et al*. 1979), and by the fact that dose-ratios calculated individually for their NMDA antagonist effects were additive when the two substances were tested together (Davies *et al*. 1979a). In contrast, dose-ratios for antagonism of NMDA-induced responses by Mg^{2+} and either DαAA or DAP were multiplicative when the inorganic and an organic antagonist were tested together, indicating that Mg^{2+} acted at a different site in the receptor-ionophore complex (Davies *et al*. 1979a).

Other divalent metal ions, for example Co^{2+}, Mn^{2+}, and Ni^{2+}, have been shown to have similar selective effects to those of Mg^{2+} when tested against the same range of amino acid excitants (Ault *et al*. 1980). In these cases, however, particularly with Mn^{2+} and Ni^{2+}, presynaptic effects of the divalent metal ions on transmitter release limit their usefulness as selective NMDA receptor antagonists. NMDA-receptor antagonism and inhibition of transmitter release can be differentiated by the relative potencies of Mg^{2+} and Mn^{2+}, the latter being much the more potent of the two ions in producing the presynaptic effect whereas Mg^{2+} is the more potent as an antagonist at NMDA receptors.

TABLE 6.2. *Relative susceptibility of responses produced by different excitatory amino acids to specific NMDA antagonists in frog and cat spinal cord*[1]

Agonist	Susceptibility
N-Methyl-D-aspartate	Very high
N-Methyl-L-aspartate	to high
D-Homocysteine sulphinate	
trans-2,3-piperidine dicarboxylate	
trans-2,4-piperidine dicarboxylate	
N-methyl-D-glutamate	
N-methyl-L-glutamate	
Ibotenate	
L-homocysteate	
D-glutamate	
cis-1-amino-1,3-dicarboxy-cyclopentane	
L-homocysteine sulphinate	Moderate
D-homocysteate	
L-aspartate	
D-aspartate	
L-cysteine sulphinate	
L-cysteate	
L-glutamate	
Quisqualate	Low to
(±)-Willardiine	undetectable
(±)-Bromowillardiine	
Kainate	
(±)-AMPA	

[1]Watkins and Evans 1981; Davies *et al.* 1982a,b; Mewett *et al.* 1983; Krogsgaard-Larsen *et al.* 1980; McLennan, 1983.

It is now known that the Mg^{2+} effect at NMDA receptors is voltage-dependent, being more effective at membrane potentials near to the resting potential or at hyperpolarized levels, and less effective at lower membrane potentials (Macdonald and Wojtowicz 1982; Nowak *et al.* 1984; Mayer *et al.* 1984). The effect appears to involve a gating mechanism for the ion-flow induced by activation of the ionophore (Nowak *et al.* 1984).

Many organic antagonists that are selective for the NMDA receptor are now known. The most potent and selective of those in current use are those in which the ω-carboxylic group of previously known NMDA antagonists has been replaced by a phosphono group, including the D(-) forms of 2-amino-5-phosphonopentanoate (AP5) (Evans *et al.* 1982;

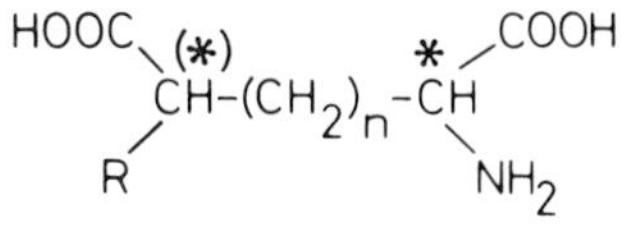

$$\underset{R}{\overset{HOOC}{\diagdown}} \underset{}{\overset{(*)}{CH}-(CH_2)_n-\overset{*}{CH}} \underset{NH_2}{\overset{COOH}{\diagup}}$$

n	R	*	(*)	Compound	Abbreviation
2	H	D	–	D-ᵅ-AMINOADIPIC ACID	D-ᵅ-AA
3	H	D	–	D-ᵅ-AMINOPIMELIC ACID	D-ᵅ-AP
4	H	D	–	D-ᵅ-AMINOSUBERIC ACID	D-ᵅ-AS
3	NH₂	D	D or L	ᵅ,ε-DIAMINOPIMELIC ACID	ᵅ,ε-DAP

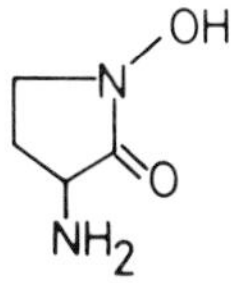

3-AMINO-1-HYDROXY-2-PYRROLIDONE
(HA-966 or HAP)

FIG. 6.6. Selective NMDA antagonists. Series 1 (Evans *et al.* 1979).

Davies and Watkins 1982) and 2-amino-7-phosphonoheptanoate (AP7) (Perkins *et al.* 1982), and two dipeptide analogues of these substances, β-D-aspartylaminomethyl phosphonate (ASP-AMP) and γ-D-glutamylaminomethyl phosphonate (GLU-AMP) (Fig. 6.7) (Jones *et al.* 1984; Davies *et al.* 1984). These substances are one or two orders of magnitude more potent than the substances of Fig. 6.6 (Evans *et al.* 1982; Jones *et al.* 1984; Davies *et al.* 1984).

Non-NMDA-receptor antagonists

Non-NMDA type excitatory amino acid receptors can be classified into different sub-types according to a number of criteria. However, the evidence underlying this sub-classification is nowhere near as definitive as that differentiating NMDA and non-NMDA receptors. Selective antagonism of quisqualate relative to kainate-induced responses was first observed in cat spinal neurones *in vivo* using L-glutamic acid diethyl ester (GDEE) (Fig. 6.8) (McLennan and Lodge 1979; Davies and Watkins 1979), although this effect is not seen in all tissues or systems (Davies *et al.* 1979b; Teichberg *et al.* 1981) On the other hand, kainate-induced responses are somewhat more susceptible to antagonism by certain dipeptides including γ-D-glutamylglycine (γDGG) (Francis *et al.* 1980; Davies and Watkins 1981) and γ-D-glutamylaminomethyl sulphonate (GAMS) (Fig. 6.8); (Jones *et al.* 1984; Davies and Watkins 1985). These results suggest the existence of different populations of non-NMDA receptors, one type perhaps activated preferentially by kainate and another type by quisqualate. As further evidence of different kainate- and

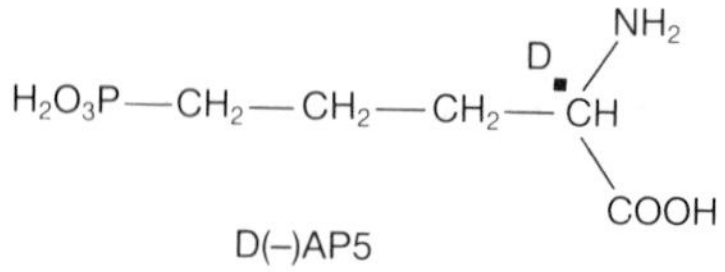

FIG. 6.7. Selective NMDA antagonists. Series 2 (Evans *et al.* 1982).

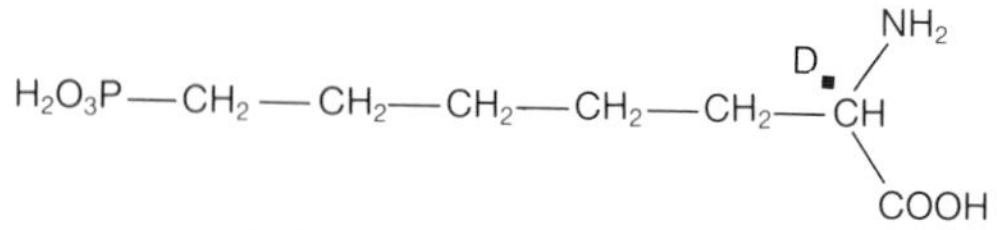

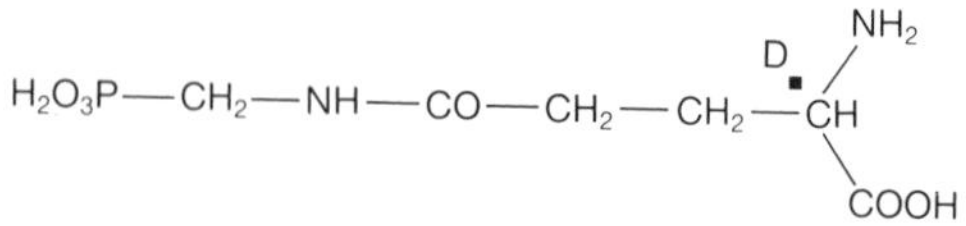

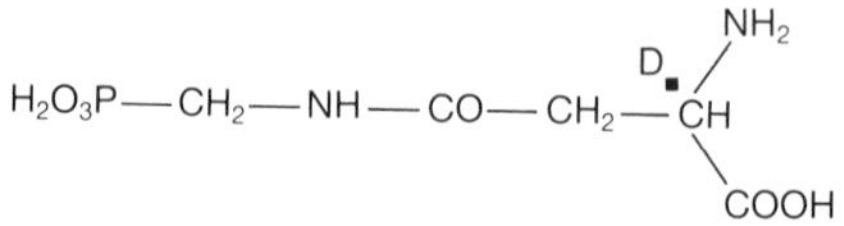

quisqualate-type receptors, one may instance the quisqualate-preferring nature of excitatory amino acid receptors at the crayfish neuromuscular junction (Shinozaki and Shibuya 1976; Takeuchi and Onodera 1975) and the kainate-preferring nature of excitatory amino acid receptors on dorsal root fibres of the 4–8-day-old rat (Davies *et al.* 1979a,b).

Other sub-type receptors and antagonists

Yet other sub-types of excitatory amino acid receptors have been proposed. Thus, responses produced by (±)-*trans*-1-amino-1,3-di-carboxycyclopentane (ADCP) in the rat spinal cord are not susceptible to antagonism by either NMDA or non-NMDA antagonists (McLennan and Liu 1982). Also, Teichberg *et al.* (1981) observed that the increase in $^{22}Na^+$-efflux from striatal slices induced by high concentrations of L-glutamate and L-aspartate were more susceptible to antagonism by GDEE than were responses to NMDA, kainate or quisqualate. On the other hand, Davies and Watkins (1985) found L-glutamate to be more resistant to antagonism by γ-D-glutamylaminomethyl

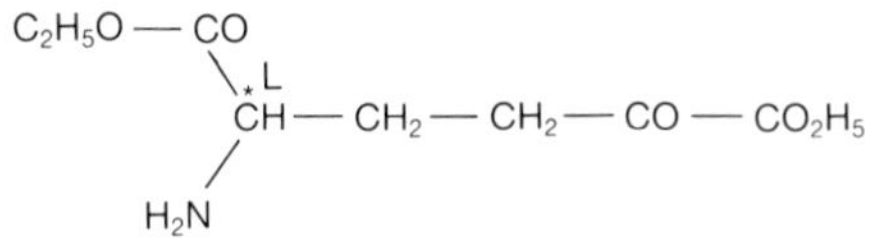

C$_2$H$_5$O — CO
    \\.L
        CH — CH$_2$ — CH$_2$ — CO — CO$_2$H$_5$
    /
H$_2$N

L–glutamic acid diethyl ester
(GDEE)

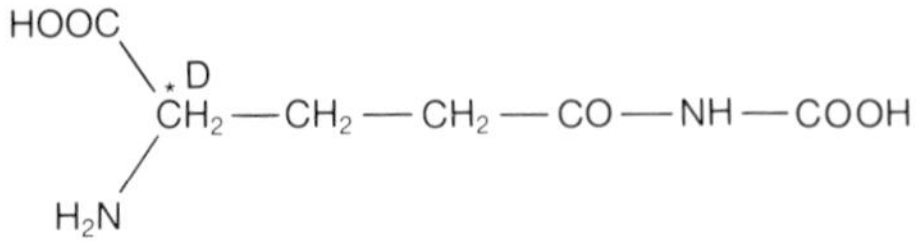

HOOC
    \\.D
        CH$_2$ — CH$_2$ — CH$_2$ — CO — NH — COOH
    /
H$_2$N

γ–D–glutamylglycine (γDGG)

HOOC
    \\.D
        CH — CH$_2$ — CH$_2$ — CO — NH — CH$_2$ — SO$_3$H
    /
H$_2$N

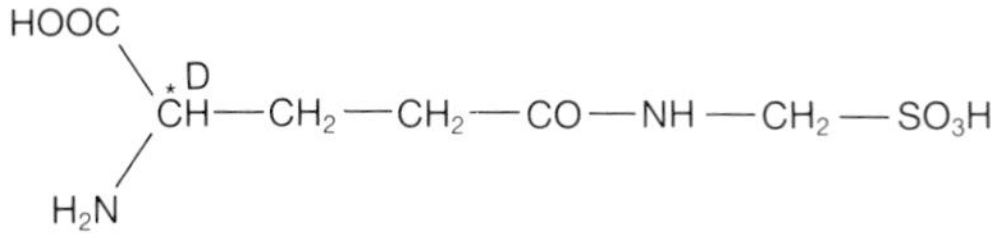

γ–D– glutamylaminomethyl sulphonic acid (GAMS)

FIG. 6.8. Structures of some antagonists at non-NMDA receptors.

sulphonate than were NMDA, kainate, or quisqualate. These two sets of observations suggest that L-glutamate acts at one or more receptor types in addition to the NMDA, kainate, or quisqualate types. Perhaps a new term—U (for classified) receptors—could be introduced for such receptors, until such time as they become more clearly characterized.

DISCUSSION

It is of interest to examine structure-activity relations for antagonism of different receptor sub-types. Table 6.3 gives relative potencies of a range of dipeptides as antagonists of different agonist-induced responses in isolated spinal cords (Jones *et al*. 1984). These results and those of related studies (Evans *et al*. 1979, 1982) lead to various conclusions as to the molecular structural features influencing potency and/or selectivity of open chain agonists in relation to their actions at NMDA and non-NMDA receptors, although it is not yet possible to define clearly those structural characteristics conducive to differential antagonism of kainate and quisqualate receptors. The following conclusions may be drawn.

TABLE 6.3. *Structure and antagonist activity of some dipeptides*

General formula: R—A—X   R = $H_2N\overset{H}{\underset{HOOC}{\overset{|}{\underset{|}{C}}}}\!\!-$ D(R)   A = Chain
X = Acidic group

No.	Structure	Feature[2]	Relative potency[1] versus		
			NMDA	K	Q
1	R—CH₂—CH₂—CO—NH—CH₂—[COOH]	b	1.0	7.8	10.7
2	R—CH₂—CH₂—CO—NH—CH₂—[PO₃H₂]		0.8	33	36
3	R—CH₂—CH₂—CO—NH—CH₂—[SO₃H]		71	9.8	7.8
2	R—CH₂—CH₂—[CO—NH—CH₂]—CH₂—PO₃H₂	c	0.8	33	36
4	R—CH₂—CH₂—[CH₂—CH₂]—CH₂—PO₃H₂		0.1	78	40
1	R—CH₂—[CH₂—CO—NH]—CH₂—COOH	d	1.0	7.8	10.7
5	R—CH₂—[CO—NH—CH₂]—CH₂—COOH		3.4	41	43
3	R—CH₂—[CH₂—CO—NH]—CH₂—SO₃H	d	71	9.8	7.8
6	R—CH₂—[CO—NH—CH₂]—CH₂—SO₃H		56	41	29

No.	Structure				
7	R—CH$_2$—CO—NH—CH$_2$—PO$_3$H$_2$	e	0.2	26	15
8	R—CH$_2$—CO—NH—CH$_2$—CH$_2$—PO$_3$H$_2$		1.2	78	63
1	R—CH$_2$—CH$_2$—CO—NH—CH$_2$—COOH	e	1.0	7.8	10.7
9	R—CH$_2$—CH$_2$—CO—NH—CH$_2$—CH$_2$—COOH		60	35	13
10	R—CH$_2$—CH$_2$—CO—NH—CH$_2$—CH$_2$—CH$_2$—COOH		167	54	24
2	R—CH$_2$—CH$_2$—CO—NH—CH$_2$—PO$_3$H$_2$	e	0.8	33	36
11	R—CH$_2$—CH$_2$—CO—NH—CH$_2$—CH$_2$—PO$_3$H$_2$		46	209	85
3	R—CH$_3$—CH$_2$—CO—NH—CH$_2$—SO$_3$H	e	71	9.8	7.8
12	R—CH$_2$—CH$_2$—CO—NH—CH$_2$—CH$_2$—SO$_3$H		202	12.3	8.5

[1] Equi-effective molar concentration ratios: γDGG (compound 1) *v.* NMDA = 1.0.
[2] See text.

(a) *Configuration of R*. All the substances of Table 6.3 have the D (R) configuration. Many of the corresponding L (S) compounds were tested, but their activity, like that of the L forms of the substances shown in Figs 6.6 and 6.7 (Evans *et al*. 1979, 1982; Davies and Watkins 1982), was usually too low to warrant quantitative comparison with the D forms. This applied to both NMDA and non-NMDA receptors (A. W. Jones, D. A. S. Smith and J. C. Watkins, unpublished observations). However, it should be emphasized that γ-L-glutamylglycine has significant antagonist activity, though considerably weaker than that of the D form (Francis *et al*. 1980; Davies and Watkins 1981).

(b) *Type of terminal acidic group X*. For NMDA receptors, the descending order of effectiveness for the terminal acidic group was $PO_3H_2 > CO_2H \gg SO_3H$, and for K/Q receptors $CO_2H \simeq SO_3H > PO_3H_2$. In dipeptides of this type the phosphonate group thus promotes NMDA selectivity and the sulphonic acid group K/Q selectivity.

(c) *Type of inter-acidic group chain A*. An all-methylene chain favours selectivity at NMDA receptors (Evans *et al*. 1979, 1982) while a peptide link in the chain (for those compounds with a terminal COOH or SO_3H group, for example, compounds 1 and 3 in Table 6.3) promotes K/Q antagonist activity.

(d) *Position of the peptide group*. Structures of the type $R\text{-}CH_2\text{-}CH_2\text{-}CO\text{-}NH\text{-}CH_2\text{-}X$ have higher K/Q antagonist potency than structures of the type $R\text{-}CH_2\text{-}CO\text{-}NH\text{-}CH_2\text{-}CH_2\text{-}X$.

(e) *Length of the inter-acidic group chain A*. The highest potency at NMDA receptors was shown by β-D-aspartylaminomethyl phosphonate (compound 7) representing a chain length of five atoms (inclusive of the carbon atom attached to the α-amino and carboxylate groups) between the acidic groups. Highest K/Q antagonist activity was seen with γ-D-glutamyl dipeptides (compounds 1, 3, and 12) in which the terminal acidic group was COOH or SO_3H and the interacidic group chain six or seven atoms long. A tentative conclusion therefore is that, for dipeptides, optimum chain length for K/Q antagonist activity is around one carbon atom longer than for NMDA antagonist activity.

Other conclusions may be drawn from the effects of some cyclic substances (Table 6.4). For example, K/Q antagonist activity increases and NMDA antagonist activity decreases or is unchanged with benzoylation of the 4-N atom of piperazine-2,3-dicarboxylic acid (PzDA). This may relate to the ability of the peptide link to increase K/Q antagonist activity in open chain compounds as discussed above and suggests the possibility that,

TABLE 6.4. *Relative potencies of some piperazine-2,3-dicarboxylic acids as excitatory amino acid antagonists*

Reference compounds	Structure	Relative potency[1] versus		
		NMDA	K	Q
γDGG[2]	see Fig. 6.8	1.0	1.0	1.0
cis-2,3-PDA		5.1[f]	1.1[f]	0.9[f]
PzDA		11.0	4.0	4.6
Piperazine derivatives				
B—PzDA	X = H	15	0.5	0.8
o—CB—PzDA	X = *o*—Cl	43	0.7	1.5
m—CB—PzDA	X = *m*—Cl	12	0.9	0.5
p—CB—PzDA	X = *p*—Cl	8.5	0.8	0.6
p—BB—PzDA	X = *p*—Br	4.3	0.6	0.5

[1]Equi-effective molar concentration ratios, γDGG = 1.0 as determined in isolated spinal cords of rat or frog (f).

[2]Abbreviations: γDGG, γ-D-glutamylglycine; PDA, piperidine dicarboxylic acid; PzDA, piperazine dicarboxylic acid; B, benzoyl; *o*-CB, *ortho*-chlorobenzoyl; *m*-CB, *meta*-chlorobenzoyl; *p*-CB, *para*-chlorobenzoyl; *p*-BB, *para*-bromobenzoyl; NMDA. N-methyl-D-aspartate; K, kainate; Q, quisqualate.

compared with the NMDA receptor, the K/Q receptor(s) contain an extra active site capable of interacting with a peptide-like linkage.

It is beyond the scope of this presentation to review the roles of these receptors to the extent that these have yet been elucidated in central nervous function. Some of the literature relating to this aspect of excitatory amino acid research has been recently summarized (Watkins & Evans 1981; Watkins 1984). In general kainate- or quisqualate-receptors appear

TABLE 6.5. *Synaptic pathways involving excitatory amino acid receptors*[1]

NMDA	NMDA/Non-NMDA	Non-NMDA
Excitatory interneurones to dorsal and ventral horn neurones in spinal cord	Schaffer-collateral fibres from CA3 to CA1 cells in hippocampus[2]	Some fast conducting primary afferents to dorsal and ventral horn spinal neurones
		Sensory afferents to neurones in cuneate n. and trigeminal n.
		Cerebral cortex to cuneate n.
		Cerebral cortex to dopaminergic caudate neurones
		Optic tract fibres to LGN neurones
		Cerebellar parallel fibres to Purkinje cells
		Perforant path to hippo-campal dentate granule cells

[1] Watkins and Evans 1981; McLennan 1983.
[2] 'Long-term potentiation' appears to be mediated by NMDA receptors and normal synaptic excitation by non-NMDA receptors (Collingridge *et al.* 1983).

to mediate transmission across synapses made by the terminals of long afferent tracts while NMDA receptors, possibly regulated by Mg^{2+}, may mediate or modulate transmission at synapses made by short interneurones with post-synaptic elements. Some pathways in which transmitter receptors have been characterized in terms of a particular type of excitatory amino acid receptor site are summarized in Table 6.5.

Much remains to be accomplished in this field. Of prime importance is the development of more potent and selective antagonists for non-NMDA receptors. Although not selective, the effectiveness of kynurenic acid (Perkins & Stone 1982; Ganong *et al.* 1983; Herrling 1984) and an analogue (Erez *et al.* 1985) as kainate/quisqualate and transmitter receptor antagonist may be a lead in this respect, as also the piperazine derivatives discussed above.

Acknowledgement

This work was supported by the Medical Research Council.

References

Ault, B., Evans, R. H., Francis, A. A., Oakes, D. J., and Watkins, J. C. (1980). Selective depression of excitatory amino acid induced depolarization by magnesium ions in isolated spinal cord preparations. *J. Physiol. (Lond.)* **307**, 413–28.

Biscoe, T. J., Davies, J., Dray, A., Evans, R. H., Francis, A. A., Martin, M. R., and Watkins, J. C. (1977). Depression of synaptic excitation and of amino acid induced excitatory responses of spinal neurones by D-α-aminoadipate, α, ε-diaminopimelic acid and HA-966. *Eur. J. Pharmacol.* **45**, 315–6.

Collingridge, G. L., Kehl, S. J., and McLennan, H. (1983). Excitatory amino acids in synaptic transmission in the Schaffer collateral-commissural pathway of the rat hippocampus. *J. Physiol. (Lond.)* **334**, 33–46.

Curtis, D. R., Phillis, J. W., and Watkins, J. C. (1961). Actions of amino acids on the isolated hemisected spinal cord of the toad. *Br. J. Pharmacol.* **16**, 262–83.

—— and Watkins, J. C. (1960). The excitation and depression of spinal neurones by structurally related amino acids. *J. Neurochem.* **6**, 117–41.

Davies, J., Evans, R. H., Francis, A. A., Jones, A. W., Smith, D. A. S., and Watkins, J. C. (1982a). Conformational aspects of the actions of some piperidine dicarboxylic acids at excitatory amino acid receptors in the mammalian and amphibian spinal cord. *Neurochem. Res.* **7**, 1119–33.

——, ——, ——, and Watkins, J.C. (1979a). Excitatory amino acids: receptor differentiation by selective antagonists and role in synaptic excitation. In *Advances in Pharmacology and Therapeutics. Vol. 2. Neurotransmitters.* (ed. P. Simon), pp. 161–70. Pergamon Press, Oxford.

——, ——, ——, and —— (1979b). Excitatory amino acid receptors and synaptic excitation in the mammalian central nervous system. *J. Physiol. (Paris),* **75**, 641–5.

——, ——, Jones, A. W., Smith, D. A. S. and Watkins, J. C. (1982b). Differential activation and blockade of excitatory amino acid receptors in the mammalian and amphibian central nervous system. *Comp. Biochem. Physiol.* **72C**, 211–24.

——, Jones, A. W., Sheardown, M. J., Smith, D. A. S., and Watkins, J. C. (1984). Phosphonodipeptides and piperazine derivatives as antagonists of amino acid-induced and synaptic excitation in mammalian and amphibian spinal cord. *Neurosci. Lett.* **52**, 79–84.

—— and Watkins, J. C. (1977). Effect of magnesium ions on the responses of spinal neurones to excitatory amino acids and acetylcholine. *Brain Res.* **130**, 364–8.

—— and —— (1979). Selective antagonism of amino acid-induced and synaptic excitation in the cat spinal cord. *J. Physiol. (Lond.)* **297**, 621–36.

—— and —— (1981). Differentiation of kainate and quisqualate receptors in the cat spinal cord by selective antagonism with γ-D(and L)-glutamylglycine. *Brain Res.* **206**, 172–7.

—— and —— (1982). Actions and D and L forms of 2-amino-5-phosphonovalerate and 2-amino-4-phosphonobutyrate in the cat spinal cord. *Brain Res.* **235**, 378–86.

—— and —— (1985) Depressant actions of γ-D-glutamylaminomethyl sulfonate (GAMS) on amino acid-induced and synaptic excitation in the cat spinal cord. *Brain Res.* **327**, 113–20.

Erez, U., Frenk, H., Goldberg, O., Cohen, A., and Teichberg, V. I. (1985). Anti-convulsant properties of 3-hydroxy-2-quinoxaline carboxylic acid, a newly found antagonist of excitatory amino acids. *Eur. J. Pharmacol.* **110**, 31–9.

Evans, R. H., Francis, A. A., Hunt, K., Oakes, D. J., and Watkins, J. C. (1979). Antagonism of excitatory amino acid-induced responses and of synaptic excitation in the isolated spinal cord of the frog. *Br. J. Pharmacol.* **67**, 591–603.

——, ——, Jones, A. W., Smith, D. A. S. and Watkins, J. C. (1982). The effects of a series of ω-phosphonic α-carboxylic amino acids on electrically evoked and amino acid induced responses in isolated spinal cord preparations. *Br. J. Pharmacol.* **75**, 65–75.

——, —— and Watkins, J. C. (1977). Selective antagonism by Mg^{2+} of amino acid-induced depolarization of spinal neurones. *Experientia (Basel),* **33**, 489–91.

——, ——, and —— (1978). Mg^{2+}-like selective antagonism of excitatory amino acid-induced responses by α, ε-diaminopimelic acid, D-α-aminoadipate and HA-966 in isolated spinal cord of frog and immature rat. *Brain Res.* **148**, 536–42.

Francis, A. A., Jones, A. W., and Watkins, J. C. (1980). Dipeptide antagonists of amino acid-induced and synaptic excitation in the frog spinal cord. *J. Neurochem.* **35**, 1458–60.

Ganong, A. H., Lanthorn, T. H., and Cotman, C. W. (1983). Kynurenic acid inhibits synaptic and acidic amino acid induced responses in the rat hippocampus and spinal cord. *Brain Res.* **273**, 170–4.

Herrling, P. L. (1984). Evidence that the cortically evoked e.p.s.p. in cat caudate neurones is mediated by non-NMDA excitatory amino acid receptors. *J. Physiol. (Lond.)* **353**, 98P.

Jones, A. W., Smith, D. A. S., and Watkins, J. C. (1984). Structure-activity relations of dipeptide antagonists of excitatory amino acids. *Neurosci.* **13**, 573–81.

Krogsgaard-Larsen, P., Honoré, T., Hansen, J. J., Curtis, D. R. and Lodge, D. (1980). New classes of glutamate agonists structurally related to ibotenic acid. *Nature (Lond.)* **284**, 64–6.

Mayer, M. L., Westbrook, G. L., and Guthrie, P. B. (1984). Voltage dependent block by Mg^{2+} of NMDA responses in spinal cord neurones. *Nature (Lond.)* **309**, 261–3.

Macdonald, J. F. and Wojtowicz, J. M. (1982). The effects of L-glutamate and its analogues upon the membrane conductance of central murine neurones in culture. *Can. J. Physiol. Pharmacol.* **60**, 282–96.

McLennan, H. (1983). Receptors for the excitatory amino acids in the mammalian central nervous system. *Prog. Neurobiol.* **20**, 251–71.

—— and Liu, J. R. (1982). The action of six antagonists of the excitatory amino acids on neurones of the rat spinal cord. *Exp. Brain Res.* **45**, 151–6.

—— and Lodge, D. (1979). The antagonism of amino acid-induced excitation of spinal neurones in the cat. *Brain Res*. **169**, 83–90.

Mewett, K. N., Oakes, D. J., Olverman, H. J. Smith, D. A. S., and Watkins, J. C. (1983). Pharmacology of the excitatory actions of sulphonic and sulphinic amino acids. In *CNS Receptors: From Molecular Pharmacology to Behaviour*. (Eds P. Mandel and F. V. DeFeudis), pp. 163–74. Raven Press, New York.

Nowak, L., Bregestowski, P., Ascher, P., Herbet, A., and Prochiantz, A. (1984). Magnesium gates glutamate-activated channels in mouse central neurones. *Nature (Lond.)* **307**, 462–5.

Perkins, M. N., Collins, J. F., and Stone, T. W. (1982). Isomers of 2-amino-7-phosphonoheptanoic acid as antagonists of neuronal excitants. *Neurosci. Lett*. **32**, 65–8.

—— and Stone, T. W. (1982). An iontophoretic investigation of the action of convulsant kynurenines and their interaction with the endogenous excitant quinolinic acid. *Brain Res*. **247**, 184–7.

Shinozaki, H. and Shibuya, I. (1976). Effects of kainic acid analogues on crayfish opener muscle. *Neuropharmacol*. **15**, 145–7.

Takeuchi, A. and Onodera, K. (1975). Effects of kainic acid on the glutamate receptors of crayfish muscle. *Neuropharmacol*. **14**, 619–25.

Teichberg, V. I., Goldberg, O., and Luini, O. (1981). The stimulation of ion fluxes in brain slices by glutamate and other excitatory amino acids. *Mol. Cell. Biochem*. **39**, 281–95.

Watkins, J. C. (1984). Excitatory amino acids and central synaptic transmission. *Trends Pharmacol. Sci*. **5**, 373–6.

—— and Evans, R. H. (1981). Excitatory amino acid transmitters. *Ann. Rev. Pharmacol. Toxicol*. **21**, 165–204.

7

Multiple benzodiazepine and related receptors as targets for psychotropic drug action

SOLOMON H. SNYDER

Since the beginnings of recorded history mankind has sought psychotropic agents that exert a calming influence both for recreational and medicinal purposes. Alcohol itself was the first widely used surgical anesthetic and has been employed more than any other substance to relieve anxiety and agitation, and to promote sleep. In the twentieth century many synthetic sedative agents emerged, which appear to act at sites like those affected by alcohol. Although exact molecular mechanisms are not yet rigorously established, it seems likely that alcohol, chloral hydrate, paraldehyde, bromides, barbiturates, meprobamates, and benzodiazepines act in similar ways. The best evidence is the cross-tolerance that occurs between alcohol and all of these drugs. Besides their sedative, hypnotic, and antianxiety effects, these agents generally are anticonvulsants and muscle relaxants. Since certain of these substances are more selective for one or another effect, such as phenobarbital as an anticonvulsant, heterogeneity of sites of actions is likely. However, the overall similarity in pharmacological effects of all these agents argues for a family of closely related target sites.

BENZODIAZEPINE-GABA RECEPTOR COMPLEX

A major advance in characterizing molecular sites of action for sedative-hypnotics was the identification of benzodiazepine receptor binding sites (Mohler and Okada 1977; Squires and Braestrup 1977). The techniques employed were essentially the same as those which had previously elucidated properties of opiate receptors (Snyder 1975). Briefly, ^{3}H-diazepam incubated with brain membranes bound with nanomolar affinity to apparently homogeneous populations of sites, and the relative potencies of benzodiazepines in competing for these binding sites paralleled their pharmacological activity in tests that predict antianxiety actions.

Continuing the analogy with opiate receptors, researchers wondered if benzodiazepine binding sites may in fact be receptors for a physiologic neurotransmitter. In initial screens no known neurotransmitters appeared to compete for the binding sites, prompting a major effort to identify the 'endogenous Valium'. More careful analysis revealed that the inhibitory neurotransmitter GABA, while not competing directly for benzodiazepine receptors, does influence benzodiazepine binding profoundly (Tallman *et ai*. 1980). Micromolar concentrations of GABA stimulate benzodiazepine receptor binding, the relative potencies of GABA derivatives in augmenting binding parallels their GABA-like synaptic effects, and GABA effects are blocked by the GABA antagonist bicuculline. Thus, GABA acts at its own synaptic receptors to allosterically influence benzodiazepine binding which presumably occurs to a separate site on the GABA receptor protein complex. Postsynaptic GABA type A receptors can be labelled with ^{3}H-GABA or the potent, rigid GABA derivative ^{3}H-muscimol (Enna and Snyder 1975; Snyder 1984; Bowery 1983). Purification of benzodiazepine receptors to apparent homogeneity results in co-purification of ^{3}H-muscimol binding sites indicating their association with the same macromolecule (Sigel and Barnard 1984). A relationship of GABA and benzodiazepine receptors is also favoured by autoradiographic investigations. The microscopic localizations of ^{3}H-muscimol labelled GABA receptors and benzodiazepine binding sites overlap, but do display a number of differences (Unnerstall *et al*. 1981). GABA receptors can be discriminated into two types, one with low and the other with high affinity. If one examines these two receptor subtypes differentially, a close relationship is apparent between low affinity GABA-A receptors and benzodiazepine receptors. This fits with the relatively low affinity of GABA in augmenting benzodiazepine binding.

One of the most striking aspects of receptor binding research has been the ability to detect interactions between recognition or binding site and second messenger events by studying the influence of second messenger related chemicals upon ligand binding. For instance, regulation of opiate receptor binding by sodium and guanine nucleotides reflects an interaction of the opiate receptor with a GTP binding protein which links the receptor to adenylate cyclase (Blume 1978; Childers and Snyder 1978, 1980). Sodium and GTP decrease the affinity of agonists but not antagonists for receptors, permitting the characterization of drugs as pure agonists, pure antagonists, and mixed agonist/antagonists. Similar differentiation of agonists and antagonists by guanine nucleotides has been demonstrated for most receptor sites that are associated with adenylate cyclase, including alpha$_2$-adrenergic, beta-adrenergic, muscarinic cholinergic, serotonin, and dopamine receptors (Snyder 1984). Amino acid neurotransmitters such as GABA and glycine are thought to act by directly affecting the permeability of ion channels. The inhibitory effects of glycine and GABA involve hyperpolarization elicited by increased chloride ion conductance. Chloride and other ions decrease ^{3}H-strychnine binding to glycine receptors, and diminish the affinity of glycine at receptor sites in close parallel with their ability to mimic chloride synaptic actions in neurophysiological studies (Young and Snyder 1974; Muller and Snyder 1978). ^{3}H-GABA binding to GABA-A receptors is similarly regulated by anions, though the correlation of binding and neurophysiologic data is not as strong as for glycine receptors (Enna and Snyder 1977). Benzodiazepine receptor binding is stimulated by chloride and anions with relative potencies like those seen at GABA receptors (Martin and Candy 1978; Costa *et al*. 1979). Thus, the benzodiazepine binding sites involve a macromolecular complex which includes the GABA receptor and an anion binding site which may be associated with chloride ion channels.

MULTIPLE BENZODIAZEPINE RECEPTORS

Benzodiazepines vary in the extent to which they elicit particular pharmacologic effects. For instance, clonazepam is more potent as an anticonvulsant than most other benzodiazepines. The ratio of anxiolytic to sedating potency also varies among benzodiazepines. Drugs which differ in chemical structure from the benzodiazepines, such as the tri-azolopyridazine CL-218872, in animal models can display an antianxiety profile without detectable sedative effects. One possible explanation for such differences is the existence of multiple benzodiazepine receptors. Heat inactivation studies first suggested a heterogeneity of benzodiazepine binding sites (Squires *et al*. 1979). CL-218872 was then found to bind with higher affinity to receptors in the cerebellum than in the frontal cortex or hippocampus (Lippa *et al*. 1982). The receptors with relatively higher affinity for CL-218872 and most concentrated in the cerebellum were termed Type I receptors in contrast to the Type II receptors which have a lower affinity for CL-218872. A variety of carbolines with considerable potency at benzodiazepine receptors also demonstrate preferential affinity for Type I sites (Braestrup and Nielsen 1981). Autoradiographic studies in the presence or absence of CL-218872 permits a differential estimate of the two benzodiazepine binding sites and reveals differences in their regional localization (Young *et al*. 1981).

Differentiations based upon drug potencies can be explained without invoking the existence of separate receptor proteins. Distinct electrophoretic mobilities for apparent Type I and II proteins have been detected following photolabeling with ^{3}H-flunitrazepam (Sieghart and Karobath 1980), though proteolysis might contribute to observed differences (Klotz *et al*. 1984). More definitive evidence comes from the physical separation of two receptor subtypes based on their differential detergent solubility (Lo *et al.* 1982). Most detergents readily solubilize benzodiazepine binding sites with a drug specificity of Type II receptors, while high salt concentrations must be combined with detergents to solubilize Type I receptors.

The study of physically separated, solubilized benzodiazepine receptor subtypes has facilitated the characterization of these two receptors. Type I and II receptors display the same affinity for most benzodiazepines, especially the ligand ^{3}H-flunitrazepam employed routinely for labelling receptor sites. Kinetic investigations of Type I and Type II solubilized receptors reveal that, while the two receptors have the same affinity for flunitrazepam, there are marked differences in kinetics. ^{3}H-flunitrazepam associates and dissociates with solubilized Type I receptors ten times more rapidly than with Type II sites (Trifiletti *et al.* 1984a) (Fig. 7.1). Since the affinity of the ligand for the receptor is dependent upon the ratio of the rate constants for association and dissociation, the affinity or dissociation constant is the same for ^{3}H-flunitrazepam at both receptors.

With solubilized Type I and Type II receptors the stimulation of benzodiazepine binding by chloride and other anions occurs exclusively at Type II receptors, while Type I sites are not affected (Lo and Snyder 1983). In the solubilized receptors anion effects are somewhat different from those observed in particulate receptors. One major difference between particulate and soluble fractions is that endogenous GABA cannot be fully removed from particulate preparations so that apparent anion effects may involve GABA receptors which in turn modulate benzodiazepine binding. With solubilized receptors only chloride, bromide, and iodine enhance binding, which does not fit the pattern observed in particulate preparations or the relative anionic abilities to permeate the chloride ion channels associated with GABA receptors. Thus, ions may affect benzodiazepine receptors in two distinct ways. By acting at presumed chloride channels they influence GABA-receptor interactions which in turn modulate benzodiazepine binding probably at both Type I and Type II receptors. On the other hand, certain anions directly regulate only Type II benzodiazepine receptors.

Variations in kinetics and ionic regulation raise questions as to the differential functions of Type I and Type II receptors. Autoradiographic studies have discriminated the localizations of these receptor subtypes in two ways. One involves assessing localization before and after treating brain slices with Type I specific drugs such as CL-218872 or a carboline (Young *et al.* 1981). A second approach is to treat brain slices with detergents in the absence of salt to solubilize selectively Type II receptors (Lo *et al.* 1983a). In the former case one assesses Type I sites by substracting Type II binding levels from total binding, while in the second

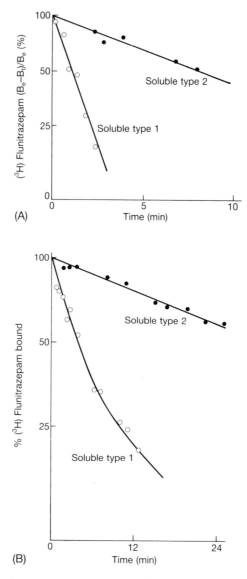

FIG. 7.1. Association and dissociation of [³H]FNZ at solubilized Type I and Type II receptors. (A) Association of [³H]FNZ to differentially solubilized benzodiazepine receptors from bovine cerebral cortex at 0°C. Benzodiazepine receptors were differentially solubilized from bovine cerebral cortex, and association kinetics to the respective soluble fractions were determined. Data for 1 per cent Triton X-100/1 M NaCl (soluble Type I) fractions are presented as a transformation appropriate for pseudo-first order conditions. Data are from a representative experiment replicated three times. (B) Dissociation of [³H]FNZ

instance one subtracts Type I sites from total binding to determine Type II receptor number. Results have been the same with both procedures. Type I receptors are selectively enriched in the cerebellum which contains four or five times more Type I than Type II sites. The most selective enrichment of Type II receptors occurs in limbic structures such as the dentate gyrus and hippocampus which contain up to 10 times more Type II than Type I receptors. Other areas are intermediate.

Autoradiographic studies have also differentiated the localization within neurons of receptor subtypes. The substantia nigra contains roughly equal numbers of Type I and Type II receptors. Lesions of the descending striatonigral pathway markedly deplete Type II receptors in the nigra while increasing numbers of Type I receptors up to 300 per cent (Lo *et al.* 1983b). These findings suggest that Type II sites occur on nerve terminals of the striatonigral pathway, while Type I sites are post-synaptic. Such a conclusion fits well with the sensitivity of Type II receptors to detergent solubilization, since nerve terminal membranes are uniquely sensitive to disruption by detergents. The post-synaptic density, a key element in synaptic transmission, is extremely resistant to detergent treatment and thus a candidate site for Type I receptors. Indeed, one of the preparative steps in procedures for isolation of postsynaptic densities is the treatment of tissue extracts with detergents. Accordingly, we conducted detailed subcellular fraction studies and demonstrated a striking enrichment of Type I receptors selectively in postsynaptic densites (Trifiletti and Snyder 1985).

SEDATIVE, CONVULSANT, AND CYCLOPYRROLONE DRUG ACTION

Evidence that drugs such as barbiturates and benzodiazepines act at the same sites stems from neurophysiologic investigations showing that both facilitate synaptic effects of GABA (Haefely *et al.* 1981). Barbiturates do not compete directly at benzodiazepine binding sites, but do stimulate benzodiazepine binding much like GABA (Leeb-Lundberg *et al.* 1980; Davis and Ticku 1981). While the GABA enhancement of benzodiazepine binding is blocked by the GABA antagonist bicuculline, barbiturate stimulation is blocked by convulsants such as picrotoxinin. In neurophysiological investigations both bicuculline and picrotoxinin

---

from differentially solubilized benzodiazepine receptors of bovine cerebral cortex at 0°C. Data for 1 per cent Triton X-100 soluble (soluble Type 2) and 1 per cent Triton X-100/1 M NaCl (soluble Type 1) fractions are presented.

Note that the ordinates in (A) and (B) are logarithmic scales.

antagonize synaptic effects of GABA, but act at different sites. Bicuculline effects are competitive with GABA, while picrotoxinin acts competitively with chloride ions and not GABA (Takeuchi and Takeuchi 1969). Thus, barbiturates may act at a receptor specific for sedatives and convulsants which is linked allosterically to the benzodiazepine-GABA receptor complex.

A direct approach to sites at which convulsant drugs act employs the binding of convulsants such as ^{3}H-dihydropicrotoxinin (Olsen 1981) or, more recently, ^{35}S-TBPS (*t*-butylbicyclophosphorothionate) which has higher affinity and binds with much less non-specific interaction to the same sites than ^{3}H-dihydropicrotoxinin (Squires *et al*. 1983). The relative potencies of barbiturates in competing for ^{35}S-TBPS parallel their potency in enhancing benzodiazepine receptor binding, indicating that the same macromolecular complex is probably involved in the convulsant benzodiazepine binding sites. Barbiturate inhibition of ^{35}S-TBPS binding is complex with decreases in number of binding sites as well as affinity changes (Trifiletti *et al*. 1984b; Fig. 7.2). Barbiturates markedly accelerate the dissociation of ^{35}S-TBPS from receptors whereas the convulsant picrotoxinin has no effect on dissociation. Thus, barbiturates do not act at the same site as picrotoxinin and TBPS but at separate sites which are allosterically linked. Similar interactions of the GABA antagonist bicuculline and ^{3}H-muscimol labelling of GABA receptors indicate an allosteric interaction.

Evidence that ^{35}S-TBPS labels sites associated with chloride ion channels includes the anion specificity of binding. In both membrane and solubilized

FIG. 7.2. Effect of barbiturates on equilibrium binding of [^{35}S]TBS. (A) Inhibition of [^{35}]TBPS binding to rat cerebral cortical membranes by DMBB (■), pentobarbital (0), amobarbital ([]), and phenobarbital (●). For clarity, similar data obtained for hexobarbital, methabarbital and barbituric acid are not shown. The results displayed are from a representative experiment replicated three times. The computed $\log_{10}$ IC$_{50}$ Values (mean ± S.E.M., three experiments) are: (±)-DMBB, -4.61 ± 0.06; (±)-pentobarbital, -4.27 ± 0.03; (±)-hexobarbital, -4.06 ± 0.05; (±)-amobarbital, -4.02 ± 0.08; (±)-phenobarbital, -3.47 ± 0.05; (±)-methabarbital, -3.31 ± 0.20. Barbituric acid gave no detectable inhibition at 10^{-3}M. (B) Correlation of relative potency of barbiturates to stimulate [^{3}H]diazepam binding (taken from Leeb-Lundberg *et al*. 1980) and the apparent IC$_{50}$ for inhibition of [^{35}S]TBPS binding. 1, methabarbital; 2, phenobarbital; 3, amobarbital; 4, hexobarbital; 5, pentobarbital; 6, DMBB. Data are from a representative experiment replicated three times. (C) Scatchard analysis of [^{35}S]TBPS binding to rat cerebral cortical membranes in the presence of various barbiturates. Data are from a representative experiment replicated two times. All Scatchard analyses shown are transformations of displacement experiments of [^{35}S]TBPS by unlabelled TBPS in the presence of barbiturates at the specific concentration. The total TBPS concentration utilized in Scatchard analyses ranged from 0.5 nM to 1.6 μM.

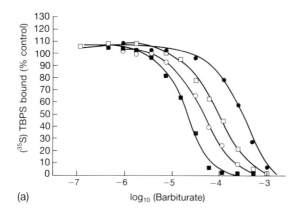

(a)

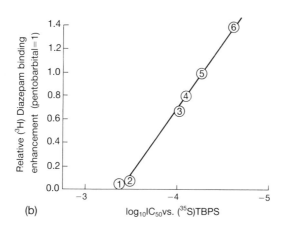

(b)

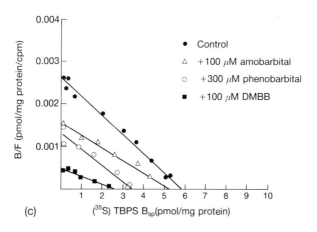

(c)

preparations ^{35}S-TBPS binding is absolutely dependent upon the presence of appropriate anions with bromide and chloride being most effective, while iodide and thiocyanate also are active. In sharp contrast, the channel-impermeant anions acetate, succinate and perchlorate do not support ^{35}S-TBPS binding. Thus, the ion specificity of these binding sites fits well with neurophysiologic properties of the chloride channel.

Identification of benzodiazepine receptors led to a search for non-benzodiazepine structures that act at the same sites. Triazolo-pyridazines such as CL-218872 compete directly at benzodiazepine binding sites and have an antianxiety profile indistinguishable from the benzodiazepines. Cyclopyrrolones also differ in structure from the benzodiazepines, but resemble their pharmacological effects. Cyclopyrrolones such as suriclone and zopiclone are among the most potent drugs known in competing for benzodiazepine binding (Blanchard and Julou 1983; Fig. 7.3). However, we recently observed that cyclopyrrolones act at sites distinct from the benzodiazepines (Trifiletti and Snyder 1984). They lower the numbers of benzodiazepine binding sites without changing affinity. Moreover, cyclopyrrolones accelerate the dissociation of ^{3}H-benzodiazepine ligands supporting their action at a novel site which is linked allosterically to benzodiazepine receptors (Fig. 7.4). Cyclopyrrolones also differ from classical benzodiazepines in that their binding is not modulated by GABA, chloride, or barbiturates. Since GABA, chloride, and barbiturates regulate the binding of benzodiazepine agonists but not benzodiazepine antagonists, cyclo-pyrrolones behave more like benzodiazepine antagonists than agonists. However, the behavioural effects of cyclopyrrolones are essentially the same as that of benzodiazepine agonists.

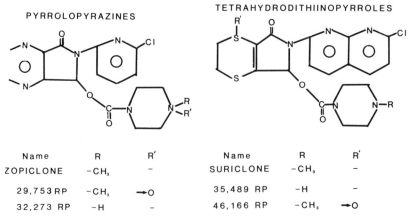

FIG. 7.3. Structures of various cyclopyrrolone drugs.

Even more striking differences between benzodiazepines and cyclopyrrolones are apparent in their interactions with ^{35}S-TBPS binding (Trifiletti *et al*. 1984c). Under conventional binding conditions, benzodiazepines are extremely weak in competing for ^{35}S-TBPS binding requiring concentrations 1000–1 000 000 times greater than those required to compete directly at benzodiazepine binding sites (Table 7.1), although in the presence of GABA, benzodiazepine becomes more potent at TBPS sites (Lawrence *et al*. 1984). By contrast, cyclopyrrolones compete for ^{35}S-TBPS binding in the low nanomolar range. Suriclone reduces the number of TBPS binding sites with little effect on affinity. Since suriclone also accelerates the dissociation of ^{35}S-TBPS, one can conclude that cyclopyrrolones act at a distinct site which is linked allosterically to the convulsant receptor. A close association of the cyclopyrrolone site to GABA receptors is indicated by the ability of the GABA antagonist bicuculline to reverse the inhibition by suriclone of ^{35}S-TBPS binding. The GABA agonist muscimol and barbiturates also influence ^{35}S-TBPS binding in a fashion similar to suriclone. However, other receptor interactions indicate that muscimol and barbiturates act at different sites than the cyclopyrrolones.

TABLE 7.1. *Inhibition of* [^{35}S]*TBPS binding to rat cerebral cortex by various drugs*

Drug	$IC_{50}(\mu m)$
Suriclone	0.003,3*
RP35,489	0.003,3*
RP46,166	0.003,3*
Zopiclone	0.01,3*
TBPS	0.07
Picrotoxinin	0.21
Muscimol	0.49
GABA	3.0
CL-218,872	4.5
Flunitrazepam	5.8
β-CCM[1]	6.2
Pentobarbital	83.2
Ro-15-1788	>100

Values are from a representative experiment replicated at least two times with less than 20 per cent variation. Each inhibition curve was determined using at least six concentrations of the appropriate drug.

*These IC_{50} values refer to the inhibition of the suriclone-sensitive (lower values) and -insensitive (higher values) components of [^{35}S]TBPS binding.

[1] β-CCM, methyl- β-carboline 3-carboxylate.

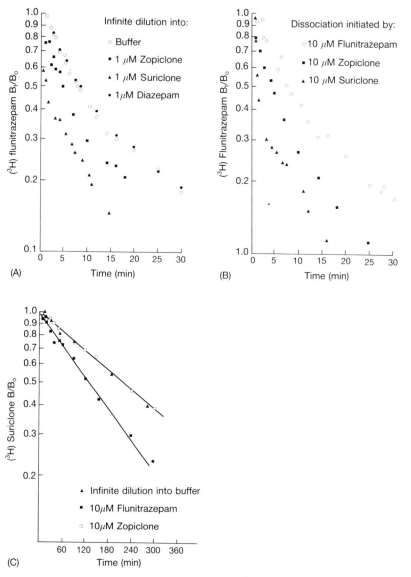

FIG. 7.4. Dissociation kinetics differentiate cyclopyrrolone and benzodiazepine binding sites. (A) Dissociation of [³H]flunitrazepam initiated by limiting dilution into various media. Rat cerebral cortical membranes were prepared and dissociation kinetics at 0°C were determined. Dissociation was initiated by 100-fold dilution into 50 nM Tris-citrate (pH 7.2) buffer or the same buffer containing 1 μM zopiclone, 1 μM suriclone, or 1 μM diazepam. Total specific binding prior to initiation of dissociation was approximately 3500 ct/min and non-specific binding was 600 ct/min. Data are from a representative experiment

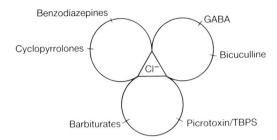

FIG. 7.5. A hypothetical model of the benzodiazepine/GABA-A/chloride ionophore complex. The complex appears to consist of three 'domains' (depicted by large circles) each associated with a chloride anionophore (depicted by the central triangle): a 'benzodiazepine' domain, a 'GABA' domain, and a 'sedative/convulsant' domain, which can be labelled with [³H]diazepam, [³H]GABA and [³⁵S]TBPS, respectively. Cyclopyrrolones, bicuculline and barbiturates appear to act at sites closely linked to (but physically distinct from) sites where benzodiazepines, GABA, and picrotoxin/TBPS, respectively, act. The above figure does not imply that the various distinct recognition sites are on distinct protein subunits.

The diverse binding properties of these drugs suggest that a single macromolecular complex with several interacting subunits is involved (Fig. 7.5). All of these components occur in the same protein complex, since the benzodiazepine binding protein purified to apparent homogeneity retains regulation by GABA, chloride ions and sedative drugs (Sigel and Barnard 1984). The receptor complex is associated with a chloride ion channel which is specific for GABA. The reason for this conclusion is that convulsants such as picrotoxinin act via chloride channels to block GABA effects, but not those of glycine which are also mediated by chlorine channels.

Antianxiety effects of drugs may involve at least three different receptor sites. Type I and Type II benzodiazepine receptors differ in their general localization throughout the brain and in their pre- and post-synaptic

replicated two times. (B) Dissociation of [³H]flunitrazepam from rat cerebral cortical membranes at 0°C initiated by addition of 10 μM flunitrazepam, 10 μM zopiclone, or 10 μM suriclone. Binding prior to initiation of dissociation was as described in A. Results are from a representative experiment replicated twice. (C) Dissociation of [³H]suriclone from rat cerebral cortical membranes at 0°C initiated by various procedures. Total specific binding prior to initiation of dissociation was approximately 700 ct/min and non-specific binding was 100 ct/min. Results are from a representative experiment replicated twice. All lines drawn are weighted linear least squares fits to the experimental data. Note that the ordinates of (A)–(C) are logarithmic scales.

occurrences. Cyclopyrrolones act via a receptor which is associated with GABA, benzodiazepine, and convulsant sites as well as the chloride channel. These extremely potent drugs have clinical activity as antianxiety agents and hypnotics. Whether or not the existing cyclopyrrolones represent a major therapeutic advance, they indicate the possibility for designing agents with unique selectivity for components of the GABA receptor complex.

References

Blanchard, J. C., and Julou, L. (1983). Suriclone, a new cyclopyrrolone derivative recognizing receptors labelled by benzodiazpines in rat hippocampus and cerebellum. *J. Neurochem*. **40**, 601–7.

Blume, A. J. (1978). Interactions of ligands with opiate receptors of brain membranes: regulation by ions and nucleotides. *Proc. Nat. Acad. Sci. USA*, **75**, 1713–7.

Bowery, N. G. (1983). Classification of GABA receptors. In *The GABA Receptors* (ed. S. J. Enna), pp. 178–213. Humana Press, New Jersey.

Braestrup, C. and Nielsen, M. (1981). ^{3}H-propyl-beta-carboline-3-carboxylate as a selective radioligand for the BZ_1 benzodiazepine receptor subclass. *J. Neurochem*. **37**, 333–41.

Childers, S. R. and Snyder, S. H. (1978). Guanine nucleotides differentiate agonist and antagonist interactions with opiate receptors. *Life Sci*. **23**, 759–62.

—— and —— (1980). Differential regulation by guanine nucleotides of opiate agonist and antagonist receptor interactions. *J. Neurochem*. **34**, 583–93.

Costa, T., Rodbard, D., and Pert, C. B. (1979). Is the benzodiazepine receptor coupled to a chloride anion channel. *Nature*, **277**, 315–7.

Davis, W. C. and Ticku, M. K. (1981). Pentobarbital enhances ^{3}H-diazepam binding to soluble receptors at the benzodiazepine-GABA-receptor-ionophore complex. *Neurosci. Lett*. **23**, 209–13.

Enna, S. J., and Snyder, S. H. (1975). Properties of gamma-aminobutyric acid (GABA) receptor binding in rat brain synaptic membrane fractions. *Brain Res*. **100**, 81–97.

—— and —— (1977). Influences of ions, enzymes, and detergents on gamma-aminobutyric acid-receptor binding in synaptic membranes of rat brain. *Mol. Pharmacol*. **13**, 442–53.

Haefely, W., Pieri, L., Polc, P., and Schaffner, R. (1981). General pharmacology and neuropharmacology of benzodiazepine derivatives. In *Handbook of Experimental Pharmacology, Part 2* (eds F. Hoffmeister and G. Stille) Vol. 55, pp. 13–262. Springer-Verlag, Berlin.

Klotz, K. L., Bocchetta, A., Neale, J., Thomas, J. W., and Tallman, J. F. (1984). Proteolytic degradation of neuronal benzodiazepine binding sites. *Life Sci*. **34**, 293–99.

Lawrence, L. J., Gee, K. W., and Yamamura, H. I. (1984). Benzodiazepine anticonvulsant action: gamma-aminobutyric acid-dependent modulation of the chloride ionophore. *Biochem. Biophys Res. Comm*. **123**, 1130–7.

Leeb-Lundberg, F., Snowman, A., and Olsen, R. W. (1980). Barbiturate receptor sites are coupled to benzodiazepine receptors. *Proc. Nat. Acad. Sci. USA*, 77, 7468–72.

Lippa, A. S., Meyerson, L. R., and Beer, B. (1982). Molecular substrates of anxiety: Clues from the heterogeneity of benzodiazepine receptors. *Life Sci.* 31, 1408–17.

Lo, M. M. S., Niehoff, D. L., Kuhar, M. J., and Snyder, S. H. (1983a). Autoradiographic differentiation of multiple benzodiazepine receptors by detergent solubilization and pharmacologic specificity. *Neurosci. Lett.* 39, 37–44.

——, ——, ——, and —— (1983b). Differential localization of Type I and Type II benzodiazepine binding sites in substantia nigra. *Nature*, 306, 57–60.

—— and Snyder, S. H. (1983). Two distinct solubilized benzodiazepine receptors: Differential modulation by ions. *J. Neurosci.* 3, 2270–9.

——, Strittmatter, S. M., and Snyder, S. H. (1982). Physical separation and characterization of two types of benzodiazepine receptors. *Proc. Nat. Acad. Sci. USA,* 79, 680–4.

Martin, I. L., and Candy, J. M. (1978). Facilitation of benzodiazepine binding by sodium chloride and GABA. *Neuropharmacol.* 17, 993–8.

Mohler, H., and Okada, T. (1977). Demonstration of benzodiazepine receptors in the central nervous system. *Science*, 198, 849–51.

Muller, W. E. and Snyder, S. H. (1978). Strychnine binding associated with synaptic glycine receptors in rat spinal cord membranes: Ionic influences. *Brain Res.* 147, 107–16.

Olsen, R. W. (1981). GABA-benzodiazepine-barbiturate-receptor interactions. *J. Neurochem.* 37, 1–13.

Sieghart, W. and Karobath, M. (1980). Molecular heterogeneity of benzodiazepine receptors. *Nature (Lond.)* 286, 285–7.

Sigel, E., and Barnard, E. A. (1984). A γ-aminobutyric acid benzodiazepine receptor complex from bovine cerebral cortex: Improved purification with preservation of regulatory sites and their interactions. *J. Biol. Chem.* 259, 7219–23.

Squires, R. F., Benson, D. I., Braestrup, C., Coupet, J., Klepner, C. A., Myers, V., and Beer, B. (1979). Some properties of brain specific benzodiazepine receptors: new evidence for multiple receptors. *Pharmacol. Biochem. Behav.* 10, 825–30.

—— and Braestrup, C. (1977). Benzodiazepine receptors in rat brain. *Nature (Lond.)* 266, 732–4.

——, Casida, S. E., Richardson, M., and Saedrup, E. (1983). [^{35}S]*tert*-butylbicyclophosphorothionate binds with high affinity to specific sites coupled to gamma-aminobutyric acid-A and ion recognition sites. *Mol. Pharmacol.* 23, 326–36.

Snyder, S. H. (1975). Opiate receptor in normal and drug altered brain function. *Nature (Lond.)* 257, 185–9.

—— (1984). Drug and neurotransmitter receptors in the brain. *Science*, 224, 22–31.

Takeuchi, A. and Takeuchi, N. (1969). A study of the action of picrotoxin on the inhibitory neuromuscular junction of the crayfish. *J. Physiol.* 205, 377–91.

Tallman, J. F., Paul, S. M., Skolnick, P., and Gallager, D. W. (1980). Receptors for the age of anxiety: pharmacology of benzodiazepines. *Science*, **207**, 247–81.

Trifiletti, R. R., Lo, M. M. S., and Snyder, S. H. (1984a). Kinetic differences between Type I and Type II benzodiazepine receptors. *Mol. Pharmacol*. **26**, 228–40.

——, Snowman, A. M., Snyder, S. H. (1984b). Barbiturate recognition site on the GABA/benzodiazepine receptor complex is distinct from the picrotoxinin/TBPS recognition site. *Eur. J. Pharmacol*. **106**, 441–7.

——, ——, and —— (1984c). Anxiolytic cyclopyrrolone drugs allosterically modulate the binding of [^{35}S]t-butylbicyclophosphorothionate to the benzodiazepine/gamma aminobutyric acid-A receptor/chloride anionophore complex. *Mol. Pharmacol*. **26**, 470–6.

——, and Snyder, S. H. (1984). Anxiolytic cyclopyrrolones zopiclone and suriclone bind to a novel site linked allosterically to benzodiazepine receptors. *Mol Pharmacol*. **26**, 458–69.

——, and —— (1985). Localization of Type I benzodiazepine receptors to postsynaptic densities in bovine brain. *J. Neurosci*. **5**, 1049–57.

Unnerstall, J. R., Kuhar, M. J., Niehoff, D. L., and Palacios, J. M. (1981). Benzodiazepine receptors are coupled to a subpopulation of gamma-aminobutyric acid (GABA) receptors: evidence from a quantiative autoradiographic study. *J. Pharmacol. Exp. Ther*. **218**, 797–804.

Young, A. B. and Snyder, S. H. (1974). The glycine synaptic receptor: Evidence that strychnine binding is associated with the ionic conductance mechanism. *Proc. Nat. Acad. Sci. USA*, **71**, 4002–5.

Young, W. S., III, Niehoff, D., Kuhar, M. J., Beer, B., and Lippa, A. (1981). Multiple benzodiazepine receptor localization by light microscopic radiohistochemistry. *J. Pharmacol. Exp. Ther*. **216**, 425–30.

8

Amino acids as fast signals: discussion

E. H. F. WONG AND J. KEMP

Discussion of the chapter by Dr J. Barker raised the question of the correlation between GABA agonist affinity in binding studies and the average Cl⁻ ion channel lifetime they evoke. Although muscimol and dihydromuscimol activate channel durations double those of other GABA agonists their receptor affinities are also correspondingly higher. This correlation suggests that the ability of agonists to keep the Cl⁻ ion channel open is, in some way, conferred by its receptor affinity. It is of interest to know how agonist efficacy is related to channel lifetime and rate of opening.

The question was raised of similarities between GABA and glycine activated Cl⁻ conductance states and the possible similarities between the protein structure of the GABA and glycine receptors. Professor Barnard commented that the purified proteins of the benzo-diazepine/GABA-linked receptor and the strychnine/glycine-linked receptor were clearly distinct, although antibodies to one polypeptide of the GABA-A receptor complex had weak cross immuno-reactivity with a polypeptide from the purified glycine/strychnine receptor. Could this possibly represent the structural determinant of the Cl⁻ ion channel?

Much of the discussion of the chapter by Dr J. S. Kelly centred on the problems of identifying the transmitter at a specific synapse by using selective receptor antagonists or the electrophysiological properties of the post-synaptic potential. This is particularly difficult when the transmitter candidates such as glutamate and aspartate are capable of acting on several receptor subtypes. Thus, whilst the antagonist may block a synaptic response and identify the receptor subtype being acted upon, it will not reveal the identify of the synaptic transmitter. Furthermore, as the electrophysiological properties of the response are a consequence of the receptor type activated, this also fails to identify the endogenous transmitter.

Several questions arising from the chapter by Dr A. M. Sillito were related to the difficulties in studying the modulatory effects of peptides, particularly in a physiologically functional system such as the cat visual cortex. For example, if the peptide transmitter was being released tonically, would the rate at which the system was activated by visual stimuli, and thus the amount of endogenous peptide released, affect the responsiveness of cells to exogenously applied peptide? It is possible that under certain circumstances the peptidergic response could already be activated maximally. In these experiments it is, however, possible to stimulate the system at different rates and examine the response over varying time scales, and several experimental paradigms had been employed. Professor K. Krnjevic commented that the fact that 'fast' neurotransmitters may also mediate longer lasting effects should be remembered. Acetylcholine is a good example of this with fast actions at the nicotinic receptor and slower muscarinic receptor mediated events. GABA may also exert more prolonged effects by actions at GABA-B receptors, which could mediate the slow i.p.s.p.'s seen in the hippocampus and cortical regions.

Discussion of the chapter by Dr P. Krogsgaard-Larsen included the analgesic actions of GABA agonists. THIP has been developed as a potent, selective and clinically active GABA agonist based on systematic modification of the GABA molecule and the introduction of rigid analogues. However, THIP produces analgesia which is atropine sensitive, potentiated by physostigmine and reported to be insensitive to blockade by bicuculline. Whilst the involvement of a cholinergic system in THIP's analgesic effects is clearly apparent, the bicuculline insensitivity must be viewed with caution as in these studies low, subconvulsant doses of bicuculline were used (Hill *et al*. 1981; Grognet *et al*. 1983). Although this may be necessary in order to measure pain thresholds effectively, it also suggests that GABA-A receptors may not be adequately blocked. All the other effects of THIP, both *in vivo* and *in vitro*, are sensitive to bicuculline blockade.

The interaction between the GABA and excitatory glutaminergic system was also discussed in view of the possible involvement of over-activity of the glutaminergic system in some major neurological disorders, such as epilepsy and Huntington's Chorea. Dr Krogsgaard-Larsen described experiments from cultures (Meier *et al*. 1984) which indicated that GABA-A receptor activation inhibited the release of glutamic acid. However, as Dr Martin pointed out, this is not a universal phenomenon since in the cortico-striatal glutamergic pathway, GABA stimulates rather than inhibits glutamate release (Mitchell, 1980).

Questions arose from the chapter by Dr J. C. Watkins concerning the identity of the endogenous ligand(s) for the so-far identified heterogeneous excitatory amino acid receptors, i.e. NMDA, kainate, and quisqualate. Dr Watkins stated that even though many of the antagonists for this system are small peptides, there is as yet no evidence to suggest that specific peptide transmitters exist for the different receptors. It is more likely that glutamate acting on different receptors, probably with different effectors/modulators, produces different effects for different purposes. The fact that some glutamate receptor antagonists have definite anticonvulsant activity suggests a possible future for NMDA antagonists in antiepileptic therapy, providing the problem of absorption and penetration into CNS can be overcome. Towards this end, Dr Watkins felt that it is generally advisable to reduce the polarity of the molecule as in the development of ring compounds; coupling antagonists to a lipophilic residue to form a labile structure should also facilitate entry and delivery.

The blockade of NMDA receptor coupled ion channels by Mg^{2+} is an interesting phenomenon and the possibility of fluctuations of extracellular free Mg^{2+} levels under physiological circumstances was raised. At present there is little evidence either for or against this and it remains a possibility which cannot be ruled out. However, to have any marked effect the levels of free Mg^{2+} would need to fall well below the 1.3–2.0 mM estimated for cerebrospinal fluid. The voltage sensitive nature of the Mg^{2+} blockade may well play a more important role in the control of NMDA receptor mediated events.

The multiplicity and complexity of the GABA/benzodiazepine/picrotoxin/barbiturate receptor system and its involvement in the central action of minor tranquillizers has prompted considerable research efforts. However, the functional significance of the so-called 'peripheral' or Ro5-4864-selective benzodiazepine receptor site is a puzzle. Dr S. H. Snyder suggested in his chapter the involvement of Ro5-4864 sites in energy metabolism as indicated by localization of these sites on mitochondrial membranes. Indeed, calcium mobilization might also be a consequence of activation of these sites. Such observations may provide some understanding of the peripheral side-effects of benzodiazepines, in particular those with high affinity for the Ro5-4864 sites, e.g. diazepam. However, as Dr Barker pointed out, this does not explain why Ro5-4864 attenuates GABA-A receptor mediated responses in the cuneate nucleus in the mammalian C.N.S. (Simmonds, 1984) and the ability to elicit convulsion in certain animal model (Weissman *et al.* 1983). Does this imply a heterogeneity in the receptor mechanism between the central Ro5-4864 sites and those in the periphery?

References

Grognet, A., Hertz, Fe., and DeFeudis, F. V. (1983). Comparison of the analgesic actions of THIP and morphine. *Gen. Pharmacol.* **14**, 585–9.

Hill, R. C., Maurer, R., Buescher, H. H., and Roemer, D. (1981). Analgesic properties of the GABA-mimetic, THIP. *Eur. J. Pharmacol.* **69**, 221–4.

Meier, E., Jongen, D., and Schousboe, A. (1984). GABA induces functionally active low-affinity GABA receptors on cultured cerebellar granule cells. *J. Neurochem.* **43**, 1737–44.

Mitchell, P. R. (1980). A novel GABA receptor modulates stimulus-induced glutamate release from cortico-striatal terminals. *Eur. J. Pharmacol.* **67**, 119–22.

Simmonds, M. A. (1984). Interactions of the benzodiazepine Ro5-4864 with the GABA-A receptor complex. *Br. J. Pharmacol.* **82**, 198P.

Weissman, B. A., Cott, J., Paul, S. M., and Skolnick, P. (1983). Ro5-4864: a potent benzodiazepine convulsant. *Eur. J. Pharmacol.* **90**, 149–50.

Part II
NEUROPEPTIDES AND MONOAMINES AS SLOW SIGNALS

9

Receptors and second messengers for neuropeptides and monoamines

G. N. WOODRUFF

This section will be devoted to neuropeptides and monoamines as slow signallers in the central nervous system. In many areas of the brain, amines and neuropeptides may be present in the same neurones. Over the last few years there have been important advances in our understanding of the amine and peptide transmitters, the receptors upon which they act and the secondary processes set in motion as a result of receptor activation. However, we are still relatively ignorant of the full physiological role of the coexisting peptides.

There is currently much interest in how clinically-active drugs affect neurotransmission. Drugs affecting central transmitter systems are of tremendous importance as potential therapeutic agents in the treatment of disorders such as schizophrenia, depression, dementia, and Parkinson's disease. However, these same drugs are also important research tools. Their use has advanced our basic understanding of the receptor mechanisms involved in the actions of central neurotransmitters and of the significance of the various secondary messengers. One example of this is the discovery of multiple receptors for various amine neurotransmitters. Obtaining evidence for the multiplicity of these receptors only became possible with the development of specific drugs. For example, Black *et al*. (1972) showed, with the development of burimamide and later cimetidine, that histamine H_2 receptors really do exist, following on from the work of Ash and Schild (1966), who had shown earlier that only some of the actions of histamine could be blocked by the then available antihistamines. The use of drugs like mepyramine and cimetidine has allowed a good understanding of the significance of these receptors in the periphery, although we are still largely ignorant about their role in the brain.

Acetylcholine receptors were long ago divided into muscarinic and nicotinic receptors. More recently, the use of pirenzipine has produced evidence suggesting that different types of muscarinic receptor, M1 and M2, might exist (Bernie *et al*. 1980), although the functional significance of these possible subtypes is again not understood and the development of new specific antagonists is required before the hypothesis can be confirmed.

Similarly, there are multiple receptors for catecholamine transmitters. The original classification of catecholamine receptors into α and β was made mainly on the basis of the rank order of potency of different agonists. This hypothesis was substantiated with the development of the β-blockers. More selective drugs have led to a further subdivision of these receptors into $\alpha1$, $\alpha2$, $\beta1$, and $\beta2$ (Phillips 1980). Furthermore, separate receptors for dopamine are now known. The dopamine system is a good example to illustrate how new, selective drugs have allowed us to discover new information on brain receptors and secondary messengers, but at the same time to emphasize our lack of understanding of the physiological significance of multiple receptors. The discovery of Greengard and his group of a dopamine-sensitive adenylate cyclase in homogenates of striatum was an important finding, suggesting that the actions of dopamine in the mammalian CNS might be mediated by increasing levels of cAMP. This finding stimulated much research interest and it soon became apparent that there were some discrepancies. The neuroleptic sulpiride has been a key compound in probing the possible involvement of cyclic AMP in mediating the actions of dopamine. Sulpiride, unlike the 'traditional neuroleptics' does not block the stimulatory effect of dopamine agonists on the dopamine-stimulated adenylate cyclase. It is, however, a potent dopamine antagonist in behavioural and electrophysiological tests of dopamine receptor function (Woodruff 1982).

These and other findings led to the concept that there were two types of dopamine receptor, a D1 receptor, linked to adenylate cyclase, and a D2 receptor, non-cyclase linked. Binding studies using [^{3}H]-sulpiride were not entirely consistent with the hypothesis. The finding that agonist affinity for [^{3}H]-sulpiride binding sites was decreased by GTP analogues suggested that sulpiride-sensitive dopamine receptors might indeed be linked to adenylate cyclase, perhaps causing inhibition of the enzyme. Support for this has come from the measurement of cyclic AMP overflow in striatal slices. Interestingly, the neuropeptide CCK, which coexists with dopamine in some neurones, also enhances cyclic AMP overflow in striatal slices and this effect is also blocked by sulpiride and by the recently-introduced SCH 22390 (Long *et al*. 1985).

The hypothesis of D1 and D2 dopamine receptors could not be fully substantiated or refuted without the availability of selective antagonists for both types of postulated receptor. Thus, the physiological role of dopamine-stimulated adenylate cyclase remained elusive in the absence of a specific antagonist for D1 receptors. SCH 22390, a close structural analogue of the D1 agonist SK&F 38393, was the first such antagonist

(Hyttel 1983). SCH 22390 blocks the action of dopamine on the dopamine-sensitive adenylate cyclase and in binding studies has a high affinity for D1 receptors, but it does not displace binding from D2 receptors as labelled, for example, with [^{3}H]-sulpiride (Holden-Dye *et al*. 1985). Although SCH 22390 is not entirely selective, since it has a relatively high affinity for 5-HT receptors, it is nevertheless an important research tool.

With the development of specific drugs over the last decade, it has become apparent that multiple receptors for slow and fast neurotransmitters appears to be the rule rather than the exception. As yet the significance of these multiple receptors, or the secondary messengers which they activate, is not fully understood. However, progress is rapid and some of the key investigators in this area will be presenting their latest results in this section.

References

Ash, A. S. F. and Schild, H. O. (1966). Receptors mediating some actions of histamine. *Br. J. Pharmacol*. **27**, 427–39.

Black, J. W., Duncan, W. A. M., Durant, G. J., Ganellin, C. R., and Parsons, M. E. (1972). Definition and antagonism of histamine H$_2$-receptors. *Nature*, **236**, 385–90.

Bernie, C. P., Birdsall, M. J. M., Burgen, A. S. U. and Hulme, E. C. (1980). Pirenzipine distinguishes between different subclasses of muscarinic receptor. *Nature*, **283**, 90–2.

Holden-Dye, L., Poat, J. A., Senior, K., and Woodruff, G. N. (1985). The characterization of [^{3}H] sulpiride binding sites in pig striatal membranes. *Biochem. Pharmacol*. **34**, 2905–9.

Hytell, J. (1983). SCH 23390—the first selective dopamine D-1 agonist. *Eur. J. Pharmac*. **91**, 153–4.

Long, S. K., O'Shaughnessy, C. T., Poat, J. A., and Turnbull, M. J. (1985). Studies on the effects of cholecystokinin on cyclic AMP efflux from rat striatal slices. *Br. J. Pharmacol*. (Proc. Suppl.) **84**, 6P.

Phillips, D. K. (1980). Chemistry of alpha- and beta-adrenoceptor agonists and antagonists. In *Adrenergic Activators and Inhibitors Part I* (ed. L. Szerkeres) Handbook Exp. Pharm. 54/1, pp. 1–61. Springer, Berlin.

Woodruff, G. N. (1982). Plenary lecture on dopamine receptors. In: *Advances in Dopamine Research* (eds) M. Kohsaka, T. Shohmori, Y. Tsukada, and G. N. Woodruff), pp. 1–24. Pergamon Press, Oxford.

10

Voltage-sensitive ion channels mediating modulatory effects of acetylcholine, amines, and peptides

DAVID A. BROWN

All that an excitatory transmitter has to do to transmit a nerve impulse speedily and efficiently across a synapse is to produce a depolarization rapidly enough to open voltage-gated Na^+ channels without inactivating them and for the depolarization to subside in time for the next impulse. This 'fast-switching' function is performed beautifully by acetylcholine at vertebrate neuromuscular and ganglionic synapses, and by glutamate and allied substances at arthropod nerve-muscle junctions and at vertebrate central synapses, through the triggered opening of discrete receptor-coupled cation channels of short lifetime. The elementary requirement for inhibiting this process is to produce a sufficient increase in membrane conductance to an ionic species whose movement is in such a direction to reduce the excitatory synaptic potential deflexion below the threshold for action potential generation; or, if presynaptic, to shunt and abbreviate the presynaptic action potential to the extent required to impair Ca^{2+}-influx. Chloride is an ideal candidate for this inhibitory ionic species because most nerve cell membranes are relatively impermeant to Cl^- in the resting state and internal Cl^- concentrations are generally low, so a large increase in membrane conductance can be generated through the activation of a relatively small number of receptors by the inhibitory transmitters, GABA and glycine.

This system of fast 'on' and 'off' devices, operated by a limited variety of rather simple chemical transmitters, may be regarded as the 'switches' in the hard-wiring of the brain's circuitry. However, over the last 15 years or so, it has become increasingly apparent that a complete extra tier of chemical control mechanisms is superimposed on this fast-switching circuitry, operating to modify the input-output function of synaptic systems. Three facets of this second tier may be noted.

(a) It is operated by an extraordinary variety of chemical agents, but including such 'classical' transmitters as acetylcholine and noradrenaline.
(b) It occurs over time-scales several orders of magnitude longer than the millisecond time required to transmit the individual 'bits' of information in the hard-wired circuit.
(c) Although some second-tier chemicals act like first-line transmitters to open new ionic channels in the nerve cell membrane, others modify those already present, and thereby change the excitable behaviour pattern of the neurone.

This article is concerned specifically with the latter group of transmitters.

VOLTAGE-GATED ION CHANNELS

As originally shown by Hodgkin and Huxley (1952), the individual signals are generated and conducted through two voltage-gated ionic currents—the fast Na^+-current (I_{Na}) and the 'delayed rectifier' K^+-current (I_K). (In mammalian myelinated fibres, this is further reduced to the Na^+-current, spike repolarization occurring through Na^+-inactivation; (Chiu *et al.* 1979). However, more recent studies have revealed a variety of additional currents in the neurone somata and dendrites whose effect is to determine the pattern of discharge activity of the neurones (see reviews by Adams *et al.* 1980; Adams 1982; Crill and Schwindt 1983). To exemplify, some of the currents present in sympathetic and hippocampal neurones are listed in Tables 10.1 and 10.2, and indicated diagrammatically in Fig. 10.1. So far, at least 4 of these currents have been shown to be modified by endogenous neurotransmitters—the K^+-currents I_M, I_A and $I_{K(Ca)}$, and the Ca-current, I_{Ca} (see Table 10.3 and below for references). As with other fields of physiological investigation, pharmacological modification of the individual currents sometimes tells us as much about the normal function of the current as about the action of the drug.

M-CURRENT, I_M

This is a non-inactivating K^+-current which is activated at membrane potentials positive to -70 mV. It may be distinguished from the delayed rectifier current by (a) its more negative activation range, (b) small amplitude (a few nA) and (c) slow kinetics (maximum time-constant $\simeq 150$ ms) (see Adams *et al.* 1982a). It is not 'Ca-activated' in the sense of depending on prior influx of Ca^{2+} ions (Adams *et al.* 1982a) though some tonic control by intracellular divalent cation cannot be excluded (Galvan *et al.* 1984; T. G. Smart, unpublished data). I_M was

TABLE 10.1. *Voltage-sensitive currents in sympathetic neurones*

	Current	Ion	Properties	V_{thr}	Inhibited by	Reference
Inward spike currents	$\{ I_{Na}$	Na	Fast, inactivating	−40	TTX	1
	$\ I_{Ca}$	Ca	Fast, slow inactivating	−30	Cd, Co NA	2,3
Ca-independent K-currents	$\{ I_k$	K	Fastish, slow inactivation	−30	TEA	1,4,5
	$\ I_A$	K	Fast, completely inactivating	−60	4-AP	4,6,7
	$\ I_M$	K	Small, slow, sustained	−70	Ba ACh LHRH SP, Ang	4,8,9,10
Ca-activated K-currents	$\{ I_C$	K	Large, fast, V-sensitive	−30	TEA	11,12
	$\ I_{AHP}$	K	Small, slow, V-insensitive	None	dTC apamin ACh	13

References (1) Belluzzi *et al.* (1985b). (2) Adams (1981). (3) Galvin and Adams (1982). (4) Adams *et al.* (1982a). (5) Galvin and Sedlmeir (1984). (6) Galvin (1982). (7) Belluzzi *et al.* (1985a). (8) Brown and Adams (1980). (9) Adams *et al.* (1982b). (10) Constanti and Brown (1981). (11) Adams *et al.* (1982c). (12) Brown *et al.* (1983). (13) Adams *et al.* (1984).

TABLE 10.2. *Voltage-sensitive current in hippocampal neurones*

Current	Ion	Properties	Inhibited by		References
I_{Na}	Na	Fast, inactivating	TTX		1
$I_{Ca(1)}$	Ca	Slow, persistent	Cd, Co, DHP		2,3
$I_{Ca(2)}$	Ca	Fast, transient	Cd		4
I_K	K	Fastish, sustained	TEA		1
I_A	K	Fast, transient	4-AP		1,5
I_M	K	Slow, sustained	Ba	ACh/mus	6,7
I_C	K	Fast, V-sensitive	TEA		7
I_{AHP}	K	Slow, V-insensitive		ACh NA Hist	8
I_Q	Mixed	Activated by hyperpoln.	Cs		6

References: (1) Segal and Barker (1984). (2) Johnston *et al.* (1980). (3) Brown and Griffith (1983b). (4) Halliwell (1983). (5) Gustaffson *et al.* (1982). (6) Halliwell and Adams (1982). (7) Brown and Griffiths (1983a). (8) Lancaster and Adams (1984), R. A. Nicoll, personal communication.

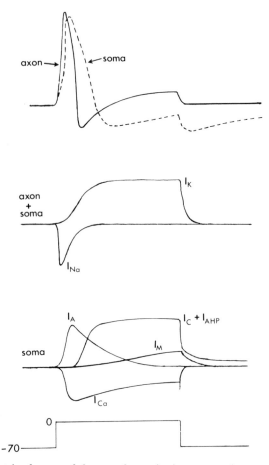

FIG. 10.1. Sketch of some of the membrane ionic currents (not to scale) activated on depolarizing an axon or soma from -70 mV to 0 mV about 50 ms (outward current upwards). Principal axonal currents are the Hodgkin-Huxley Na and K currents designated I_{Na} and I_K. Additional somatic currents depicted in this sketch are the transient K-current I_A, the inward Ca-current I_{Ca}, two Ca-activated K-currents I_C and I_{AHP}, and the slow K-current I_M. See Tables 10.1 and 10.2 for source material.

originally detected in frog sympathetic neurones (Brown and Adams 1980) and has since been identified in mammalian sympathetic neurones (Constanti and Brown 1981; Hashiguchi *et al.* 1982; Freschi 1983), hippocampal pyramidal cells (Halliwell and Adams 1982), olfactory cortex cells (Constanti and Galvan 1983), spinal cord neurones (Nowak and MacDonald 1983), neuroblastoma cells (D. A. Brown, H. Higashida, and G. Rougon, unpublished data) and even certain toad smooth muscle cells (Sims *et al.* 1984).

TABLE 10.3. *Transmitter-sensitivities of some endogenous membranes currents*

Tissue	I_M	I_{AHP}	I_A	I_{Ca}
Frog symp. gn.	ACh/mus[1] LHRH[2] SP[3,4]	ACh/mus[7,8]		
Rat symp. gn.	ACh/mus[5] Ang[5]			NA[17,18]
Hippocampal neurones	ACh/mus[6]	ACh/mus[9] NA[10,11] Hist[11] CRF[12]		
Sensory neurones				Dynorphin[19]
Locus coeruleus			5-HT[16]	
Myenteric neurones		ACh/mus[13] SP[14] 5-HT[15] Histamine[21]		GABA(?)[20]

References: (1) Brown and Adams (1980). (2) Adams and Brown (1980). (3) Adams *et al.* (1983). (4) Akasu *et al.* (1984). (5) Constani and Brown (1981). (6) Halliwell and Adams (1982). (7) Tokimasa (1984). (8) Adams *et al.* (1984). (9) Cole and Nicoll (1983). (10) Madison and Nicoll (1982). (11) Haas and Konnerth (1983). (12) Aldenhoff *et al.* (1983). (13) Morita *et al.* (1982). (14) North *et al.* (1983). (15) Wood and Mayer (1979b). (16) Aghajanian (1985). (17) Horn and McAfee (1980). (18) Galvin and Adams (1982). (19) Werz and MacDonald (1984). (20) Cherubini and North (1984). (21) Nemeth *et al.* (1984).

Pharmacological sensitivity of I_M

The most interesting aspect of I_M lies in its susceptibility to inhibition by slow transmitter substances. Thus, in most locations it is readily inhibited by muscarinic acetylcholine-receptor agonists. Indeed, it was this response which originally led to its identification (Brown and Adams 1980). However, I_M is neither uniquely nor ubiquitously sensitive to such agents. Not unique, because in frog sympathetic neurones I_M can be inhibited by a variety of other agents, acting through independent receptors, including mammalian and teleost LHRH (Adams and Brown, 1980; Katayama and Nishi 1982; Kuffler and Sejnowski 1983; Jones *et al.* 1984), substance P (Adams *et al.* 1983; Akasu *et al.* 1983b), and uridine and adenosine nucleotides (Adams *et al.* 1982b; Akasu *et al.* 1983a). Since these agents are mutually occlusive, one must suppose their effects on independent surface recognition sites to be transduced by a common messenger, but to date, the nature of this transduction step has not been unequivocally identified (see Adams *et al.* 1982b).

Regarding ubiquity: in the small 'C' subclass of frog sympathetic neurones, I_M is inhibited by LHRH, but only infrequently and weakly by acetylcholine (Jones 1984). This accords with the observation that synaptically-released acetylcholine produces a hyperpolarizing inhibitory post-synaptic potential (i.p.s.p.) in these cells, rather than a depolarizing excitatory post-synaptic potential (e.p.s.p.: see below), by *increasing* K^+-conductance (Dodd and Horn 1983). In hippocampal and cortical neurones also I_M appears to be less sensitive to acetylcholine than another, Ca^{2+}-activated, K^+-current (see below; B. Lancaster, D. V. Madison and R. A. Nicoll, unpublished data; Constanti and Sim 1985).

Role of I_M in slow synaptic excitation

Inhibition of I_M plays a major role in two forms of slow excitatory synaptic transmission in sympathetic ganglia—the cholinergic (muscarinic) slow e.p.s.p. observed in frog lumbar sympathetic 'B' cells (Adams & Brown 1982; Akasu *et al*. 1984) and in rat superior cervical ganglion cells (Brown and Selyanko 1985b); and the peptidergic (*t*LHRH-mediated) late slow e.p.s.p. (Jan and Jan 1982) in frog lumbar B- and C-cells (Katayama and Nishi 1982; Adams *et al*. 1984). I_M-inhibition may also contribute, in conjunction with inhibition of Ca^{2+}-dependent currents (Cole and Nicoll 1984a,b), to the cholinergic slow e.p.s.p. in hippocampal pyramidal neurones generated by septal afferents (Gähwiler and Brown 1985a and unpublished data).

Examination of the consequences of I_M-inhibition in sympathetic ganglia reveals two important facets to this form of slow synaptic excitation: the development of a *voltage-dependent* inward (depolarizing) current; and a change in the *excitable properties* of the neurone.

Voltage-dependent current

Although I_M is small, it operates over a membrane potential range—from rest potential to spike threshold—where other voltage-dependent currents are absent, and hence contributes the major outward component of the steady-state current. Inhibition of I_M therefore reduces the amount of outward current and thereby induces an inward current. Because the amplitude of I_M increases steeply with membrane depolarization, the slope factor for the underlying conductance G_M being $10\,mV$ for an e-fold increase in G_M (Adams *et al*. 1982a), so the inward current arising from a constant amount of I_M-inhibition also increases steeply with membrane potential. This has the functional consequence that the membrane depolarization may be quite small at hyperpolarized potentials, where I_M is minimal, but will be increased when the cell is under another depolarizing influence. Calculations suggest (A. Constanti and D. A. Brown unpublished data), and experiments confirm (Tosaka *et al*. 1983), that a significant amount of I_M-activation accompanies a single nicotinic fast e.p.s.p. in the frog ganglion, so the slow e.p.s.p. may be enhanced during tonic excitatory activity. Conversely, and notwithstanding the reduced driving force, the nicotinic fast e.p.s.p. is augmented and prolonged during the peptidergic late slow e.p.s.p. when I_M is inhibited (Schulman and Weight 1976). This pin-points one characteristic of this form of voltage-dependent slow synaptic potential, that it is an *interactive* synaptic event which may have minimal function significance on its own but serves to reinforce, and is reinforced by, adjuvent synaptic activity.

Excitability changes

Notwithstanding the aforementioned paragraphs, and although strong peptidergic or cholinergic stimulation can evoke spike discharges in the absence of nicotinic synaptic activity, it seems unlikely that the principal function of the slow and late slow synaptic potentials in sympathetic ganglia is to induce a spike discharge. I_M-inhibition is a relatively inefficient mechanism for such an action, since the anticipated relationship between the inhibition of the M-conductance and the resultant depolarization is highly sublinear (see Fig. 26 in Adams *et al*. 1982b): this is, a very large fraction ($\geqslant 80$ per cent) of M-channels would need to be closed to produce a suprathreshold depolarization, because of the corrective effect of the outward current through the residual, open M-channels on the membrane potential change. Instead, the most striking effect of I_M-inhibition is to increase the excitability of the neurones and, in particular, to facilitate repetitive firing (Adams *et al*. 1982b; see also Fig. 10.3a below). This is readily observed following preganglionic stimulation or the application of I_M-inhibitors at strengths or dosages below those eliciting overt spike discharges *per se*. It appears that cells with well-developed M-currents are essentially 'phasic' in firing behaviour; that is, they respond well to intermittent stimulation, but are unable to sustain trains of spikes during prolonged depolarization. Adams *et al*. (1982b)) attributed this to the fact that the increased M-conductance generated by prolonged depolarizations (and by repeated spikes) both counteracts the depolarization and raises spike threshold. When I_M is inhibited, this increase in conductance no longer occurs and the cell is free to sustain a discharge at a rate of up to 20 Hz or so, subject to the normal accommodative processes. Indeed, under conditions of intense I_M-inhibition, mammalian ganglion cells can fire spontaneously for many minutes on end if excessive I_{Na}-inactivation is prevented by occasional brief hyperpolarizing pulses (see Brown and Selyanko 1985a). The cells then undergo a radical change in behaviour from a state where they only respond phasically to external stimuli to a tonically active state reminiscent of pacemaker cells. The same effect has been reported in frog sympathetic neurones following inhibition of I_M by intense peptidergic stimulation (Dodd & Horn 1983). For these reasons, Adams *et al.* (1982b) referred to I_M as a 'braking' current, and to synaptic inhibition of I_M as 'removing the brake'.

A corollary to this is that tonically-discharging cells, or cells capable of sustained, high-frequency repetitive discharges during prolonged membrane depolarization are unlikely to have well-developed M-currents or other slow adapting current. Such cells might include certain parasympathetic neurones in the gut and bladder walls (Nishi & North 1973; Wood and Mayer 1979a; Griffith *et al.* 1980). Although direct evidence for this proposition is lacking, it is interesting that these neurones show a rather different form of slow e.p.s.p., resulting from the inhibition of a voltage-*independent* species of K^+-conductance such that the e.p.s.p. may be readily reversed by membrane hyperpolarization (Johnson *et al.* 1980; Grafe *et al.* 1980; Gallagher *et al.* 1982). This would provide an appropriate mechanism for increasing the rate of spike-discharge without altering spike threshold (as may be likely if the synaptic depolarization were accompanied by an increased conductance). Conversely, the discharge rate can be reduced by increasing K^+-conductance, the prevalent form of i.p.s.p. in such cells (Gallagher *et al.* 1982; see also Dodd & Horn 1983).

Ca^{2+}-ACTIVATED K^+-CURRENTS

There appear to be at least two species of Ca^{2+}-activated K^+-current in vertebrate neurones: a large rapid, voltage-dependent current sometimes termed I_C (see Brown *et al.* 1983); and a smaller, slower, voltage-insensitive current termed I_{AHP} (Pennefather *et al.* 1985b). Where present, I_C probably plays a major role in spike-repolarization and in the early phase of the spike after-hyperpolarization. It is readily blocked by tetraethylammonium (TEA), but so far has not been shown to be controlled by neurotransmitters. I_{AHP} generates a more prolonged after-hyperpolarization. It is generally insensitive to TEA but, unlike I_C, is exquisitely sensitive to a variety of neurotransmitters. Thus, in myenteric neurones, the after-hyperpolarization is inhibited by 5-hydroxytryptamine (Wood and Mayer 1979b) substance P (North *et al.* 1983) and muscarinic agonists (North and Tokimasa 1983) and by stimulating the nerves releasing the corresponding transmitter substance (Wood and Mayer 1979a; Grafe *et al.* 1980; North and Tokimasa 1983; Morita and North, 1985). These agents also reduce the *resting* K^+-conductance in these cells, in a voltage-independent manner. This raises the possibility that the component of resting membrane conductance inhibited by transmitters is tonically activated by intracellular Ca^{2+}. Evidence in favour of this view has been provided by Grafe *et al.* (1980), who showed that the effects of neural stimulation and 5-hydroxytryptamine were replicated by Mn^{2+} ion (which inhibits Ca^{2+} entry), and by North and Tokimasa (1983) using Ba^{2+} ion to replicate acetylcholine action [which inhibits $I_{K(Ca)}$]. The latter suggest that acetylcholine modifies the amount of Ca available to activate the K^+ current.

An interesting situation arises where I_M and I_{AHP} co-present in the same neurones: both appear to be sensitive to inhibition by acetylcholine, but with varying differential sensitivities and the two have different sensitivities to other transmitters. Thus, in bullfrog sympathetic ganglia, the long I_{AHP}-generated spike after-hyperpolarization is sensitive to muscarinic agonists (Tokimasa 1984), but direct recording under voltage-clamp revealed a smaller effect of muscarine on I_{AHP} than on I_M (Adams *et al.* 1984). Likewise, I_{AHP} in rat sympathetic ganglia is inhibited much less than I_M by muscarine (Fig. 10.2). In contrast, in hippocampal and olfactory pyramidal cells, I_{AHP} (and spike after-hyperpolarization) appears to be more sensitive to acetylcholine than I_M (Cole and Nicoll 1983; D. V. Madison and R. A. Nicoll, unpublished data; Constanti and Sim 1985), and is strongly suppressed following cholinergic afferent stimulation (Cole and Nicoll 1983). Like that in myenteric neurones, the long hippocampal after-hyperpolarization is sensitive to a variety of agents, including acetylcholine, noradrenaline, histamine, and CRF (Madison and Nicoll, 1982; Haas and Konnerth 1983; Cole and Nicoll 1983; Aldenhoff *et al.* 1983). The effect of noradrenaline is mediated by β-receptors and, like other β-mediated effects, appears to be transduced by cAMP (Madison and Nicoll 1982); the action of acetylcholine is imitated more closely by cGMP (Cole and Nicoll 1984b).

It is important to recognize that I_{AHP} and I_M are quite different currents. This is indicated not only by their very different kinetics, but also by a number of sharp pharmacological distinctions. Nevertheless, to a considerable extent they share the same functional role of 'adapting currents', since the principal effect of selectively inhibiting hippocampal I_{AHP} with noradrenaline is to reduce spike adaptation during long depolarizations and to facilitate repetitive discharge (Madison and Nicoll 1984).

In bullfrog ganglion cells, where both currents co-exist, inhibition of the two is synergistic in that the repetitive discharge produced by combined inhibition of I_M and I_{AHP} is more intense than that achieved by inhibition of either alone (Pennefather *et al.* 1985a). In rat sympathetic ganglia I_M appears to exert a more dominant adapting role, since inhibition of I_M alone induces a pronounced repetitive discharge (Fig. 10.3a) whereas

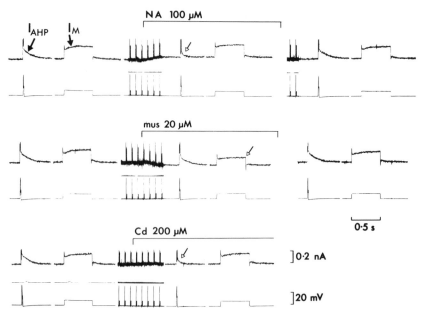

Fig. 10.2. Differential effects of noradrenaline (NA), muscarine (mus) and $CdCl_2$ ions (Cd) on the two outward K^+-currents, I_M and I_{AHP}, in a rat superior cervical sympathetic neurone. The cell was clamped at -50 mV and depolarized alternately to 0 mV for 20 ms or -35 mV for 500 ms. The former generated a spike and an inward Ca current (incompletely clamped): the resultant Ca-activated K-current I_{AHP} is recorded as a slowly-declining outward after-current (see Brown *et al*. 1982; Pennefather *et al*. 1985). The 10 mV, 500 ms command activates an additional component of I_M, seen as an outward relaxation during the command (see Adams *et al*. 1982a). Noradrenaline and Cd inhibit I_{AHP}, by reducing the priming Ca-current (see Galvan and Adams, 1982), but do not reduce I_M or affect steady current or conductance at the holding potential. (The small outward current produced by noradrenaline is insensitive to Cd and may reflect the small α_2-mediated hyperpolarization seen in these cells: Brown and Caulfield 1979). Muscarine selectively reduces I_M with little (<20 per cent) inhibition of I_{AHP}. Muscarine also reduces the resting conductance and produces an inward current at the holding potential, reflecting depression of that component of I_M already activated at -45 mV. (D. A. Brown and S. J. Marsh, unpublished experiment; see Brown and Selyanko 1985a for technical details).

indirect inhibition of I_{AHP} with noradrenaline (which inhibits I_{Ca} in these cells: Horn and McAfee 1980; Galvan and Adams 1982) has little effect on spike frequency adaptation (Fig. 10.3b), even after I_M is suppressed.* This accords with Barrett and Barrett's (1976) studies on Ca-dependent

---

*Some recent studies using apamin and hexamethonium to directly inhibit I_{AHP} suggest that I_{AHP} does influence spike train adaptation in rat ganglia in a similar manner to that seen in frog ganglia (Kawai *et al*. 1985, Galvan and Behrends 1985).

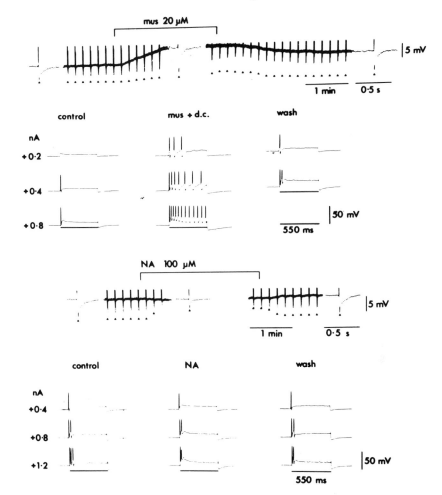

FIG. 10.3 Dissection of the roles of I_M and I_{Ca}/I_{AHP} in spike train adaptation in a rat superior cervical ganglion cell, using muscarine and noradrenaline. Upper (continuous) records show membrane potential and spike after-hyperpolarizations following single spikes generated by short (5 ms) depolarizing current injections. Lower records show spike trains evoked by long (550 ms) depolarizing current injections recorded at a fixed (−53 mv) membrane potential before, during and after drug application. Muscarine (mus) depolarized the cell through inhibition of I_M and induced repetitive spikes but did not inhibit the Ca-activated after-hyperpolarization (see Fig. 10.2); noradrenaline (NA) inhibited the after-hyperpolarization but did not inhibit I_M (see Fig. 10.2) and did not induce repetitive firing.

spike after-hyperpolarizations in frog motoneurones, in which it was concluded that the after-hyperpolarization acted as a 'spacer', maintaining rather than limiting low-frequency repetitive discharge. Notwithstanding, neurotransmitter-induced suppression of I_{AHP} affords another clear example of a slow synaptic event whose result is to change the behaviour pattern of cell excitability rather than to directly excite the cell (Nicoll 1982).

TRANSIENT K⁺-CURRENT, I_A

I_A has a venerable history as the first of the non-spiking, rhythm controlling neuronal currents to receive detailed study (in snail neurones: Conner and Stevens 1971; Neher 1971). It is a fast transient K^+-current which is activated below spike-threshold (from -60 mV upwards), but then undergoes rapid and usually complete inactivation with a time-constant in the 50–100 ms range. This means that I_A can only be activated by depolarization from hyperpolarized ($\geqslant -60$ mV) membrane potentials.

I_A, or an equivalent transient K^+ current, is present in a wide variety of vertebrate neurones, including sympathetic ganglion cells (Adams *et al.* 1982a; Galvan 1982; Galvan and Sedlmeir 1984; Belluzzi *et al.* 1985a,b), hippocampal pyramidal cells (Gustafsson *et al.* 1982; Segal and Barker 1984) and locus coeruleus neurones (Williams *et al.* 1984). Several functional effects of I_A have been identified. (a) In repetitively-firing invertebrate neurones and axons it acts as a 'spacer' current between action potentials, enabling low-frequency tonic discharges by delaying the spike and reducing I_{Na}-inactivation (Connor 1978). (b) It is activated sufficiently rapidly to reduce the amplitude of synaptic potentials evoked from hyperpolarized membrane potentials (Daut 1973; Williams *et al.* 1984). (c) It can accelerate spike repolarization when triggered from negative membrane potentials (Belluzzi *et al.* 1985a,b).

Since I_A (rather than the delayed rectifier current I_K) appears to be the prime target for the aminopyridine drugs (Thompson 1977; Gustafsson *et al.* 1982) the widespread pharmacological effects of 4-aminopyridine in enhancing transmitter release and producing convulsions suggest that I_A plays a very important role in controlling neuronal excitability. Until recently, it has proved singularly resistant to neurotransmitters. However Aghajanian (1985) has very recently reported that I_A in locus coeruleus neurones is inhibited by α_1-adrenoceptor agonists, and has suggested that this may be the mechanism whereby noradrenergic autosynapses in this region accelerate neuronal firing rate.

CALCIUM CURRENT, I_{Ca}

Calcium currents appear to be present in all neurones and play a number of crucial roles in triggering transmitter release, in dendritic signalling and in controlling neuronal spike patterns via their primary effect on Ca^{2+}-activated K^+-currents (see Llinas & Walton 1980). Possibly with the known sensitivity of cardiac calcium currents to neurotransmitters in mind (see Reuter 1983), I_{Ca} has been frequently probed for its response to neurotransmitters. In particular, depression of I_{Ca} has been put forward as the mechanism for presynaptic inhibition by a variety of neurotransmitters on the basis that such agents (including noradrenaline, opioid peptides and GABA) reduce the duration of the Ca^{2+}-component of the somal action potential in sensory ganglion cells (e.g. Dunlap and Fischbach, 1981). However, with some exceptions, this hypothesis has not withstood close examination of isolated Ca^{2+}-currents using voltage-clamp techniques. Instead, these studies indicate the principal cause of the reduced Ca^{2+} influx, in the soma at least, is more usually the initiation of a countervailing increase in K^+-conductance. This has been shown for enkephalin acting at μ receptors on sensory ganglion cells (Werz and Macdonald 1983; MacDonald and Werz 1985), for noradrenaline acting at α_2-receptors on locus coeruleus neurones (Williams and North 1985a) and for GABA acting at GABA-B (baclofen) receptors in hippocampal pyramidal cells (Newberry and Nicoll 1984, 1985; Gähwiler and Brown 1985b), all well-known inhibitors of transmitter release.

There remain three examples of direct inhibition of the Ca^{2+}-current. (a) Noradrenaline clearly inhibits I_{Ca} in rat sympathetic ganglion cells under conditions where K^+-currents are suppressed with Cs^+ (Galvan and Adams 1982; cf. Horn and McAfee 1980), but it also hyperpolarizes these cells under normal conditions (Brown and Caulfield 1979) so there may be two effects: the hyperpolarization is clearly mediated by α_2-receptors but the nature of the receptors inhibiting I_{Ca} is less well-documented. Noradrenaline has also recently been shown to inhibit I_{Ca} in locus coeruleus neurones, but by a different (and unclassified) receptor from the α_2-receptor increasing K^+-conductance (Williams and North 1985b). (b) Dynorphin, acting at κ-receptors, inhibits I_{Ca} in Cs-loaded sensory ganglion cells (Macdonald and Werz 1984, 1986), in contrast to enkephalin. (c) Adenosine has been reported to inhibit I_{Ca} in voltage-clamped rat sympathetic neurones (Henon and MacAfee 1984), but increases K^+-conductance in hippocampal (Halliwell and Scholfield 1984) and cardiac (Hartzell 1979) cells.

This resistance of neuronal Ca^{2+}-currents to neurotransmitters is rather surprising in view of the sensitivity of cardiac currents to adrenaline and acetylcholine, and in view of the fact that, like the cardiac currents, Ca-currents in isolated internally-perfused sensory neurones seemed to be regulated by intracellular cAMP (Kostyuk 1981). Since inhibition of adenylate cyclase has been implicated in the action of adenosine, of noradrenaline on α_2-receptors, and of GABA on GABA-B receptors, some inhibition of I_{Ca} might have been expected as a consequence of this effect. However, it does seem as though the neuronal Ca-current may not be that closely regulated by cAMP after all under normal conditions since, as pointed out above, activation of β-receptors or injection of cAMP in hippocampal neurones appears to inhibit directly the Ca^{2+}-mediated K^+-currents and not to alter the Ca^{2+}-current (Madison and Nicoll 1982, 1985). Here one can only note a close parallel to the best-studied invertebrate system where the facilitatory action of 5-hydroxytryptamine on Ca-spike in Aplysia neurones (Klein and Kandel 1978) turned out to result from a cAMP-mediated inhibition of a K^+-current (the 'S-current': Klein and Kandel 1980) rather than an enhancement of I_{Ca} *per se*. Teleologically, this may not be too surprising since it might be more efficient to regulate the amount of Ca^{2+}-entry during an action potential by adjusting countervailing K^+-currents than by modifying the Ca^{2+}-current at source, to judge from the often dramatic effects of K^+-channel blockers on the duration of Ca^{2+}-spikes.

CONCLUSIONS

This is obviously a rather limited survey; for example, peptides appear under-represented in Table 10.3. This may simply reflect the limited extent of current studies: as voltage-clamp studies are extended to a greater variety of neurones and transmitter candidates, more examples of neurotransmitter control of endogenous neuronal currents may be anticipated. The evidence so far available suggests that small, subthreshold currents controlling the rate and rhythm of nerve cell discharges, such as I_M and I_{AHP}, are likely to be the prime target for transmitter control.

One aspect which clearly warrants further attention is the tendency of several different transmitters to modify the same ionic current. Such duplication might most easily be accommodated by assuming some form of common intermediary biochemical step(s). A particularly clear example is the serotonin-induced closure of the S-channel in Aplysia neurones, mediated by cyclic AMP-dependent protein phosphorylation (Siegelbaum and Tsien, 1983). Studies on this system were enormously facilitated by the identification of individual S-channel and direct recording of S-channel closure. The identification of the corresponding channels underlying I_M and I_{AHP} is clearly a major prerequisite for further analysis.

At another level, observations on the effect of transmitters on voltage-sensitive currents directs attention to the control of neuronal excitability as mediated by the interplay of membrane currents. A precise understanding of the effect of reducing (say) I_M or I_{AHP} is not trivial, in view of the number and variety of other ionic currents present in nerve cell somata.

Acknowledgements

The work of the author referred to in this article was supported by the Medical Research Council. I wish to thank Mrs C. Pybus-Johnson for manuscript preparation.

References

Adams, D. J., Smith, S. J., and Thomson, S. H. (1980). Ionic currents in molluscan somata. *Ann. Rev. Neurosci.* **3**, 141–67.

Adams, P. R. (1981). The calcium current of a vertebrate neurone. In *Advances in Physiological Sciences* (ed. J. Salanki) Vol. 4, pp. 135–8. Akademiai Kaido, Budapest.

—— (1982). Voltage-dependent conductances of vertebrate neurones. *Trends Neurosci.* **5**, 116–9.

—— and Brown, D. A. (1980). Luteinizing hormone-releasing factor and muscarinic agonists act on the same voltage-sensitive K$^+$-current in bullfrog sympathetic neurons. *Br. J. Pharmacol.* **68**, 353–5.

——, and —— (1982). Synaptic inhibition of the M-current: slow excitatory post-synaptic potential mechanism in bullfrog sympathetic neurones. *J. Physiol. Lond.* **332**, 263–72.

——, ——, and Constanti, A. (1982a). M-currents and other potassium currents in bullfrog sympathetic neurones. *J. Physiol. Lond.* **330**, 537–72.

——, ——, and —— (1982b). Pharmacological inhibition of the M-current. *J. Physiol. Lond.* **332**, 223–62.

——, ——, and Jones S. W. (1983). Substance P inhibits the M-current in bullfrog sympathetic neurones. *Br. J. Pharmacol.* **79**, 330–3.

——, Constanti, A., Brown, D. A., and Clark, R. B. (1982c). Intracellular Ca^{2+} activates a fast voltage-sensitive K$^+$-current in vertebrate sympathetic neurones. *Nature*, **296**, 746–9.

——, Pennefather, P., Lancaster, B., Nicoll, R. A. (1984). Hybrid current/voltage clamp of the bullfrog sympathetic neuron slow after hyperpolarization. *Neurosci. Abs.* **10**, 147.

Aghajanian, G. (1985). Modulation of a transient outward current in serotonergic neurones by α_1-adrenoceptors. *Nature* **315**, 501–3.

Akasu, T., Gallagher, J. P., Koketsu, K., and Shinnick-Gallagher, P. (1984). Slow excitatory post-synaptic currents in sympathetic neurones of *Rana catesbiana*. *J. Physiol.* **351**, 583–93.

——, Hirai, K., and Koketsu, K. (1983a). Modulatory actions of ATP on membrane potentials of bullfrog ganglion cells. *Brain Res.* **258**, 313–7.

——, Nishimura, T., and Koketsu, K. (1983b). Substance P inhibits the action potentials in bullfrog sympathetic ganglion cells. *Neurosci. Lett.* **41**, 161–6.

Aldenhoff, J. B., Gruol, D. L., Rivier, J., Vale, W., and Siggins, G. R. (1983). Corticotrophin releasing factor decrease postburst hyperpolarizations and excites hippocampal neurons. *Scince,* **221**, 875–7.

Barratt, E. F. and Barratt, J. N. (1976). Separation of two voltage-sensitive potassium currents, and demonstration of a tetrodotoxin-resistant calcium current in frog motoneurones. *J. Physiol.* **255**, 737–74.

Belluzzi, O., Sacchi, O., and Wanke, E. (1985a). A fast transient outward current in the rat sympathetic neurone studied under voltage clamp. *J. Physiol.* **358**, 91–108.

——, —— and —— (1985b). Identification of delayed potassium and calcium currents in the rat sympathetic neurone under voltage clamp. *J. Physiol.* **358**, 109–30.

Brown, D. A. (1983). Slow cholinergic excitation—a mechanism for increasing neuronal excitability. *Trends Neurosci.* **6**, 302–6.

——, and Adams, P. R. (1980). Muscarinic suppression of a novel voltage sensitive K^+-current in a vertebrate neurone. *Nature (Lond.)* **283**, 673–6.

——, ——, and Constanti, A. (1982). Voltage-sensitive K-currents in sympathetic neurons and their modulation by neurotransmitters. *J. Autonomic Nerv. Syst.* **6**, 23–35.

——, and Caulfield, M. P. (1979). Hyperpolarizing α_2-adrenceptors in rat sympathetic ganglia. *Br. J. Pharmacol.* **65**, 435–45.

——, Constanti, A., and Adams, P. R. (1983). Ca-activated potassium current in vertebrate sympathetic neurones. *Cell Calcium*, **4**, 407–20.

——, and Griffith, W. H. (1983a). Calcium-activated outward in voltage-clamped hippocampal neurones of the guinea-pig. *J. Physiol.* **337**, 287–301.

——, and —— (1983b). Persistent slow inward current in voltage-clamped hippocampal neurones of the guinea-pig. *J. Physiol.* **337**, 303–20.

——, and Selyanko, A. A. (1985a). Two components of muscarine-sensitive membrane current in rat sympathetic neurones. *J. Physiol.* **358**, 335–64.

——, and —— (1985b). Membrane currents underlying the cholinergic slow excitatory post-synaptic potential in the rat superior cervical sympathetic ganglion. *J. Physiol.* **365**, 365–87.

Cherubini, E. and North, R. A. (1984). Actions of α-aminobutyric acid on neurones of guinea-pig myenteric plexus. *Br. J. Pharmacol.* **82**, 93–100.

Chiu, S. Y., Ritchie, J. M., Rogart, R. D., and Stagg, D. (1979). A quantitative description of membrane currents in rabbit myelinated nerve. *J. Physiol.* **292**, 149–66.

Cole, A. E. and Nicoll, R. A. (1983). Acetylcholine mediates a slow synaptic potential in hippocampal pyramidal cells. *Science*, **221**, 1299–301.

—— and —— (1984a). Characterization of a slow cholinergic post-synaptic potential recorded *in vitro* from rat hippocampal pyramidal cells. *J. Physiol.* **352**, 173–88.

——, and —— (1984b). The pharmacology of cholinergic excitatory responses in hippocampal pyramidal cells. *Brain Res.* **305**, 283–90.

Connor, J. A. (1978). Slow repetitive activity from fast conductance changes in neurones. *Fed. Proc.* **37**, 2139–45.

——, and Stevens, C. F. (1971). Voltage-clamp studies of a transient outward membrane current in gastropod neural somata. *J. Physiol.* **213**, 21–30.

Constanti, A. and Brown, D. A. (1981). M-currents in voltage-clamped mammalian sympathetic neurones. *Neurosci. Lett.* **24**, 289–94.

——, and Galvan, M. (1983). M-current in voltaged-clamped olfactory cortex neurones. *Neurosci. Lett.* **24**, 289–94.

——, and Sim, J. A. (1985). A slow, muscarine-sensitive Ca-dependent K current in guinea-pig olfactory cortex neurones *in vitro*. *J. Physiol* **365**, 47**P**.

Crill, W. E. and Schwindt, P. C. (1983). Active currents in mammalian central neurons. *Trends Neurosci.* **6**, 236–40.

Daut, J. (1973). Modulation of the excitatory synaptic response by fast transient K⁺-current in snail neurones. *Nature*, **246**, 193–6.

Dodd, J. and Horn, J. P. (1983). Muscarinic inhibition of sympathetic C neurones in the bullfrog. *J. Physiol. Lond.* **334**, 271–91.

Dunlap, K. and Fischbach, G. D. (1981). Neurotransmitters decrease the calcium conductance activated by depolarization of embryonic chick sensory neurones. *J. Physiol.* **317**, 519–35.

Freschi, J. (1983). Membrane currents of cultured rat sympathetic neurons under voltage clamp. *J. Neurophysiol.* **50**, 1460–78.

Gähwiler, B. and Brown, D. A. (1985a). Functional innervation of cultured hippocampal neurones by cholinergic afferents from co-cultured septal explants. *Nature*, **313**, 577–9.

——, —— (1985b). GABA$_B$-receptor-activated K⁺ current in voltage-clamped CA$_3$ pyramidal cells in hippocampal cultures. *Proc. Nat. Acad. Sci. USA*, **82**, 1558–62.

Gallagher, J. P., Griffith, W. H., and Shinnick-Gallagher, P. (1982). Cholinergic transmission in cat parasympathetic ganglia. *J. Physiol.* **332**, 473–86.

Galvan, M. (1982). A transient outward current in rat sympathetic neurones. *Neurosci. Lett.* **31**, 295–300.

——, and Adams, P. R. (1982). Control of calcium current in rat sympathetic neurones by noradrenaline. *Brain Res.* **244**, 135–44.

——, and Behrends (1985). Apamin blocks calcium-dependent spike after-hyperpolarization in rat sympathetic neurones. *Pflüg. Arch.* **403**, (suppl.), R50.

——, Satin, L. S., and Adams, P. R. (1984). Comparison of conventional microelectrode and whole-cell patch clamp recordings from cultured bullfrog ganglion cells. *Neurosci.* Abs. **10**, 146.

——, and Sedlmeir, C. (1984). Outward currents in voltage-clamped rat sympathetic neurones. *J. Physiol.* **356**, 115–33.

Grafe, P., Mayer, C. J., and Wood, J. D. (1980). Synaptic modulation of calcium-dependent potassium conductance in myenteric neurones in the guinea-pig. *J. Physiol.* **305**, 235–48.

Griffith, W. H., Gallagher, J. P., and Shinnick-Gallagher, P. (1980). An intracellular investigation of cat vesical pelvic ganglia. *J. Neurophysiol.* **43**, 343–54.

Gustafsson, B., Galvan, M., Grafe, P., and Wigstrom, H. (1982). A transient outward current in a mammalian central neurone blocked by 4-aminopyridine. *Nature*, **299**, 252–4.

Haas, H. L. and Konnerth, A. (1983). Histamine and noradrenaline decreases calcium-activated potassium conductance in hippocampal pyramidal cells. *Nature*, **302**, 432–4.

Halliwell, J. V. (1983). Caesium-loading reveals two distinct Ca-currents in voltage-clamped guinea-pig hippocampal neurones *in vitro*. *J. Physiol.* **341**, 10–1*P*.

——, and Adams, P. R. (1982). Voltage-clamp analysis of muscarinic excitation in hippocampal neurons. *Brain Res.* **250**, 71–92.

——, and Scholfield, C. N. (1984). Somatically recorded Ca-currents in guinea-pig hippocampal and olfactory cortex neurones are resistant to adenosine action. *Neurosci. Lett.* **50**, 13–8.

Hartzell, H. C. (1979). Adenosine receptors in the frog sinus venosus: slow inhibitory potentials produced by adenine compounds and acetylcholine. *J. Physiol.* **293**, 23–49.

Hashiguchi, T., Kobayashi, H., Tosaka, T., and Libet, B. (1982). Two muscarinic depolarizing mechanisms in mammalian sympathetic neurons. *Brain Res.* **242**, 378–82.

Henon, B. K. and McAfee, D. A. (1984). Modulation of calcium currents by adenosine receptors on mammalian sympathetic neurons. In *Regulatory Function of Adenosine* (eds R. M. Berne, T. W. Rall, and R. Rubio), pp. 455–66. Martinus Nijhoff, The Hague.

Hodgkin, A. L. and Huxley, A. F. (1952). A quantitative description of membrane current and its application to conduction and excitation in nerve. *J. Physiol.* **117**, 500–44.

Horn, J. P. and McAfee, D. A. (1980). Alpha-adrenergic inhibition of calcium-dependent potentials in rat sympathetic neurones. *J. Physiol.* **301**, 191–204.

Jan, L. Y. and Jan, Y. N. (1982). Peptidergic transmission in sympathetic ganglia of the frog. *J. Physiol.* **327**, 219–46.

Johnson, S. M., Katayama, Y., and North, R. A. (1980). Slow synaptic potentials in neurones of the myenteric plexus. *J. Physiol.* **301**, 505–16.

Johnston, D., Hablitz, J. J., and Wilson, W. A. (1980). Voltage-clamp discloses slow inward current in hippocampal burst-firing neurones. *Nature*, **286**, 391–3.

Jones, S. W. (1984). Muscarinic and peptidergic action on C cells of bullfrog sympathetic ganglia. *Neurosci. Abs.* **10**, 207.

——, Adams, P. R., Brownstein, M. J., and Rivier, J. E. (1984). Telcost luteinizing hormone-releasing hormone: action on bullfrog sympathetic ganglia is consistent with role as nuerotransmitter. *J. Neurosci.* **4**, 420–9.

Katayama, Y. and Nishi, S. (1982). Voltage-clamp analysis of peptidergic slow deploarizations in bullfrog sympathetic ganglion cells. *J. Physiol. Lond.* **333**, 305–13.

Kawai, *et al.* (1985). Hexamethonium increases the excitability of sympathetic neurons by the blockade of the Ca^{2+}-activated K^+ channels. *Life Sci.* **36**, 2339–46.

Klein, M. and Kandel, E. R. (1978). Presynaptic modulation of voltage-dependent Ca^{2+}-current. *Proc. Nat. Acad. Sci.* **75**, 3512–6.

Klein, M. and Kandel, E. R. (1980). Mechanism of calcium current modulation underlying presynaptic facilitation on behavioural sensitization in *Aplysia*. *Proc. Nat. Acad. Sci.* **77**, 6912–6.

Kostyuk, P. G. (1981). Calcium channels in the neuronal membrane. *Biochim. Biophys. Acta*, **650**, 128–50.

Kuffler, S. W. and Sejnowski, T. J. (1983). Peptidergic and muscarinic excitation at amphibian sympathetic synapses. *J. Physiol. Lond.* **341**, 257–78.

Lancaster, B. and Adams, P. R. (1984). Single electrode voltage-clamp of the slow AHP current in rat hippocampal pyramidal cells. *Neurosci. Abs.* **10**, 872.

Llinas, R. and Walton, K. (1980). Voltage-dependent calcium conductances in neurones. In *The Cell Surface and Neuronal Function* (eds C. W. Cobman, G. Poste, and G. L. Nicolson), pp. 87–118. Elsevier/North Holland: Amsterdam.

MacDonald, J. F. and Werz, M. A. (1986). *J. Physiol.* (in press).

Madison, D. V. and Nicoll, R. A. (1982). Noradrenaline blocks accommodation of pyramidal cell discharge in the hippocampus. *Nature*, **299**, 636–8.

——, and Nicoll, R. A. (1984). Control of the repetitive discharge of rat CA1 pyramidal cells *in vitro*. *J. Physiol.* **354**, 319–31.

——, and —— (1986). *J. Physiol.* (in press).

Morita, K. and North, R. A. (1985). Significance of slow synaptic potentials for transmission of excitation in guinea-pig myenteric plexus. *Neurosci.* **14**, 661–72.

——, ——, and Tokimasa, T. (1982). Muscarinic agonists inactivate potassium conductance in guinea-pig myenteric neurons. *J. Physiol.* **321**, 471–91.

Neher, E. (1971). Two fast transient current components during voltage clamp on snail neurons. *J. Gen. Physiol.* **58**, 36–53.

Nemeth, P. R., Ort, C. A., and Wood, J. D. (1984). Intracellular study of effects of histamine on electrical behaviour of myenteric neurones in guinea-pig small intestine. *J. Physiol.* **355**, 411–25.

Newberry, N. R. and Nicoll, R. A. (1984). Direct hyperpolarizing action of baclofen on hippocampal pyramidal cells. *Nature*, **308**, 450–2.

——, and —— (1985). Comparison of the action of baclofen with gamma-aminobutyric acid on rat hippocampal pyramidal cell *in vitro*. *J. Physiol.* **360**, 161–86.

Nicoll, R. A. (1982). Neurotransmitters can say more than just 'yes' or 'no'. *Trends Neurosci.* **5**, 369.

Nishi, S. and North, R. A. (1973). Intracellular recording from the myenteric plexus of the guinea-pig ileum. *J. Physiol.* **231**, 471–91.

North, R. A., Morita, K., and Tokimasa, T. (1983). Peptide actions on autonomic nerves. In *Systemic Role of Regulatory Peptides* (eds S. R. Bloom, and J. Polak pp. 77–88. Shattauer, New York.

——, and Tokimasa, T. (1983). Depression of calcium-dependent potassium conductance of guinea-pig myenteric neurones by muscarinic agonists. *J. Physiol.* **342**, 253–66.

Nowak, L. & MacDonald, R. L. (1983). Muscarine-sensitive voltage-dependent potassium current in cultured murine spinal cord neurons. *Neurosci. Lett.* **35**, 85–91.

Pennefather, P., Jones, S. W., and Adams, P. R. (1985a). Modulation of repetitive firing in bullfrog sympathetic ganglion cells by two distinct K currents, I_{AHP} and I_M. *Neurosci. Abs.* **11**, 148.

——, Lancaster, R., Adams, P. R., and Nicoll, R. A. (1985b). Two distinct Ca-dependent K-currents in bullfrog sympathetic ganglion cells. *Proc. Nat. Acad. Sci.* **11**, 787.

Reuter, H. (1983). Calcium channel modulation by neurotransmitters, enzymes and drugs. *Nature*, **301**, 569–74.

Schulman, J. A. and Weight, F. F. (1976). Synaptic transmission: long-lasting potentiation by a postsynaptic mechanism. *Science* NY, **194**, 1437–9.

Segal, M. and Barker, J. L. (1984). Rat hippocampal neurons in culture: potassium conductances. *J. Neurophysiol.* **51**, 1409–33.

Siegelbaum, S. A. and Tsien, R. W. (1983). Modulation of gated ion channels as a mode of transmitter action. *Trends Neurosci.* **6**, 307–313.

Sims, S., Singer, J., and Walsh, J. (1984). Identification of a muscarine-sensitive

K^+-current in single smooth muscle cells. *Abs. IUPHAR 9th Int. Congr. Pharmacol.* p. 1246.

Thompson, S. H. (1977). Three pharmacologically distinct potassium channels in molluscan neurones. *J. Physiol.* **265**, 465–88.

Tokimasa, T. (1984). Muscarinic agonists depress calcium-dependent g_k in bullfrog sympathetic neurons. *J. Autonomic Nerv. Syst.* **10**, 107–16.

Tosaka, T., Tasaka, J., Miyazaki, T., and Libet, B. (1983). Hyperpolarization following activation of K^+ channels by excitatory postsynaptic potentials. *Nature*, **305**, 148–150.

Werz, M. A. and MacDonald, J. F. (1983). Opioid peptides selective for mu- and delta-opiate receptors reduce calcium-dependent action potential duration by increasing potassium conductance. *Neurosci. Lett.* **42**, 173–8.

——, —— (1984). Dynorphin reduces calcium-dependent action potential duration by decreasing voltage-dependent calcium conductance. *Neurosci. Lett.* **46**, 185–90.

Williams, J. T. and North, R. A. (1985a). Characterization of α_2-adrenoceptors which increase potassium conductance in rat locus coeruleus neurones. *Neurosci.* **14**, 93–101.

——, and —— (1985b). Catecholamine inhibition of calcium action potentials in rat locus coeruleus neurones. *Neurosci.* **14**, 103–9.

——, ——, Shefner, S. A., Nishi, S., and Egan, T. M. (1984). Membrane properties of rat locus coeruleus neurones. *Neurosci.* **13**, 137–56.

Wood, J. D. and Mayer, C. J. (1979a). Intracellular study of tonic-type enteric neurones in guinea-pig small intestine. *J. Neurophysiol.* **43**, 569–81.

——, and —— (1979b). Serotonergic activation of tonic-type enteric neurones in the guinea-pig small bowel. *J. Neurophysiol.* **42**, 582–93.

11

First messengers, second messengers, and protein phosphorylation in CNS

S. IVAR WALAAS, JAMES K.-T. WANG, AND PAUL GREENGARD

INTRODUCTION

There is now strong evidence that a wide variety of extracellular signals produce many of their biological effects by regulating protein phosphorylation systems. Protein phosphorylation is a final common pathway through which many neurotransmitters exert their effects on target neurons (Greengard 1978; Nestler *et al.* 1984; Walaas and Nairn 1985). The present chapter briefly reviews some general properties of the protein phosphorylation systems in brain, and describes in more detail the properties of synapsin I and DARPP-32, two brain phosphoproteins which appear to be involved in regulation of synaptic transmission.

PROTEIN PHOSPHORYLATION PATHWAYS IN BRAIN

A large number of extracellular signals have been shown to regulate protein phosphorylation in the nervous system. The molecular pathways underlying these reactions are shown in Table 11.1.

First messengers in the nervous system include neurotransmitters and hormones as well as the nerve impulse itself. The first messengers regulate the intracellular concentrations of second messengers such as cyclic AMP (cAMP), cyclic GMP (cGMP), calcium (Ca^{2+}), diacylglycerol, and inositol trisphosphate (Downes 1982; Berridge 1984; Nestler and Greengard 1984a; Nishizuka 1984). Present evidence suggests that most of the second messenger actions of cAMP, cGMP, and diacylglycerol are mediated by the activation of specific protein kinases. The mechanisms of action of Ca^{2+} [and of inositol trisphosphate, which appears to achieve most or all of its effects by releasing Ca^{2+} from intracellular stores (Berridge 1984)] are

TABLE 11.1. *Regulation of protein phosphorylation systems in brain by first and second messengers**

Type of first messenger[†]	Second messenger	Cofactor	Protein kinase	Substrate specificity[‡]
Neurotransmitters; Hormones; Drugs;	cyclic AMP		cyclic AMP-dependent protein kinase, Type I and II	Broad
Neurotransmitters; Nerve impulse; Drugs;	cyclic GMP		cyclic GMP-dependent protein kinase	Narrow
Nerve impulse; Neurotransmitters;	calcium	calmodulin	Ca^{2+}/calmodulin-dependent protein kinases I and II	Broad
			Myosin light chain kinase	Narrow
			Phosphorylase kinase	Narrow
Neurotransmitters; Hormones;	diacylglycerol, calcium (inositol trisphosphate)	phospholipid	Ca^{2+}/phospholipid-dependent protein kinase	Broad

*Modified from Walaas and Greengard (1986).
[†] Recent reviews on signal molecules which regulate the levels of second messengers in brain can be found in Drummond (1983), Downes (1982), Nishizuka (1984), and Michell (1975).
[‡] Indicates the number and diversity of known proteins which are specific substrates for the different protein kinases.

more diverse (Marme 1985), although recent evidence suggests that many Ca^{2+} effects also involve the activation of specific protein kinases (see, for example, Walaas and Nairn 1985). There is virtually one type of cAMP-dependent protein kinase and one type of cGMP-dependent protein kinase in mammalian brain, but there are at least two classes of Ca^{2+}-dependent protein kinases (Nairn *et al.* 1985). One of these is activated in the combined presence of Ca^{2+} and the Ca^{2+}-binding protein calmodulin, and the other is activated by the combined presence of Ca^{2+}, diacylglycerol, and membrane phospholipids such as phosphatidylserine (Nishizuka 1984). At present, five distinct Ca^{2+}/calmodulin-dependent protein kinases have been found in brain (Nairn *et al.* 1985), while only one calcium/phospholipid-dependent protein kinase (also known as protein kinase C) has been identified (Kikkawa *et al.* 1982; Nairn *et al.* 1985).

Activation by second messengers of the various protein kinases results in the phosphorylation of specific substrate proteins and such phosphorylation leads, through one or more steps, to the production of specific biological responses in target neurons. Extensive evidence suggests that a wide range of physiological events related to synaptic transmission are, indeed, mediated or modulated by such protein modifications (see Rosen and Krebs 1981, for review).

DIRECT EVIDENCE FOR A ROLE OF PROTEIN PHOSPHORYLATION IN NEURONAL FUNCTION

Recent studies, using intracellular injection of components of various protein phosphorylation systems, have provided direct evidence for a causal relationship between protein phosphorylation (cAMP-dependent, cGMP-dependent, calcium/calmodulin-dependent and calcium/phospho-lipid-dependent) and the ensuing physiological responses in various excitable cells. Injection of the catalytic subunit of cAMP-dependent protein kinase has been shown to control the action potential firing frequency of the bag cell neurons of the mollusc *Aplysia* by regulating K^+ channels (Kaczmarek *et al.* 1980, 1984), and to affect transmitter release from a sensory neuron in *Aplysia* by regulating what appears to be a different set of K^+ channels (Castellucci *et al.* 1980, 1982; Siegelbaum *et al.* 1982). Other studies have shown that the diacylglycerol-activated, Ca^{2+}/phospholipid-dependent protein kinase may regulate voltage-dependent Ca^{2+} channels in the bag cell neurons of *Aplysia* (DeRiemer *et al.* 1985). Intracellular injection of cGMP-dependent protein kinase appears to regulate input resistance in mammalian cortical pyramidal cells, thereby mimicking the effect of acetylcholine on these cells (Bartfai *et al.* 1985). Studies with Ca^{2+}/calmodulin-dependent protein kinase II have shown that neurotransmitter release from the squid giant axon can be directly modulated by intracellular injection of this enzyme without any apparent effect on ion channel properties (Llinas *et al*. 1985). These and other studies indicate that detection and characterization of components of various neuronal protein phosphorylation systems will eventually give us a more detailed understanding of the molecular mechanisms involved in the generation or modulation of fast and slow responses in nerve cells (Nestler and Greengard 1984a; Walaas and Greengard 1986).

SPECIFIC BRAIN PHOSPHOPROTEINS

This section describes in more detail the properties of two brain phosphoproteins, synapsin I and DARPP-32, that appear to be involved in neuronal signal transduction. Some of the properties of these two proteins are summarized in Table 11.2.

Synapsin I, a synaptic vesicle-associated phosphoprotein

Synapsin I (previously termed Protein I) was first identified as a phosphoprotein in particulate synaptic fractions of brain (Johnson *et al*. 1972). It has since been purified and extensively characterized (Ueda and Greengard 1977; Huttner *et al*. 1981). Synapsin I, which consists of two polypeptides of M_r 86 000 and 80 000 (designated synapsin Ia and Ib, respectively), is a major brain substrate for at least three distinct protein kinases, and undergoes multisite phosphorylation. One serine residue (site 1) of the protein is phosphorylated by both cAMP-dependent protein kinase and Ca^{2+}/calmodulin-dependent protein kinase I (Huttner *et al*. 1981; Nairn and Greengard 1983), while one or two other serine residues (the number apparently depending on the species), collectively termed site 2, are phosphorylated by Ca^{2+}/calmodulin-dependent protein kinase II (Huttner *et al*. 1981; Kennedy and Greengard 1981; Nairn *et al*. 1985).

Synapsin I is present only in neurons, and appears to be associated with synaptic vesicles in virtually all nerve terminals (De Camilli *et al*. 1983a,b). It represents a major neuron-specific protein, comprising 1 per cent of neuronal protein, and 6 per cent of synaptic vesicle protein (Goelz *et al*. 1981). It is an extrinsic protein of the synaptic vesicle membranes, being located at the outer, cytoplasmic surface (De Camilli *et al*. 1983b; Huttner, *et al*. 1983). Synaptic vesicles contain a specific, saturable and high-affinity binding site for synapsin I (Schiebler *et al*. 1983). Recent studies have shown that the domain of synapsin I which contains phosphorylation site 2 (the so-called 'tail region') is the region of the molecule that binds to synaptic vesicles (Ueda 1981; Huttner *et al*. 1983), and that phosphorylation of site 2 decreases this binding (Huttner *et al*. 1983;

TABLE 11.2. *Some properties of Synapsin I and DARPP-32*[*]

	Synapsin I	DARPP-32
Molecular size	Ia, 86 000 dalton; Ib, 80 000 dalton (by SDS gel electrophoresis)	32 000 dalton (by SDS-gel electrophoresis) 22 951 dalton (by amino acid sequencing)[†]
Conformation	elongated	elongated
Isoelectric point	very basic	acidic
Distribution		
Regional	all neurons	Dopaminoceptive neurons which have D-1 receptors
Subcellular	synaptic vesicles	Cytosol
First messengers regulating phosphorylation	nerve impulses (site 1–2) serotonin, dopamine, norepinephrine (site 1)	Dopamine
Protein kinases catalyzing phosphorylation	cyclic AMP-dependent (site 1) Ca²⁺/calmodulin-dependent I (site 1) Ca²⁺/calmodulin-dependent II (site 2)	cyclic AMP-dependent
Effect of phosphorylation	Decreases affinity for synaptic vesicles (site 2)	Increases potency as phosphatase inhibitor
Effect of neuronal injection	Regulates neurotransmitter release (squid giant axon)	—

First messengers regulating phosphorylation: nerve impulses (site 1–2) serotonin, dopamine, norepinephrine (site 1)

Protein kinases catalyzing phosphorylation: cyclic AMP-dependent (site 1) Ca²⁺/calmodulin-dependent I (site 1) Ca²⁺/calmodulin-dependent II (site 2)

[*] Modified from Nestler and Greengard (1984b); see text for references.
[†] Williams *et al.* (1985).

Schiebler *et al*. 1983). Phosphorylation of site 1 does not influence binding to vesicles (Schiebler *et al*. 1983); however, it is possible that phosphorylation of this site of synapsin I may affect its interaction with some other component of the nerve terminal.

The state of phosphorylation of site 1 of synapsin I has been shown to be specifically regulated by various neurotransmitters, including serotonin, dopamine, and norepinephrine, in a number of well-defined, relatively homogeneous regions of the central and peripheral nervous system (Nestler and Greengard 1980; Dolphin and Greengard 1981; Tsou and Greengard 1982; Mobley and Greengard, 1985). In each of these regions, the respective neurotransmitter stimulates the phosphorylation of synapsin I in presynaptic terminals, presumably by activating presynaptic receptors (Mobley and Greengard 1985) which increase cAMP levels in these terminals.

In contrast, the state of phosphorylation of both sites 1 and 2 of synapsin I has been found to be regulated by depolarization-induced Ca^{2+}-influx into nerve terminals (Krueger *et al*. 1977; Forn and Greengard 1978; Huttner and Greengard 1979) or, in the presence of extracellular Ca^{2+}, by brief periods of nerve impulse conduction at physiological frequencies in presynaptic fibers (for example, the superior cervical ganglion (Nestler and Greengard 1982, 1984b) and the posterior pituitary (Tsou and Greengard 1982)). The studies of the superior cervical ganglion showed that as few as 20 nerve impulses significantly stimulated synapsin I phosphorylation, with approximately 2 per cent of the total presynaptic synapsin I becoming phosphorylated per nerve impulse, and up to 80 per cent of presynaptic synapsin I becoming converted from dephospho- to phospho-synapsin I (Nestler and Greengard 1984b). The incorporation of ^{32}P into both sites 1 and 2 was found to be rapid, and reversible within seconds following termination of the stimulus.

The localization of synapsin I to the cytoplasmic surface of synaptic vesicles in virtually all nerve terminals, and evidence for the regulation of its state of phosphorylation by first messengers that increase neuronal Ca^{2+} or cyclic AMP levels, led to the hypothesis that synapsin I plays some role in regulating the process by which neurotransmitters are released from the nerve terminal (Greengard 1981). An elevation in either Ca^{2+} or cyclic AMP has been shown to potentiate the release of neurotransmitters from various nerve terminals under different experimental conditions (see, for example, Dunwiddie and Hoffer 1982, for review). For example, brief periods of impulse conduction, apparently acting through Ca^{2+}, increase the amount of neurotransmitter released in response to a single nerve impulse in numerous types of neuronal preparations in a process called 'post-tetanic potentiation' (Rosenthal 1969; Zengel *et al*. 1980). Recent studies with the squid giant synapse have provided direct evidence that synapsin I and Ca^{2+}/calmodulin-dependent protein kinase II regulate neurotransmitter release (Llinas *et al.* 1985). In these studies, the dephosphorylated form of synapsin I, or the holoenzyme of Ca^{2+}/calmodulin-dependent protein kinase II, was injected into the presynaptic terminal digit of the giant synapse and the effects on presynaptic calcium influx (as measured under presynaptic voltage-clamp conditions) and on neurotransmitter release (as measured by the post-synaptic potential) were determined. The results indicated that injection of dephospho-synapsin I decreased neurotransmitter release, while injection of Ca^{2+}/calmodulin-dependent protein kinase II potentiated neurotransmitter release from these terminals. Neither of these treatments changed the depolarization-induced presynaptic Ca^{2+} influx. These results support the hypothesis that the dephospho-form of synapsin I, which is tightly bound to the synaptic vesicles (see above), limits the availability of synaptic vesicles for neurotransmitter release, and that its phosphorylation by Ca^{2+}/calmodulin-dependent protein kinase II on site 2 (which decreases the affinity of synaptic vesicles for synapsin I, see above) relieves this inhibition (Llinas *et al.* 1985). Such a mechanism may act synergistically or antagonistically with other protein phosphorylation systems which regulate neurotransmitter release (Kandel and Schwartz 1982; Levitan *et al.* 1983).

DARPP-32, a dopamine-regulated phosphatase inhibitor

DARPP-32 (dopamine- and cAMP-regulated phosphoprotein, $M_r = 32\,000$) is a protein which was first detected in homogenates of rat neostriatum as an endogenous substrate for cAMP-dependent protein kinase (Walaas *et al.* 1983a,b; Walaas and Greengard 1984). The protein has since been purified and extensively characterized (Hemmings *et al.* 1984b,c,d; Ouimet *et al.* 1984; Walaas and Greengard 1984). Purified DARPP-32 is an effective substrate for cAMP-dependent protein kinase, which rapidly and specifically phosphorylates a single threonine residue in the protein (Hemmings *et al.* 1984b,c). In contrast to synapsin I, DARPP-32 is not an effective substrate for Ca^{2+}-dependent protein kinases.

The distribution of DARPP-32 has been studied with a variety of techniques, including microdissection and biochemical assays (Walaas and Greengard 1984), subcellular fractionation (Walaas and Greengard 1984), immunological assays (H. C. Hemmings and P. Greengard, unpublished data) and immunocytochemistry (Ouimet *et al.* 1984). These studies have shown that DARPP-32 is highly enriched in the nervous system, where it is concentrated in brain regions that are densely innervated by dopaminergic neurons (Walaas *et al.* 1983a; Walaas and Greengard 1984; Ouimet *et al.*

1984). Within these brain regions, DARPP-32 appears to be specifically enriched only in neurons that receive dopaminergic innervation (i.e. dopaminoceptive neurons), but not in the dopaminergic neurons themselves (Walaas and Greengard 1984). Furthermore, DARPP-32 is found specifically in those dopaminoceptive neurons that possess dopamine receptors linked to the stimulation of adenylate cyclase activity [the so-called D-1 receptor (Kebabian and Calne 1979)]. One major example of this class of neuron is the so-called medium-sized spiny neuron of the neostriatum which represents the majority of cells in this brain region. These cells receive the most important inputs to the neostriatum, i.e. the putative glutamate corticostriatal fibres, the thalamostriatal fibres, and the dopaminergic nigrostriatal fibres (see Divac and Öberg 1979, for review). Furthermore, they represent the major origin for the outflow from the neostriatum, since they generate most of the striatopallidal and striatonigral fibres, which terminate in the globus pallidus and substantia nigra, respectively (Divac and Öberg 1979). Within the medium-sized spiny neurons, DARPP-32 is distributed as a cytosolic protein, being present diffusely throughout the neuronal cytoplasm in the cell bodies, dendrites, axons, and nerve terminals (Ouimet *et al*. 1984).

Dopamine (but not serotonin or norepinephrine) and cAMP analogues have been shown to increase the state of phosphorylation of DARPP-32 in slices of rat neostriatum (Walaas *et al*. 1983a; Walaas and Greengard 1984). In contrast, depolarizing agents, which induce Ca^{2+} fluxes into neurons, did not alter the state of phosphorylation of the protein (Walaas and Greengard 1984). These results are consistent with observations made on broken cell and purified preparations that DARPP-32 is a specific substrate for cAMP-dependent but not Ca^{2+}-dependent protein kinases (Hemmings *et al*. 1984c; Walaas *et al*. 1983b). They are also consistent with a functional model whereby dopamine stimulates DARPP-32 phosphorylation through the activation of D-1 receptors, and the resultant activation of dopamine-sensitive adenylate cyclase and cAMP-dependent protein kinase.

Recent studies have indicated one possible biological function for DARPP-32. Biochemical characterization showed that the phosphorylated single threonine residue is surrounded by an amino acid sequence (Hemmings *et al*. 1984d) very similar to that surrounding the phosphothreonyl residue of the well-characterized protein phosphatase inhibitor-1 (Nimmo and Cohen 1978). Furthermore, similar to phosphate inhibitor-1, phospho-DARPP-32 (but not dephospho-DARPP-32) is a potent inhibitor of protein phosphatase-1, one of the major protein phosphatases involved in cellular regulation (Hemmings *et al*. 1984a). In addition, neither phospho-DARPP-32 nor phosphate inhibitor-1 are active towards protein phosphatase-2A, -2B and -2C, other major protein phosphatases involved in cellular regulation (Ingebritsen and Cohen 1983). Therefore, one possible biological role for DARPP-32 phosphorylation, mediated through dopamine-induced D-1 receptor activation, might be to inhibit protein phosphatase-1 in dopaminoceptive target cells. This could lead to changes in the state of phosphorylation of other proteins, some of which may be substrates for cAMP-dependent protein kinase and hence also dopamine-regulated, and some may be substrates for other protein kinases which are regulated by other first messengers influencing these neurons (Nestler *et al*. 1984). This hypothetical scheme could therefore represent a molecular mechanism by which a classical 'slow-acting' synaptic transmitter such as dopamine could influence the response of a neuron to a 'fast-acting' transmitter without primarily involving specific ion channels. One example of where this mechanism might be operative is the well-known interaction between the effects of dopamine and glutamate on the medium-sized spiny neurons of the neostriatum: glutamate will rapidly increase the firing frequency of these neurons, and this effect is potently prevented or reversed by dopamine (Moore and Bloom 1978; Bunney 1979). We have hypothesized that this effect might be caused by changes in the state of phosphorylation of some modulator protein regulated by phospho-DARPP-32. This modulator protein could thereby change the responsiveness of the post-synaptic membrane to glutamate (Nestler *et al*. 1984). This as yet unproven scheme might also be extended to other cells in the nervous system as a general molecular mechanism by which the interactions of two or more neurotransmitters might be regulated at the level of protein phosphatase inhibition (Nestler *et al*. 1984).

CONCLUSION

This chapter has briefly reviewed recent data which indicate that protein phosphorylation is a major molecular mechanism involved in many aspects of brain function. For example, injections of various components of cAMP-dependent, cGMP-dependent or Ca^{2+}-dependent protein phosphorylation systems into neurons have provided direct evidence that neurotransmitter release and many ion channels may be regulated by reversible protein phosphorylation. Such a biochemical mechanism might primarily be expected to represent a response mechanism utilized by 'slow-acting' first messengers. However, the data on synapsin I and on DARPP-32 reviewed above indicate that fast-acting transmitters and their effects may also be influenced by protein phosphorylation, either at the presynaptic and/or at the post-synaptic level. We believe that further analysis of brain protein phosphorylation systems, including the detection and characterization of neuronal substrates for various protein kinases and protein phosphatases, will lead to a more complete understanding of the molecular pathways through which both fast-acting and slow-acting extracellular signals produce their specific physiological responses in neurons.

References

Bartfai, T., Woody, C. D., Gruen, E., Nairn, A., and Greengard, P. (1985). Intracellular injection of cGMP-dependent protein kinase results in increased input resistance in neurons of the mammalian motor cortex. *Soc. Neurosci. Abs.* **11**, 1093.

Berridge, M. J. (1984). Inositol trisphosphate and diacylglycerol as second messengers. *Biochem. J.* **220**, 345–60.

Bunney, B. S. (1979). Electrophysiology of the midbrain dopamine systems. In *The Neurobiology of Dopamine* (eds A. S. Horn, J. Korf, and B. H. C. Westerink), pp. 417–52. Academic Press, New York.

Castellucci, V. F., Kandel, E. R., Schwartz, J. H., Wilson, F. D., Nairn, A. C., and Greengard, P. (1980). Intracellular injection of the catalytic subunit of cyclic AMP-dependent protein kinase simulates facilitation of transmitter release underlying behavioral sensitization in *Aplysia*. *Proc. Nat. Acad. Sci. USA*, **77**, 7492–6.

——, Nairn, A., Greengard, P., Schwartz, J. H., and Kandel, E. R. (1982). Inhibitor of adenosine 3': 5'-monophosphate-dependent protein kinase blocks presynaptic facilitation in *Aplysia*. *J. Neurosci.* **2**, 1673–81.

De Camilli, P., Cameron, R., and Greengard, P. (1983a). Synapsin I (Protein I), a nerve terminal-specific phosphoprotein: I. Its general distribution in synapses of the central and peripheral nervous system demonstrated by immunofluorescence in frozen and plastic sections. *J. Cell Biol.* **96**, 1337–54.

——, Harris, S. M., Huttner, W. B., and Greengard, P. (1983b). Synapsin I (Protein I), a nerve terminal-specific phosphoprotein: II. Its specific association with synaptic vesicles demonstrated by immunocytochemistry in agarose-embedded synaptosomes. *J. Cell Biol.* **96**, 1355–73.

DeRiemer, S. A., Strong, J. A., Albert, K. A., Greengard, P., and Kaczmarek, L. K. (1985). Enhancement of calcium current in *Aplysia* neurones by phorbol ester and protein kinase C. *Nature*, **313**, 313–6.

Divac, I. and Öberg, G. E. (eds) (1979). *The Neostriatum*. Pergamon Press, London.

Dolphin, A. C., and Greengard, P. (1981). Serotonin stimulates phosphorylation of Protein I in the facial motor nucleus of rat brain. *Nature*, **289**, 76–9.

Downes, C. P. (1982). Receptor-stimulated inositol phospholipid metabolism in the central nervous system. *Cell Calcium*, **3**, 413–28.

Drummond, G. I. (1983). Cyclic nucleotides in the nervous system. *Adv. Cycl. Nucleotide Res.* **15**, 373–494.

Dunwiddie, T. V., and Hoffer, B. G. (1982). The role of cyclic nucleotides in the nervous system. *Handb. Exp. Pharmacol.* **58** (Part II), 389–463.

Forn, J., and Greengard, P. (1978). Depolarizing agents and cyclic nucleotides regulate the phosphorylation of specific neuronal proteins in rat cerebral cortex slices. *Proc. Nat. Acad. Sci. USA*, **75**, 5195–9.

Goelz, S. E., Nestler, E. J., Chehrazi, B., and Greengard, P. (1981). Distribution of Protein I in mammalian brain as determined by a detergent-based radioimmunoassay. *Proc. Nat. Acad. Sci. USA*, **78**, 2130–4.

Greengard, P. (1978). Phosphorylated proteins as physiological effectors. *Science*, **199**, 146–52.

—— (1981). Intracellular signals in the brain. In: *The Harvey Lectures, Series 75*, pp. 277–331, Academic Press, New York.

Hemmings, H. C., Jr, Greengard, P., Tung, H. Y. L., and Cohen, P. (1984a). DARPP-32, a dopamine-regulated neuronal phosphoprotein, is a potent inhibitor of protein phosphatase-1. *Nature*, **310**, 503–5.

——, Nairn, A. C., Aswad, D. W., and Greengard, P. (1984b). DARPP-32, a dopamine and adenosine 3′: 5′-monophosphate regulated phosphoprotein enriched in dopamine-innervated brain regions. II. Purification and characterization of the phosphoprotein from bovine caudate nucleus. *J. Neurosci.* **4**, 99–110.

——, ——, and Greengard, P. (1984c). DARPP-32, a dopamine- and adenosine 3′:5′-monophosphate-regulated neuronal phosphoprotein: II. Comparison of the kinetics of phosphorylation of DARPP-32 and phosphatase inhibitor-1. *J. Biol. Chem.* **259**, 14491–7.

——, Williams, K. R., Konigsberg, W. H., and Greengard, P. (1984d). DARPP-32, A dopamine- and adenosine 3′:5′-monophosphate-regulated neuronal phosphoprotein: I. Amino acid sequence around the phosphorylated threonine. *J. Biol. Chem.* **259**, 14486–90.

Huttner, W. B., DeGennaro, L. J., and Greengard, P. (1981). Differential phosphorylation of multiple sites in purified Protein I by cyclic AMP-dependent and calcium-dependent protein kinases. *J. Biol.* Chem. **256**, 1482–8.

——, and Greengard, P. (1979). Multiple phosphorylation sites in Protein I and their differential regulation by cyclic AMP and calcium. *Proc. Nat. Acad. Sci. USA*, **76**, 5402–6.

——, Schiebler, W., Greengard, P., and De Camilli, P. (1983). Synapsin I (Protein I), a nerve terminal-specific phosphoprotein. III. Its association with synaptic vesicles studied in a highly-purified synaptic vesicle preparation. *J. Cell. Biol.* **96**, 1374–88.

Ingebritsen, T. S., and Cohen, P. (1983). Protein phosphatases: Properties and role in cellular regulation. *Science*, **221**, 331–8.

Johnson, E. M., Ueda, T., Maeno, H., and Greengard, P. (1972). Adenosine 3′:5′-monophosphate-dependent phosphorylation of a specific protein in synaptic membrane fractions from rat cerebrum. *J. Biol. Chem.* **247**, 5650–2.

Kaczmarek, L. K., Jennings, K. R., Strumwasser, F., Nairn, A. C., Walter, U., Wilson, F. D., and Greengard, P. (1980). Microinjection of catalytic subunit of cyclic AMP-dependent protein kinase enhances calcium action potentials of bag cell neurons in cell culture. *Proc. Nat. Acad. Sci. USA*, **77**, 7487–91.

——, Nairn, A. C., and Greengard, P. (1984). Microinjection of protein kinase inhibitor prevents enhancement of action potentials in peptidergic neurons of *Aplysia. Soc. Neurosci Abs.* **10**, 895.

Kandel, E. R., and Schwartz, J. H. (1982). Molecular biology of learning: modulation of transmitter release. *Science*, **218**, 433–43.

Kebabian, J. W., and Calne, D. B. (1979). Multiple receptors for dopamine. *Nature*, **277**, 93–6.

Kennedy, M. B., and Greengard, P. (1981). Two calcium/calmodulin-dependent protein kinases, which are highly concentrated in brain, phosphorylate Protein I at distinct sites. *Proc. Nat. Acad. Sci. USA*, **78**, 1293–7.

Kikkawa, U., Takai, Y., Minakuchi, R., Inohara, S., and Nishizuka, Y. (1982). Calcium-activated, phospholipid-dependent protein kinase from rat brain. Subcellular distribution, purification and properties. *J. Biol. Chem.* **257**, 13341–8.

Krueger, B. K., Forn, J., and Greengard, P. (1977). Depolarization-induced phosphorylation of specific proteins, mediated by calcium ion influx, in rat brain synaptosomes. *J. Biol. Chem.* **252**, 2764–73.

Levitan, I. B., Lemos, J. R., and Novak-Hofer, I. (1983). Protein phosphorylation and the regulation of ion channels. *Trends Neurosci.* **6**, 496–9.

Llinas, R., McGuinness, T., Leonard, C. S., Sugimori, M., and Greengard, P. (1985). Intraterminal injection of synapsin I or calcium/calmodulin-dependent protein kinase II alters neurotransmitter release at the squid giant synapse. *Proc. Nat. Acad. Sci. USA*, **82**, 3035–9.

Marme, D. (ed.) (1985). *Calcium and Cell Physiology*. Springer, Heidelberg.

Michell, R. H. (1975). Inositol phospholipids and cell surface receptor function. *Biochim. Biophys. Acta*, **415**, 81–147.

Mobley, P. M., and Greengard, P. (1985). Evidence for widespread effects of noradrenaline on axon terminals in the rat frontal cortex. *Proc. Nat. Acad. Sci. USA*, **82**, 945–7.

Moore, R. Y., and Bloom, F. E. (1978). Central catecholamine neuron systems: anatomy and physiology of the dopamine systems. *Ann. Rev. Neurosci.* **1**, 129–69.

Nairn, A. C., and Greengard, P. (1983). Purification and characterization of brain Ca^{2+}/calmodulin-dependent protein kinase I that phosphorylates Synapsin I. *Soc. Neurosci. Abs.* **9**, 1029.

——, Hemmings, H. C., Jr, and Greengard, P. (1985). Protein kinases in the brain. *Ann. Rev. Biochem.* **54**, 931–76.

Nestler, E. J., and Greengard, P. (1980). Dopamine and depolarizing agents regulate the state of phosphorylation of Protein I in the mammalian superior cervical sympathetic ganglion. *Proc. Nat. Acad. Sci. USA*, **77**, 7479–83.

——, and —— (1982). Distribution of Protein I and regulation of its state of phosphorylation in the rabbit superior cervical ganglion. *J. Neurosci*, **2**, 1011–23.

——, and —— (1984a). *Protein Phosphorylation in the Nervous System*. John Wiley and Sons, Inc. New York.

——, and —— (1984b). Protein phosphorylation in nervous tissue. In *Catecholamines: Basic and Peripheral Mechanisms*, pp. 9–22. Alan Liss, New York.

——, Walaas, S. I., and Greengard, P. (1984). Neuronal phosphoproteins: physiological and clinical implications. *Science*, **225**, 1357–64.

Nimmo, G. A., and Cohen, P. (1978). The regulation of glycogen metabolism: purification and characterisation of protein phosphatase inhibitor-1 from rabbit skeletal muscle. *Eur. J. Biochem.* **87**, 341–51.

Nishizuka, Y. (1984). Turnover of inositol phospholipids and signal transduction. *Science* **255**, 1365–70.

Ouimet, C. C., Miller, P. E., Hemmings, H. C., Jr, Walaas, S. I., and Greengard, P. (1984). DARPP-32, a dopamine and adenosine 3′:5′-monophosphate-regulated phosphoprotein enriched in dopamine-innervated bran regions. III. Immunocytochemical localization. *J. Neurosci.* **4**, 111–24. *Neurosci.* **4**, 111–24.

Rosen, O. M., and Krebs, E. G. (eds) (1981). *Protein Phosphorylation*. Cold Spring Harbor Press, Cold Spring Harbor, New York.

Rosenthal, J. (1969). Post-tetanic potentiation at the neuromuscular junction of the frog. *J. Physiol.* **203**, 121–33.

Schiebler, W., Rothlein, J., Jahn, R., Doucet, J. P., and Greengard, P. (1983). Synapsin I (Protein I) binds specifically and with high affinity to highly purified synaptic vesicles from rat brain. *Soc. Neurosci. Abs.* **9**, 582.

Siegelbaum, S. A., Camardo, J. S., and Kandel, E. R. (1982). Serotonin and cyclic AMP close single K^+ channels in *Aplysia* sensory neurons. *Nature*, **299**, 413–7.

Tsou, K., and Greengard, P. (1982). Regulation of phosphorylation of Proteins I, III_a and III_b in rat neurohypophysis *in vitro* by electrical stimulation and by neuroactive agents. *Proc. Nat. Acad. Sci. USA*, **79**, 6075–9.

Ueda, T. (1981). Attachment of the synapse-specific phosphoprotein Protein I to the synaptic membrane: a possible role of the collagenase-sensitive region of Protein I. *J. Neurochem.* **36**, 297–300.

——, and Greengard, P. (1977). Adenosine 3′:5′-monophosphate-regulated phosphoprotein system of neuronal membranes. I. Solubilization, purification, and some properties of an endogenous phosphoprotein. *J. Biol. Chem.* **252**, 5155–63.

Walaas, S. I., Aswad, D. W., and Greengard, P. (1983a). A dopamine- and cyclic AMP-regulated phosphoprotein enriched in dopamine-innervated brain regions. *Nature*, **301**, 69–71.

——, and Greengard, P. (1984). DARPP-32, a dopamine and adenosine 3′:5′-monophosphate regulated phosphoprotein enriched in dopamine-innervated brain regions. I. Regional and cellular distributin in the rat brain. *J. Neurosci.* **4**, 84–98.

——, and —— (1986). Phosphorylation of brain proteins. In *The Enzymes, Vol. 18: Enzyme regulation by protein phosphorylation* (eds P. D. Boyer and E. G. Krebs). Academic, New York (in press).

——, and Nairn, A. C. (1985). Calcium-regulated protein phosphorylation in mammalian brain. In *Calcium and Cell Physiology* (ed. D. Marme), pp. 238–64. Springer, Heidelberg.

——, ——, and Greengard, P. (1983b). Regional distribution of calcium- and cyclic adenosine 3′5′-monophosphate-regulated protein phosphorylation systems in mammalian brain. II. Soluble systems. *J. Neurosci.* **3**, 302–11.

Williams, K. R., Hemmings, H. C., Jr, LoPresti, M. B., Konigsberg, W. H., and Greengard, P. (1985). Primary structure of bovine brain DARPP-32 and its homology with protein phosphatase inhibitor-1. *J. Biol. Chem.* (in press).

Zengel, J. E., Magleby, K. L., Horn, J. P., McAfee, D. A., and Yarowsky, P. J. (1980). Facilitation, augmentation and potentiation of synaptic transmission at the superior cervical ganglion of the rabbit. *J. Gen. Physiol.* **76**, 213–31.

12

Synaptic interactions between dorsal root ganglion and dorsal horn neurons in cell culture: amino acids, nucleotides, and peptides as possible fast and slow excitatory transmitters

T. M. JESSELL AND C. E. JAHR

The organization of afferent pathways that convey cutaneous sensory information to the central nervous system has been analyzed in recent years by using physiological and anatomical techniques (Willis and Coggeshall 1978; Brown 1982; Perl 1983). Cutaneous sensory receptors that are activated by noxious stimuli exist as free endings in the skin and give rise to unmyelinated C and thinly myelinated A δ fibres whereas innocuous mechanical stimuli predominantly activate large calibre myelinated A β fibres. The central terminals of most cutaneous sensory afferents are located in the dorsal horn of the spinal cord and form synaptic contact with the dendrites of dorsal horn neurons. Physiological analysis of dorsal horn neurons *in situ* has revealed that the response properties of some neurons in the superficial dorsal horn faithfully reflect the peripheral receptive properties of specific fibre classes (Perl 1983). However, there is also considerable evidence for a convergence of input onto single dorsal horn neurons from afferent fibres conveying different sensory modalities. The integration of synaptic inputs from different classes of sensory neurons is therefore likely to play a fundamental role in the processing of incoming sensory information. Elucidation of the chemical identity and post-synaptic action of neurotransmitters that mediate the transfer of sensory information from dorsal root ganglion (DRG) neurons to dorsal horn neurons represents a first step in determining the molecular mechanisms that underlie sensory processing. In this chapter we review anatomical and physiological studies indicating that DRG neurons can release synaptic transmitters with both fast and slow excitatory actions on dorsal horn neurons. In particular we discuss recent physiological studies in cell culture that provide information on the chemical nature of fast transmitters at sensory synapses.

ANATOMICAL AND FUNCTIONAL ORGANIZATION OF SENSORY INPUT TO THE DORSAL HORN

Intracellular injection of horseradish peroxidase into functionally characterized afferent fibres has provided definitive information on the central termination of afferents conveying specific sensory modalities. Strikingly, each class of afferent fibre appears to possess a stereotyped terminal arborization pattern in the dorsal horn of the spinal cord (Brown 1982; Perl 1983). Low-threshold myelinated afferents that innervate hair follicles, and both slowly and rapidly adapting mechanoreceptors terminate in laminae III and IV of the dorsal horn (Bannatyne *et al*. 1984; Ralston *et al*. 1984). Although many classes of low-threshold myelinated afferents terminate within laminae III and IV, fibres conveying specific sensory modalities can be distinguished by the mediolateral extent of their terminal arborization and by the rostrocaudal spacing of collateral branches that give rise to the terminal fields (Brown 1982). Myelinated afferents that respond to high-threshold mechanical stimuli project to lamina I and also have collaterals that terminate in lamina V (Light and Perl, 1979; Rethelyi *et al*. 1982).

The majority of unmyelinated afferents, of which between 50 and 70 per cent in the rat correspond to polymodal nociceptors (Lynn and Carpenter 1982), terminate within laminae I and II (Rethelyi 1977). Lamina II has been subdivided according to morphological and ultrastructral features of the neuropil (Ralston 1979; Ribiero Da Silva and Coimbra 1982). Polymodal nociceptors appear to project to the outer region of lamina II whereas C fibres innervating low-threshold mechanoreceptors project to inner lamina II.

Morphological studies in the dorsal horn of rodents, cats and primates have revealed that all functional classes primary afferent terminals appear to contain small, clear synaptic vesicles (Ralston 1979). Ultrastructural analysis of single identified HRP-filled afferents has revealed that both low- and high-threshold myelinated afferents appear to possess, exclusively small clear vesicles in their synaptic terminals (Bannatyne *et al* 1984; Ralston *et al*. 1984; Rethelyi *et al*. 1982). However, many afferent terminals in the superficial dorsal horn also contain a second class of larger vesicles with electron-dense cores (Ralston 1979; Bresnahan *et al*. 1984; DiFiglia *et al*. 1982). In other regions of the CNS the presence of dense-core synaptic vesicles has been associated with storage of transmitters that are chemically distinct from those contained within small clear vesicles. Some primary afferents that project to the superficial dorsal horn may therefore store and release more than one class of sensory transmitter.

Physiological studies have provided evidence that synaptic transmitters released from sensory afferents may produce qualitatively different responses in a single post-synaptic dorsal horn neuron. Intracellular recording from neurons in the superficial dorsal horn of rat spinal cord slices maintained *in vitro* has demonstrated that activation of dorsal root fibres elicits fast excitatory post-synaptic potentials (e.p.s.p.s) that can be detected in most dorsal horn neurons (Urban and Randic 1984). By increasing the frequency of dorsal root stimulation, it is possible to evoke a slow depolarizing post-synaptic potential that persists for seconds or minutes. Although the cellular origin and chemical identity of the transmitter(s) mediating the slow synaptic response have not been established conclusively these observations suggest that primary sensory neurons release both fast and slow excitatory transmitters.

Several of the neuropeptides present within subsets of cutaneous sensory neurons appear to act as transmitters that mediate slow synaptic potentials at primary afferent synapses (Hokfelt *et al.* 1976; Otsuka *et al.* 1982; Jessell 1983a). Of the peptides localized in sensory ganglia the role of substance P has been studied in greatest detail (Otsuka *et al.* 1982; Jessell 1982, 1983b). Substance P immunoreactivity is associated with large dense-core vesicles in the central terminals of sensory neurons in laminae I and II of the dorsal horn (DiFiglia *et al.* 1982; Bresnahan *et al.* 1984). These terminals also contain small clear vesicles suggesting that substance P and other peptides may be present in afferent terminals that release fast sensory transmitters. Substance P release from sensory terminals in the dorsal horn has been demonstrated *in vitro* (Otsuka *et al.* 1982) and *in vivo* in response to activation of sensory fibres with conduction velocities in the A δ—and C-fibre range (Yaksh *et al.* 1980). Intracellular recording from dorsal horn neurons in rat spinal cord slices has demonstrated that the slow, synaptically-mediated depolarization recorded following high-frequency dorsal root stimulation can be mimicked by application of substance P (Urban and Randic 1984). Moreover, synthetic peptide substance P antagonists block the slow depolarization of dorsal horn neurons evoked by substance P and dorsal root stimulation (Urban and Randic 1984; Konishi *et al.* 1983). Collectively, these findings indicate that substance P and perhaps other peptides found within primary sensory neurons may be released as slow synaptic transmitters in the dorsal horn. Peptides released from specific subclasses of afferent fibres are likely to have important, but as yet undefined physiological roles in regulating the excitability of subsets of dorsal horn neurons that receive synaptic input mediated by fast sensory transmitters.

Somewhat surprisingly, the identity of the transmitters that mediate fast post-synaptic responses at afferent synapses has been difficult to establish. The approach we and several other groups have taken to address this question has been to study synaptic transmission between DRG and spinal cord neurons maintained in cell culture. The access to both pre- and post-synaptic neurons has made it possible to examine the chemosensitivity of dorsal horn neurons and the pharmacological characteristics of monosynaptic excitatory connections formed by DRG neurons grown in co-culture with dorsal horn neurons. By using histological and immunocytochemical methods we have also begun to identify subclasses of primary sensory neurons and dorsal horn neurons maintained in dissociated cell culture.

CYTOCHEMICAL PROPERTIES OF RAT DRG AND DORSAL HORN NEURONS *IN SITU* AND IN CELL CULTURE

Morphological and cytochemical properties that delineate different classes of spinal sensory neurons have been identified (Table 12.1). In the rat several types of dorsal root ganglion (DRG) neuron can be distinguished on the basis of cell body diameter and ultrastructural features of the neuronal cytoplasm (Lawson *et al*. 1974; Rambourg *et al*. 1983). Virtually all large-diameter DRG neurons express a 200-kD neurofilament protein that is absent from the soma of small sensory neurons (Anderton *et al*. 1982) and from some unmyelinated afferent terminals in lamina II (Ribiero Da Silva and Coimbra 1982).

Three essentially distinct populations of DRG neurons with central terminal arbors in the superficial dorsal horn can be defined by their expression of the peptides substance P and somatostatin and a fluoride-resistant acid phosphatase enzyme (FRAP) (Table 12.1) (Hokfelt *et al*. 1976; Nagy and Hunt 1982; Dodd *et al*. 1983). Other populations of DRG neurons can be classified by the presence of several other neuropeptides, isoenzymes, cytoskeletal proteins, and immunoglobulin binding sites (Dodd *et al*. 1983). More recently functional subclasses of DRG neurons have been identified by the cytoplasmic and cell surface expression of defined carbohydrate differentiation antigens (Dodd *et al*. 1984 Jessell and Dodd 1985; Dodd and Jesell 1985).

The availability of cell markers (Table 12.1) has made it possible to examine whether DRG neurons maintained in dissociated cell culture retain differentiated chemical properties (Dodd *et al*. 1983). Newborn rat DRG neurons maintained in culture for 4–6 weeks exhibit cell body diameters ranging from 10–40 μm. Between 65 and 70 per cent of DRG neurons possess cell body diameters of less than 20 μm, indicating that the size distribution of these neurons in culture is similar to that observed in adult DRG neurons *in situ*. By analyzing cytoplasmic antigens in cultured DRG neurons, we have observed a differential expression of many specific markers. The 200-kD neurofilament protein is detected in the cell bodies of about one-quarter of neurons in DRG cultures. Neuropeptides are also

TABLE 12.1 *Subclasses of DRG neurons identified by peripheral receptive properties, laminar termination, synaptic vesicle morphology, and cytochemical markers*

Sensory afferent modality	Spinal termination	Synaptic vesicle morphology	Cytochemical markers	Carbohydrate phenotype	Transmitter candidates
1a/1b afferents	Ventral horn	Small clear	200 K NF+	?	Glutamate
Hair Follicle Afferents	Laminae III, IV	Small clear	200 K NF+	Globoseries	Glutamate
Rapidly & Slowly Adapting Mechano-receptors	Laminae III, IV	Small clear	200 K NF+	Globoseries?	Glutamate
High Threshold Mechanoreceptor	Laminae I, V	Small clear	200 K NF+	Globoseries	Glutamate
C:Nociceptor Thermoreceptor?	Laminae I, IIo	Small clear Large granular	200 K NF− Substance P+ CGRP+, CCK+	Lactoseries	Glutamate Peptides
C:Nociceptor	Lamina IIo	Small clear Large granular	200 K NF− Somatostatin+	Lactoseries	Glutamate Peptides
C:Mechanoreceptor	Lamina IIi	Small clear	FRAP+	Lactoseries	Glutamate ATP

The correlation between functional and cytochemical properties is tentative. The cytochemical markers listed can be used to identify subsets of DRG neurons maintained in cell culture. For a more detailed description see Ralston (1979); Dodd *et al*. (1983); Dodd and Jessell (1985); this chapter.

200 K NF: 200 kilodalton neurofilament protein reactive with monoclonal antibody RT97; CGRP: calcitonin gene related peptide; CCK: cholecystokinin.

expressed by subpopulations of DRG neurons in culture. Substance P-like immunoreactivity was present in 15–20 per cent of DRG neurons. Both small- and large-diameter neurons express substance P immunoreactivity, although the most intensely labelled cells were usually of small diameter. Somatostatin-like imunoreativity was present in about 15 per cent of all neurons. Between 30 and 50 per cent of DRG neurons maintained in culture also exhibited intense FRAP staining. No significant overlap between FRAP containing and substance P- or somatostatin-immunoreactive neurons is observed in culture (Dodd *et al.* 1983) or *in situ* (Nagy and Hunt 1982). The carbohydrate differentiation antigens that delineate functional subsets of cutaneous afferent neurons *in situ* are also expressed on the surface of cultured DRG neurons (Dodd *et al.* 1984; Dodd and Jessell 1985). Moreover, the relationship between populations of neurons that express individual carbohydrate antigens is maintained in culture. These observations indicate that DRG neurons in culture retain many of their differentiated properties and establish that a representative cross-section of sensory neuron subclasses survive under our culture conditions.

In preliminary studies to characterize dorsal horn neurons growing in culture, we have found that up to 40 per cent of these neurons exhibit a high-affinity uptake of ^{3}H-labelled γ-aminobutyric acid (GABA) (C. E. Jahr; unpublished observations). Numerous processes and occasional cell bodies also display enkephalin-like immunoreactivity. Enkephalin-immunoreactive neurons are highly concentrated in the superficial dorsal horn of the spinal cord *in situ*, providing some evidence that these cultures are enriched in neurons from this region of the spinal cord.

SYNAPTIC TRANSMISSION BETWEEN DRG AND DORSAL HORN NEURONS IN CULTURE

Several groups have used dissociated cultures of DRG and spinal cord neurons to examine synaptic transmission at DRG-spinal cord, spinal cord-spinal cord, and spinal cord-DRG synapses (Ransom *et al.* 1977; Choi and Fischbach 1981; MacDonald *et al.* 1983). We have found that one major difficulty in studying primary affect transmission between DRG and dorsal horn neurons in dissociated cell co-cultures is the low incidence of detecting synaptically coupled pairs of neurons. To overcome this problem we have grown explants of DRG neurons in co-culture with dissociated dorsal horn neurons. By stimulating many DRG neurons simultaneously, we have been able to increase, dramatically, the probability of recording from dorsal horn neurons that can be excited monosynaptically (Jahr and Jessell 1985).

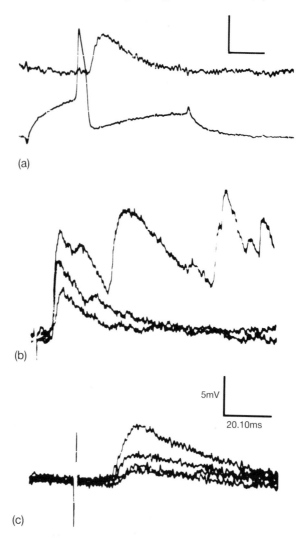

FIG. 12.1. Synaptic transmission between dorsal root ganglion and dorsal horn neurons grown in cell culture. (a). Monosynaptic e.p.s.p. recorded intracellularly from a dorsal horn neuron following intracellular stimulation of a single DRG neuron. Lower trace: action potential evoked in the presynaptic DRG neuron by intracellular current injection. Upper trace: e.p.s.p. recorded from postsynaptic dorsal horn neuron. Calibration in A; lower: 10 ms, 30 mV; upper: 10 ms, 5 mV. (b) Effect of stimulation frequency and divalent cation concentration on mono- and polysynaptic e.p.s.p.s recorded in dorsal horn neurons after extracellular stimulation of DRG explants grown in coculture. e.p.s.p.s evoked in a dorsal horn neuron by electrical stimulation of a dorsal root ganglion explant. Three stimuli were applied at 1 Hz. Note

Intracellular recordings have been obtained from dorsal horn neurons that received monosynaptic DRG input. In low concentrations of divalent cations (3 mM Ca^{2+}, 0.9 mM Mg^{2+}), spontaneous postsynaptic potentials are detectable in more than 95 per cent of dorsal horn neurons. Since no spontaneous synaptic activity or action potentials are detectable when recording from DRG neurons in the same cultures, the spontaneous e.p.s.p.s probably reflect input from other dorsal horn neurons. Electrical stimulation of DRG explants evokes e.p.s.p.s in a high proportion of dorsal horn neurons located near the explant (Fig 12.1). Post-synaptic responses in recording medium containing 3 mM Ca^{2+} and 0.9 mM Mg^{2+}, are multiphasic and prolonged, suggesting that at least part of the response is mediated by polysynaptic circuits (Fig. 12.1b). Superfusion of cultures with medium containing high divalent cations (5 mM Ca^{2+} and 3 mM Mg^{2+}) decreases or blocks all but the earliest monophasic e.p.s.p. (Fig. 12.1c).

Several characteristics of the synaptic potentials elicited in dorsal horn neurons by extracellular stimulation of DRG indicated that they are monosynaptic. As previously described (MacDonald *et al.* 1983), the divalent cation concentration in the recording medium can be increased to a level sufficient to block spontaneous synaptic input to dorsal horn neurons. In this condition, monosynaptic e.p.s.p.s generated in dorsal horn neurons by DRG stimulation can be studied without contamination from dorsal horn spinal neuron interactions The observation that evoked e.p.s.p.s follow repetitive DRG stimulation at 10 Hz and appear monophasic with a constant latency, also indicates that they are monosynaptic.

The high proportion of dorsal horn neurons that receive DRG input under these recording conditions has enabled us to compare the pharmacology of monosynaptic e.p.s.p.s with that of the potentials evoked by various excitatory transmitter candidates on the same dorsal horn neurons. Examination of the chemosensitivity of cultured dorsal horn neurons has thus far identified only two classes of compounds, excitatory amino acids and nucleotides, that have excitatory actions consistent with their roles as fast sensory transmitters (Jahr and Jessell 1983, 1985).

---

that the later components disappeared with the second and third stimuli. The recording medium contained 3 mM Ca^{2+} and 0.9 mM Mg^{2+}. Calibration in B, 5 mV × 20 ms (c) e.p.s.p.s' evoked in the same cell by four stimuli (1 Hz) after switching to medium containing 5 mM Ca^{2+} and 3 mM Mg^{2+}. Resting potential = −65 mV. Calibration in (c), 5 mv × 10 ms.

ACTIONS OF ATP ON RAT DORSAL HORN NEURONS IN CULTURE

Holton provided evidence several years ago that ATP is released from the peripheral terminals of unmyelinated sensory fibres (Holton and Holton 1954; Holton 1959) and suggested that release of ATP might also occur from central sensory terminals. To investigate the possibility that ATP acts as a central sensory transmitter we have examined the actions of nucleotides on rat dorsal horn neurons maintained in dissociated cell culture (Jahr and Jessell 1983). Intracellular recordings were made from rat dorsal horn neurons grown separately or in co-culture with DRG neurons (Yamamoto *et al.* 1981). Nucleotides and other drugs were applied to neuronal somata by pressure ejection from micropipettes with tip diameters of 4–6 μm, or by iontophoresis.

Pressure application of ATP (10^{-6}–10^{-5}M) produces a rapid and marked depolarization in about a quarter of dorsal horn neurons tested (Fig. 12.2). The response to ATP is often accompanied by action potentials superimposed on the depolarization. Other dorsal horn neurons are unaffected by ATP even at a concentration of 10^{-4}M (Fig. 12.2c). Although the majority of dorsal horn neurons receive spontaneous synaptic input, the ATP response does not depend on intact synaptic transmission. Superfusion of cultures with $CdCl_2$ completely abolishes all spontaneous synaptic activity but does not decrease the response to ATP (Fig. 12.2b).

The depolarization of dorsal horn neurons is elicited only by ATP and closely related nucleotide analogs. Adenosine, AMP, GTP, and UTP produce no effect on dorsal horn neurons, whereas ADP and CTP are less than 1/10 as active as ATP (Fig. 12.2d). Adenosine tetraphosphate, which is not an effective substrate for ATPases, and the slowly hydrolyzable ATP analogs AMP-PNP and β-δ-methylyene ATP also depolarize dorsal horn neurons indicating that the ATP-induced excitation of dorsal horn neurons is unlikely to require hydrolysis of the triphosphate chain. In contrast, α-β-methylene ATP was not an agonist on dorsal horn neurons. Comparison of the potency of nucleosides and nucleotides on dorsal horn neurons makes it unlikely that the actions of ATP are mediated by adenosine receptors (Burnstock 1978). In agreement with this, superfusion of dorsal horn neurons with 8-phenyltheophylline at a concentration sufficient to block adenosine receptors (Furshpan *et al.* 1982) does not decrease the depolarization produced by ATP.

Previous studies have suggested that Ca^{2+} chelation is responsible for the excitatory action of ATP on cuneate neurons *in vivo* (Galindo *et al.* 1967). However, we found that the disodium, magnesium, calcium, and Tris salts of ATP are equally effective in depolarizing dorsal horn neurons.

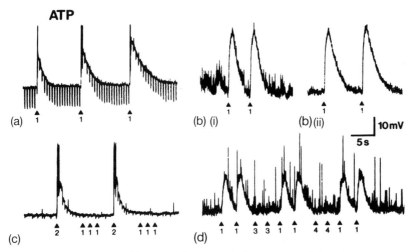

ATP

(a)

(b) (i)

(b) (ii)

10 mV

5 s

(c)

(d)

FIG.12.2. Chart records of intracellular recorded responses of dorsal horn neurons to ATP. (a) Disodium ATP (10^{-5}M) was ejected (▲) by pressure (1.5 psi) from a micropipette positioned 15 μm from the soma of the recorded neuron in pulses of 50, 100, and 200 ms duration. The fast downward deflections were produced by injecting constant current hyperpolarizing pulses of 100 ms at 80 pA through the recording electrode and provide a measure of input resistance. Action potentials are truncated in this and all subsequent figures. Resting potential = -62 mV. (b) (i) Depolarizations of another dorsal horn neuron caused by pressure ejection of 10^{-5} M ATP (0.5 s 1 psi) in control medium. (ii) After the addition of 100 μM CdSO$_4$ which blocked all spontaneous post-synaptic potentials [see in B(i) as random fast upward deflections]. Resting potential = -64 mV. (c) A dorsal horn neuron which was unresponsive to ATP (10^{-5} M; 1 s, 1.5 psi; applied at triangles labelled $'1'$) was strongly excited by L-glutamate (10^{-4} M; 0.1 s, 1.5 psi; applied at $'2'$). Fast downward deflections were spontaneous postsynaptic potentials. Resting potential = -66 mV. (d) A dorsal horn neuron which was responsive to ATP (10^{-5} M; 1 s, 1.5 psi; applied at triangles labelled $'1'$) but not adenosine (10^{-4} M; 1 s. 1.5 psi; applied at $'3'$) or GTP (10^{-4} M; 1 s, 1.5 psi; applied at $'4'$). Fast upward deflections were spontaneous postsynaptic potentials and action potentials. Resting potential = -59 mV. Calibration bar applies to all records.

Furthermore, the chelators EDTA and inorganic pyrophosphate do not affect ATP-sensitive dorsal horn neurons.

The depolarization of dorsal horn neurons by ATP is accompanied by an increase in membrane conductance that persists when the membrane potential is maintained at the resting potential (Jahr and Jessell 1983). The increase in conductance therefore reflects the action of ATP itself and does not result from voltage-dependent membrane rectification. Although hyperpolarization of dorsal horn neurons by intracellular injection of current decreases the input resistance of most neurons, it greatly increases the depolarization elicited by ATP (Fig. 12.3a). The reversal potential of

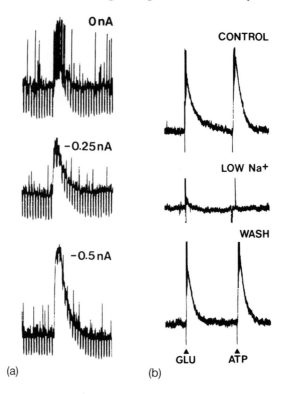

FIG. 12.3. The voltage sensitivity and ionic dependence of the ATP-evoked depolarization. (a) Three responses of the same neuron to identical pulses of ATP (10^{-5} M; 1 s × 2 psi) at the resting potential (-55 mV) during the injection of 0.25 nA of hyperpolarizing DC current (resting potential = -70 mV) and during injection of 0.5 nA (resting potential = -80 mV) Calibration: 5 s, 15 mV. (b) Responses of dorsal horn neuron to iontophoresis of Tris glutamate (100 mM; pH 8; 0.3 s, 5 nA) and Mg^{2+} $-$ATP (0.4 M; pH 5; 0.5 s, 5 nA except in the middle trace where the duration was 0.8 s) in control medium (control), after switching to medium containing 7.5 mM Na^+ (Low Na^+; sodium chloride was replaced by sucrose, 260 mM) and after returning to control medium (Wash). $CdCl_2$ (200 μM) was present in all conditions in order to block synaptic transmission. Identical results were obtained when NaCl was replaced by equimolar choline chloride.

the ATP response is therefore positive to the resting potential. Chloride ions probably do not play a major role in the ATP response. The depolarization evoked by GABA, which results from an increase in chloride conductance, can easily be converted to hyperpolarization by injecting the cell with depolarizing current. The contribution of Na^+ ions to the ATP response has been tested by altering the extracellular Na^+

concentration. Reducing extracellular Na^+ substantially decreases the depolarization evoked by ATP (Fig. 12.3b). Additional studies using patch-clamp recording techniques suggest that the depolarization of dorsal horn neurons by ATP is likely to be due to an increase in conductance to both Na^+ and K^+ (C. E. Jahr, unpublished observation).

These results provide evidence that subpopulations of dorsal horn neurons involved in the processing and regulation of sensory stimuli are potently and selectively excited by ATP. In dissociated cell culture, the identity of dorsal horn neurons that respond to ATP is unclear. *In vivo*, ATP has been reported to increase the firing rate of cuneate neurons that have cutaneous and proprioceptive input and to activate both nociceptive and non-nociceptive units in the dorsal horn of the medulla (Galindo *et al.* 1967). Central neurons with sensory input are therefore excited by ATP *in vivo* (Salt and Hill 1983b). In the spinal cord, the excitatory action of ATP may be restricted to neurons in the dorsal horn since ATP does not appear to excite neurons in the ventral horn of newborn rat (K. Yoshioka, unpublished observations). There is also considerable information on the possible transmitter actions of ATP in the peripheral nervous system (Sneddon *et al.* 1982). Evidence for a sensory transmitter role of ATP was originally obtained by Holton and her colleagues. ATP was shown to mimic the vasodilation that follows antidromic activation of nociceptive sensory fibres (Holton and Holton 1954). Moreover ATP may be released into the circulation following peripheral nerve stimulation (Holton 1959).

Several important criteria remain to be fulfilled before a transmitter role for ATP in the spinal cord can be established. While biochemical studies have demonstrated a selective release of ATP from the dorsal horn of the spinal cord (Yoshioka and Jessell 1984), the neuronal elements in the dorsal horn that release ATP remain to be identified. It will also be necessary to identify specific receptor antagonists in order to provide a more critical assessment of the role of ATP in sensory transmission. Recent studies by Fyffe and Perl (1984) demonstrate that cat dorsal horn neurons with A and C fibre mechanoreceptor input are selectively excited by ATP and thus provide additional physiological evidence for a role of ATP in sensory transmission.

EXCITATORY AMINO ACIDS AS TRANSMITTERS AT DRG-DORSAL HORN SYNAPSES IN CULTURE

Several lines of evidence suggest that the fast e.p.s.p.'s recorded from spinal cord neurons after primary afferent stimulation are mediated by L-glutamate or by compounds with similar post-synaptic actions. L-glutamate depolarizes the majority of mammalian spinal neurons *in vivo* and *in vitro* (Ransom *et al.* 1977; Watkins and Evans 1981; Salt and Hill

1983a) with a reversal potential (Mayer and Westbrook 1984) similar to that of the e.p.s.p. evoked by dorsal root ganglion (DRG) neuron stimulation (Engberg and Marshall 1979; Finkel and Redman 1983; MacDonald *et al*. 1983). Biochemical analysis has demonstrated a higher concentration of L-glutamate in dorsal than in ventral roots (Roberts *et al*. 1973) and has provided evidence for release of endogenous L-glutamate from regions of the CNS containing primary afferent terminals (Roberts 1974; Takeuchi *et al*. 1983). Moreover, L-glutamate-binding sites are found in high density in the superficial dorsal horn of rat spinal cord (Greenamyre *et al*. 1984) in regions coincident with the location of high threshold cutaneous afferent terminals.

Direct confirmation of the role of acidic amino acids as primary sensory transmitters is still lacking. In studies with intact spinal cord preparations it has been difficult to distinguish direct and indirect effects of exogenously applied amino acid transmitter candidates. A second problem has been the lack of pharmacological probes for excitatory amino acid receptors. However, recent studies have provided evidence for at least three amino acid receptor subtypes, each of which can be activated by L-glutamate (Watkins and Evans 1981; Foster and Fagg 1984). Selective ligands are available for one of the receptor subtypes: N-methyl-D-aspartate (NMDA) is a selective agonist at the NMDA subclass of receptor (Davies *et al*. 1981). In addition, the ion channel opened by NMDA receptor occupation is subject to a voltage-dependent blockade in the presence of micromolar concentrations of magnesium ions (Ault *et al*. 1980; Mayer *et al*. 1984; Nowack *et al*. 1984) thus providing a second means of distinguishing this receptor subtype. The existence of two additional classes of amino acid receptors has been proposed on the basis of pharmacological studies with the rigid amino acid analogues quisqualic acid and kainic acid, which appear to act as preferential ligands at these sites (Watkins and Evans 1981; Foster and Fagg 1984) (see Watkins, Chapter 6).

To provide information on the role of glutamate and other excitatory amino acids as primary afferent transmitters, we have studied the actions of amino acid agonists and antagonists at monosynaptic sensory neuron-dorsal horn neuron synapses formed in culture (Jahr and Jessell 1985). The sensitivity to excitatory amino acids and related compounds was tested in dorsal horn neurons that received monosynaptic sensory input, in the presence of high divalent cations. The effect of excitatory amino acids and their analogues on dorsal horn neurons were consistent in all platings. L-glutamate (10–20 μM), quisqualate (1–10 μM), and kainate (10–20 μM) were potent excitants. Kainate and L-glutamate were approximately equipotent, whereas quisqualate was 10–20 times more potent. These compounds caused rapid depolarizations which were often suprathreshold. In contrast, NMDA, at concentrations up to 200 μM, had no effect on spinal neurons and L-aspartate (100–200 μM) had either no effect or elicited only small depolarizations after repeated application. Importantly, although most dorsal horn neurons were completely insensitive to L-aspartate they clearly received a monosynpatic input from DRG neurons (Fig. 12.4). The failure of NMDA and L-aspartate to excite dorsal horn neurons could reflect the absence of NMDA receptors on most dorsal horn neurons or the inactivation of the NMDA receptor/channel complex in the presence of high Mg^{2+} concentrations (Mayer *et al*. 1984; Nowak *et al*. 1984). Studies on chick spinal cord cells grown in low density cultures have demonstrated that identified spinal motoneurons do not express NMDA receptors in the absence of innervation by interneurons (O'Brien 1985). The selective depolarization of dorsal horn neurons by L-glutamate in the presence of high divalent cations therefore suggests that L-glutamate, but not L-aspartate may mediate the fast DRG-evoked e.p.s.p.s. At present, however, we cannot exclude the possibility that L-aspartate is a primary sensory transmitter at synapses that are not detected under our recording conditions.

To determine whether amino acids, or compounds acting at amino acid receptors on dorsal horn neurons, might be transmitters at sensory synapses, we examined the effect of several antagonists of excitatory amino acid-evoked responses on the DRG-evoked monosynaptic e.p.s.p. recorded from dorsal horn neurons. Kynurenate was the most potent antagonist of both the e.p.s.p. and the depolarization evoked by pressure ejection of L-glutamate (Fig. 12.5), quisqualate, and kainate (Jahr and Jessell 1985). Kynurenate at a concentration of 1 mM completely blocks

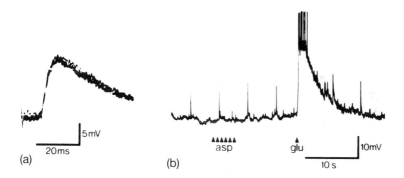

FIG. 12.4. Monosynaptic sensory neuron input in the absence of L-aspartate sensitivity. The first record shows oscilloscope traces of four superimposed e.p.s.p.s elicited by DRG explant stimulation. The second record is a chart record obtained from the same neuron, showing the responses to pressure application of L-aspartate (asp: 1 s × 200 μM: six applications) and of L-glutamate (glu: 1 s × 20 μM). The fast upward deflections occurring at 5 s intervals are the evoked e.p.s.p.s shown at a faster time base in A. Resting potential = −64 mV.

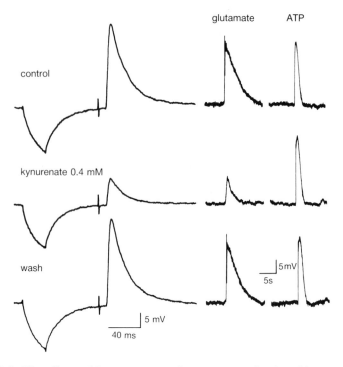

FIG. 12.5. The effects of kynurenate on the responses of a dorsal horn neuron to DRG-explant stimulation and pressure application of L-glutamate and ATP. The first traces in each row show the responses to an intracellular current pulse followed by stimulation of a nearby DRG explant. The next two traces are chart records of the responses to pressure ejection of L-glutamate (1 s × 20 μM) and ATP (1 s × 20 μM). The responses in the second row were recorded in the presence of 0.4 mM kynurenate. The responses in the third row were recorded after washing out kynurenate. Each e.p.s.p. trace is the average of 16 records. Resting potential = −68 mV.

the monosynaptic e.p.s.p. and the amino acid-evoked responses recorded from dorsal horn neurons. The reduction of e.p.s.p. amplitude and of amino acid-evoked depolarizations by kynurenate is not associated with changes in the input resistance of dorsal horn neurons or with changes in the threshold, amplitude, or duration of the action potential recorded intracellularly from the cell bodies of DRG neurons. We have found that all e.p.s.ps recorded from dorsal horn neurons, whether spontaneous or evoked in response to DRG stimulation, are antagonized by kynurenate (Jahr and Jessell 1985). In contrast, neither 2-amino-4-phosphonobutyrate nor L-glutamate diethylester at concentrations up to 1 mM significantly affected the DRG-evoked e.p.s.p. or the responses of dorsal horn neurons to L-glutamate, kainate, or quisqualate. Kynurenate has been shown previously to block all three classes of amino acid receptor. However, the selective NMDA receptor antagonist, 2-aminophosphorovaleric acid (APV) only slightly reduced the amplitude of the e.p.s.p. and of the L-glutamate-evoked depolarization.

The weak antagonist action of APV on DRG-evoked e.p.s.p.s suggests that if L-glutamate is the transmitter at these synapses, it is unlikely to mediate the initial excitation of dorsal horn neurons via the NMDA receptor, at least at a potential range near rest. It is possible, however, that if NMDA receptors are expressed on dorsal horn neurons, they could contribute to afferent transmission at depolarized potentials, under conditions in which Mg^{2+} dependent channel block is relieved.

The blockade of glutamate-evoked depolarization of dorsal horn neurons clearly demonstrates a postsynaptic site of action of kynurenate. However, kynurenate could antagonize the e.p.s.p. by blocking action potential propagation in the axons of DRG neurons. Most of the e.p.s.p.'s studied were composed of smaller units which could be separated by a continuous gradation in the stimulus strength. Graded stimulation revealed discrete steps in the amplitude of the e.p.s.p. and suggested that several DRG neurons can form functional synapses on the same dorsal horn neuron (Jahr and Jessell 1985). In the presence of kynurenate the same number of discrete steps could be evoked, although the amplitude of each was reduced to a similar extent. Moreover, recruitment of each step in the e.p.s.p. occurred at a stimulus strength identical to that required before addition of kynurenate. These findings make it unlikely that kynurenate blocks spike propagation in DRG axons and provide further evidence of the post-synaptic site of inhibition of the DRG-evoked e.p.s.p. (Jahr and Jessell 1985). The blockade by kynurenate of L-glutamate responses of dorsal horn neurons at concentrations that also reduced or abolished the DRG-evoked monosynaptic e.p.s.p. is consistent with the idea that L-glutamate is a primary sensory transmitter.

Although kynurenate antagonized the response of dorsal horn neurons to L-glutamate, it did not inhibit the response of dorsal horn neurons to other excitatory transmitter candidates. We have shown above that ATP excites a subpopulation of dorsal horn neurons by activating a membrane conductance similar to that evoked by L-glutamate (Jahr and Jessell 1983). The reversal potential of both L-glutamate (Mayer and Westbrook 1984) and ATP-evoked (C. E. Jahr, unpublished observations) responses on spinal neurons is near O mV. Superfusion of dorsal horn neurons with a concentration of kynurenate sufficient to antagonize both the DRG-evoked e.p.s.p. and the response to L-glutamate had no effect on the response of the same neurons to ATP (Fig. 12.5). The selectivity of kynurenate provides further evidence that L-glutamate or a closely related compound is the transmitter at the sensory neuron synapses examined in these experiments.

From our studies in culture, the proportion of DRG neurons that release L-glutamate-like compounds as primary afferent transmitters is not clear. It is possible that culture conditions have in some way restricted our analysis to a subpopulation of primary afferent synapses. Studies on newborn rat spinal cord preparations performed in parallel with the experiments described here, however, have demonstrated the ability of kynurenate to antagonize both Ia e.p.s.p.s and cutaneous input to dorsal horn neurons (Jahr and Yoshioka 1985) suggesting that primary afferents conveying diverse sensory modalities release L-glutamate-like compounds as primary sensory transmitters in the mammalian spinal cord. From our studies on the chemosensitivity of cultured dorsal horn neurons to putative sensory transmitters, the only compounds known to have post-synaptic actions that are consistent with a fast excitatory transmitter role are the acidic amino acids and the nucleotide ATP (Jahr and Jessell 1983, 1985). Although kynurenate is an effective antagonist of DRG evoked e.p.s.p.s our results do not rule out the possibility that ATP may be a sensory transmitter released from a small proportion of primary afferent fibres. Alternatively, the sensitivity of dorsal horn neurons to ATP may be associated with the presence of dorsal horn interneurons that use nucleotides as synaptic transmitters.

CONCLUSIONS

The studies outlined in this chapter suggest that there are likely to be two major classes of synaptic transmitters released by primary sensory neurons in the mammalian spinal cord. Many, and possibly all, primary sensory neurons appear to release transmitters with fast excitatory post-synaptic actions on spinal neurons. Although the identity of fast sensory transmitters is still not established, L-glutamate, or a closely related compound, is likely to be a sensory transmitter at synapses formed between DRG and dorsal horn neurons maintained in culture and also at Ia afferent synapses. Pharmacological studies in culture and *in vivo* have provided evidence that is consistent with the concept that ATP may be a fast transmitter at some cutaneous afferent synapses, possibly those mediating low-threshold mechanical stimuli conveyed by C fibres.

In addition to the fast excitatory sensory transmitters, several peptides have been shown to exist in subsets of primary sensory neurons that project to the superficial dorsal horn. Substance P and related tachykinins are likely to be slow excitatory transmitters released following activation of high-threshold afferent fibres. The physiological functions of tachykinins and other peptides in the processing of sensory information are still uncertain. However, it seems likely that the release of peptides may modify the responses of specific classes of dorsal horn neurons to other rapidly acting sensory transmitters. Identification of dorsal horn neurons that express receptors for amino acids and peptides will be important in defining more clearly, the functional interactions of fast and slow sensory transmitters.

Future studies to correlate the sensory fibre types that use amino acids, nucleotides, and peptides at sensory transmitters should be possible with the availability of monoclonal antibodies that label carbohydrate differentiation antigens expressed by functional subpopulations of DRG neurons *in situ* and in cell culture (Dodd *et al*. 1984; Jessell and Dodd 1985; Dodd and Jessell 1985).

Acknowledgements

This work was supported by grants from the National Institutes of Health, the McKnight Foundation, The Muscular Dystrophy Association, and the Rita Allen Foundation.

References

Anderton, B., Coakhan, H. B., Garson, J. A., Harper, A. A., Harper, E. I., and Lawson, S. N. (1982). A monoclonal antibody against neurofilament protein specifically labels the large light cell population in rat dorsal root ganglia. *J. Physiol*. **323**, 97P.

Ault, B., Evans, H., Francis, A. A., Oakes, D. J., and Watkins, J. C. (1980). Selective depression of excitatory amino acid induced depolarizations by magnesium ions in isolated spinal cord preparations. *J. Physiol. (Lond.)* **307**, 413–28.

Bannatyne, B. A., Maxwell, D. J., Fyffe, R. E. W., and Brown, A. G. (1984). Fine structure of primary afferent axon terminals of slowly adapting cutaneous receptors in the cat. *Q. J. Exp. Physiol*. **69**, 547–57.

Bresnahan, J. C., Ho, R. H., and Beattie, M. S. (1984). A comparison of the ultrastructure of substance P and enkephalin-immunoreactive elements in the nucleus of the dorsal lateral funiculus and laminae I and II of the rat spinal cord. *J. Comp. Neurol*. **229**, 497–511.

Brown, A. G. (1982). The dorsal horn of the spinal cord. *Q. J. Exp. Physiol*. **67**, 193–212.

Burnstock, G. (1978). Purinergic receptors. *J. Theor. Biol*. **62**, 491–9.

Choi, D. W., and Fischbach, G. D. (1981). GABA-mediated synaptic potentials in chick spinal cord and sensory neurons. *J. Neurophysiol*. **45**, 632–43.

Davies, J., Francis, A. A., Jones, A. W., and Watkins, J. C. (1981). 2-Amino-phosphonovalerate (2APV), a potent and selective antagonist of amino acid-induced and synaptic excitation. *Neurosci. Lett*. **21**, 77–81.

DiFiglia, M., Aronin, N., and Leeman, S. (1982). Light microscopic and ultrastructural localization of immunoreactive substance P in the dorsal horn of the monkey spinal cord. *Neurosci*. **7**, 1127–39.

Dodd, J., Jahr, C. E., Hamilton, P., Heath, M., Matthew, W. D., and Jessell, T. M. (1983). Cytochemical and physiological properties of sensory and dorsal horn neurons that transmit cutaneous sensation. *Cold Spring Harbor Symp. Quant. Biol.* **48**, 685–95.

——, and Jessell, T. M. (1985). Lactoseries carbohydrates specify subsets of dorsal root ganglia neurons projecting to the superficial dorsal horn of the spinal cord. *J. Neurosci.* (in press).

——, Solter, D., and Jessell, T. M. (1984). Monoclonal antibodies against carbohydrate differentiation antigens identify subsets of primary sensory neurons. *Nature*, **311**, 469–72.

Engberg, I., and Marshall, K. C. (1979). Reversal potential for Ia excitatory postsynaptic potentials in spinal motoneurons of cats. *Neurosci.* **4**, 1583–91.

Finkel, A. S., and Redman, S. J. (1983). The synaptic current evoked in cat spinal motoneurons by impulses in single group Ia axons. *J. Physiol. (Lond.)* **342**, 615–32.

Foster, A. C., and Fagg, G. E. (1984). Acidic amino acid binding sites in mammalian neuronal membranes: The characteristics and relationship to synaptic receptors. *Brian Res. Rev.* **7**, 103–64.

Furshpan, E. J., Potter, D. D., and Landis, S. C. (1982). On the transmitter repertoire of sympathetic neurons in culture. *Harvey Lect.* **76**, 149.

Fyffe, R. E. W., and Perl, E. R. (1984). Is ATP a central synpatic mediator for certain primary afferent fibers from mammalian brain? *Proc. Nat. Acad. Sci. USA*, **81**, 6890–3.

Galindo, A., Krnjevic, K., and Schwartz, S. (1967). Microiontophoretic studies on neurons in the cuneate nucleus. *J. Physiol.* **192**, 359.

Greenamyre, J. T., Young, A. B., and Penny, J. B. (1984). Quantitative autoradiographic distribution of L-[^{3}H]glutamate-binding sites in rat central nervous system. *J. Neurosci.* **4**, 2133–44.

Hokfelt, T., Elde, R., Johansson, O., Luft, R., Nilsson, G., and Arimura, A. (1976). Immunohistochemical evidence for separate populations of somatostatin-containing and substance P-containing primary afferent neurons in the rat. *Neurosci.* **1**, 131.

Holton, F. A., and Holton, P. (1954). The capillary dilator substances in dry powders of spinal roots: A possible role of adenosine triphosphate in chemical transmission from nerve endings. *J. Physiol.* **126**, 124.

Holton, P. (1959). The liberation of adenosine triphosphate on antidromic stimulation of sensory nerves. *J. Physiol.* **145**, 494.

Jahr, C. E., and Jessell, T. M. (1983). ATP excites a subpopulation of rat dorsal horn neurons. *Nature* **304**, 730–3.

—— and Jessell, T. M. (1985). Synaptic transmission between dorsal root ganglion and dorsal horn neurons in culture: antagonism of e.p.s.p.s and glutamate excitation by kynurenate. *J. Neurosci.* **5**, 2281–9.

——, and Yoshioka, K. (1985). Ia afferent excitation of motoneurons in the newborn rat spinal cord is selectively antagonized by kynurenate. *J. Physiol.* (in press).

Jessell, T. M. (1982). CNS neurotransmitters: Pain. *Lancet*, **ii**, 1084–8.

—— (1983a). Nociception. In *Brian Peptides* (eds D. Krieger, M. Brownstein, and J. B. Martin), pp. 315–32. Wiley, NY.

—— (1983b). Substance P in the nervous system. In *Handbook of Psycho-*

pharmacology (eds L. L. Iversen, S. D. Iversen, and S. H. Snyder) Vol. 16, p. 1. Plenum Press, NY.

——, and Dodd, J. (1985) Structure and expression of differentiation antigens on functional subsets of primary sensory neurons. *Phil. Trans. Roy. Soc. Lond. (Biol.)* **208**, 271–81.

Konishi, S., Akagi, H., Yanagisawa, M., and Otsuka, M. (1983). Enkephalinergic inhibition of slow transmission in the rat spinal cord. *Neurosci. Lett.* Suppl. 13, S107.

Lawson, S. N., Caddy, K. W. T., and Biscoe, T. J. (1974). Development of rat dorsal root ganglion neurons. Studies of cell birthdays and changes in mean cell diameter. *Cell Tissue Res.* **153**, 399–413.

Light, A. R., and Perl, E. R. (1979). Spinal termination of functionally identified primary afferent fibers with slowly conducting myelinated fibers. *J. Comp. Neurol.* **186**, 133–50.

——, Trevino, D. L., and Perl, E. R. (1979). Morphological features of functionally defined neurons in the marginal zone and substantia gelatinosa of the spinal dorsal horn. *J. Comp. Neurol.* **186**, 151–72.

Lynn, B., and Carpenter, S. E. (1982). Primary afferent units from the hairy skin of the rat hind limb. *Brain Res.* **238**, 29–43.

MacDonald, R. L., Pun, R. Y. K., Neale, E. A., and Nelson, P. G. (1983). Synaptic interactions between mammalian central neurons in cell culture. I. Reversal potential for excitatory postsynaptic potentials. *J. Neurophysiol.* **49**, 1428–41.

Mayer, M. L., and Westbrook, G. L. (1984). Mixed agonist action of excitatory amino acids on mouse spinal cord neurons under voltage clamp. *J. Physiol. (Lond.)* **354**, 29–53.

——, ——, and Guthrie, P. B. (1984). Voltage-dependent block by Mg^{2+} of NMDA responses in spinal cord neurons. *Nature*, **309**, 261–3.

Nagy, J. I., and Hunt, S. P. (1982). Fluoride-resistant acid phosphatase-containing neurons in dorsal rat ganglia are separate from those containing substance P or somatostatin. *Neurosci.* **7**, 89.

Nowak, L., Bregestovski, P., Ascher, P., Herbert, A., and Prochiantz, A. (1984). Magnesium gates glutamate-activated channels in mouse central neurons. *Nature*, **307**, 462–5.

O'Brien, R. J. (1985). Physiological properties of identified motoneurons grown in dissociated cell culture. PhD Thesis, Harvard University.

Otsuka, M., Konishi, S., Yanagisawa, M., Tsunoo, A., and Akagi, H. (1982). Role of substance P as a sensory transmitter in spinal cord and sympathetic ganglia. *Ciba Found. Symp.* **91**, 13–34.

Perl, E. R. (1983). Characterization of nociceptors and their activation of neurons in the superficial dorsal horn: first steps for the sensation of pain. *Adv. Pain Res. Ther.* **6**, 23–51.

Ralston, J. H., III (1979). Distribution of dorsal root axons in laminae I, II and III of the macaque spinal cord. *J. Comp. Neurol.* **184**, 643–84.

——, Light, A. R., Ralston, D. D., and Perl, E. R. (1984). Morphology and synaptic relationships of physiologically identified low-threshold dorsal root axons stained with intra-axonal horseradish peroxidase in the cat and monkey. *J. Neurophysiol.* **51**, 777–92.

Rambourg, A., Clermont, Y., and Beaudet, A. (1983). Ultrastructural features of six types of neurons in rat dorsal root ganglia. *J. Neurocytol.* **12**, 47.

Ransom, B. R., Bullock, P. N., and Nelson, P. G. (1977). Mouse spinal cord in cell culture. III: Neuronal chemosensitivity and its relationship to synaptic activity. *J. Neurophysiol.* **40**, 1163–77.

Rethelyi, M. (1977). Preterminal and terminal axon arborizations in the substantia gelatinosa of cat's spinal cord. *J. Comp. Neurol.* **172**, 511–28.

Rethelyi, M., Light, A. R., and Perl, E. R. (1982). Synaptic complexes formed by functionally defined primary afferent units with fine myelinated fibers. *J. Comp. Neurol.* **207**, 381–93.

Ribiero Da Silva, A., and Coimbra, A. (1982). Two types of synaptic glomeruli and their distribution in laminae I-III of the rat spinal cord. *J. Comp. Neurol.* **209**, 176–89.

Roberts, P. J. (1974). The release of amino acids with proposed transmitter function from the cuneate and gracile nuclei of the rat, *in vivo*. *Brain Res.* **67**, 419–28.

——, Keen, P., and Mitchell, J. F. (1973). The distribution and axonal transport of free amino acids and related compounds in the dorsal sensory neuron of the rat, as determined by the dansyl reaction. *J. Neurochem.* **21**, 199–209.

Salt, T. E., and Hill, R. G. (1983a). Neurotransmitter candidates of somatosensory primary afferent fibres. *Neurosci.* **10**, 1083–103.

——, and —— (1983b). Excitation of single sensory neurons in the rat caudal trigeminal nucleus by iontophoretically applied adenosine 5' triphosphate. *Neurosci. Lett.* **35**, 53.

Sneddon, P., Westfall, D. P., and Fedan, J. S. (1982). Co-transmitters in the motor nerves of the guinea pig was deferens: electrophysiological evidence. *Science*, **218**, 693.

Takeuchi, A., Onodera, K., and Kawagoe, R. (1983). The effects of dorsal root stimulation on the release of endogenous glutamate from the frog spinal cord. *Proc. Jpn Acad.* **59**, 88–92.

Urban, L., and Randic, M. (1984). Slow excitatory transmission in rat dorsal horn. Possible mediation by peptides. *Brain Res.* **290**, 336–41.

Watkins, J. C., and Evans, R. H. (1981). Excitatory amino acid transmitters. *Annu. Rev. Pharmacol. Toxicol.* **21**, 165–204.

Willis, W. D., and Coggeshall, R. E. (1978). *Sensory Mechanisms in the Spinal Cord*. Plenus Press, NY.

Yaksh, T. L., Jessell, T. M., Gamse, R., Mudge, A. W., and Leeman, S. E. (1980). Intrathecal morphine inhibits substance P release from mammalian spinal cord *in vivo*. *Nature*, **286**, 155–7.

Yamamoto, M., Steinbusch, M. W., and Jessell, T. M. (1981). Differentiated properties of identified serotonin neurons in dissociated cultures of embryonic rat brain stem. *J. Cell Biol.* **91**, 142–52.

Yoshioka, K., and Jessell, T. M. (1984). ATP release from the dorsal horn of rat spinal cord. *Soc. Neurosci. Abs.* **10**, 993.

13

Inositide metabolism in the brain: its potential role in complex neuronal pathways

R. F. IRVINE AND M. J. BERRIDGE

INTRODUCTION

Recent advances in our understanding of brain function have emphasized the role played by slow chemical signals in the integrative and cognitive aspects of this organ. The contrast has been drawn between the rapid point-to-point transmission mediated by, for example, γ-aminobutyric acid and L-glutamate, and the slow information transmission between neurons mediated by a myriad of chemical signals, peptides, amines, etc. (Schmitt 1984; Iversen 1984). The latter group of signals appear to modulate the faster pathways and may also interact with each other. Furthermore, in the light of the discovery that a single axon can release more than one substance (Viveros *et al.* 1983, Hökfelt *et al.* 1986), an almost infinite variety of chemical signals from single neurones is possible (Iversen 1984). It is these slow pathways that are considered most likely to be responsible for the extraordinary complexity of integrated circuits and memory functions which our brains must possess.

However, in the region of a synapse in which a neurotransmitter is released and then binds to its receptor, the diffusion of that neurotransmitter is not free, and so the range of such chemical compounds to which each neurone is exposed is limited. Also, in contrast to the rich variety of extracellular signals, the intracellular messenger systems which they employ are much less diverse (Iversen 1984), and indeed are very few.

So, although some potential for the complex control systems necessary for higher brain function is to be found in the diversity of information substances and in the complex anatomy ('wiring') of the brain, the subtlety of chemical controls, and so the true potential for regulation and thus integration, is limited ultimately by the flexibility of the intracellular signalling systems. How well do the different signalling systems interact with each other? And at how many levels? And within just one signalling system what is the potential for increases and decreases not only in the amounts of the individual signals, but of the degree to which they are interpreted by the cell? It is our aim to draw attention to the enormous potential for flexibility of the inositol lipid signalling pathway in this regard, and to discuss some of the molecular mechanisms involved.

MULTIFUNCTIONAL NATURE OF INOSITIDE FUNCTION

Although our main subject here will be on the role of inositides in the transmission of signals from the outside of the cell (receptor) to the inside, in particular by the generation of second messengers, it should be emphasized that this may be only one aspect of the many potential functions of inositides (Irvine and Dawson 1980; Berridge 1981; Irvine *et al.* 1982). For example, the enzymes responsible for polyphosphoinositide metabolism are not confined to the plasma membrane. Phosphatidylinositol (Ptd Ins) kinase has been reported in golgi (Jergil and Sundler 1983), lysosomes (Collins and Wells 1983) and nuclear membranes (Smith and Wells 1983). The latter membrane also has a phosphatidylinositol-4-phosphate (Ptd Ins P) phosphomonoesterase activity (Smith and Wells 1984). If we accept that Ptd Ins P kinase is a predominantly soluble enzyme (see Irvine 1982a) then whenever Ptd Ins P occurs, some phosphatidylinositol-4, 5-bisphosphate (Ptd Ins P_2) is likely to be formed. Thus, other cell membranes possess the capacity to synthesize and degrade polyphosphoinositides, and if Ptd Ins P_2 phosphodiesterase is also present in them (or alternatively they possess the ability to activate the soluble enzyme), then the possibility emerges of, for example, intranuclear second messenger formation. The nuclear membrane is in fact rich in many lipid-metabolizing enzymes, especially in neuronal tissue (for example Baker and Chang 1981), and so here is a whole new concept of regulation of nuclear activity via effects on inositide metabolism.

In addition to the capacity for inositides to generate second messengers (discussed below) one must also not forget the possible functions of the lipids themselves. Earlier documented reports of effects of Ptd Ins and the polyphosphoinositides on cellular enzymes can be found in reviews by Irvine *et al.* (1982) and Hawthorne (1983), and these have recently been added to by, for example, effects of Ptd Ins on a proteinase (Coolican and Hathaway 1984 and Ptd Ins P_2 on a plasma membrane Ca^{2+} ATP-ase (Choquette *et al.* 1984) or profilactin (Lassing and Lindberg 1985). Thus, any extracellular messengers or metabolic changes which are likely to induce modifications in the relative proportions of the various inositides, could exert influences on more than just the levels of the second messengers.

In summary, the clearly multifunctional nature of inositide metabolism in turn suggests that subtle long-term modifications of neuronal properties could be caused by changes in the levels of the various inositides over and above the short-term second messenger levels to be discussed here.

SECOND MESSENGERS GENERATED FROM INOSITIDES AND THEIR ROLES IN THE NERVOUS SYSTEM

The range of second messengers that can be derived from the inositides is a further facet of their multifunctional nature. The first such compound shown unequivocally to possess messenger functions was diacylglycerol (Takai *et al.* 1979; Nishizuka 1984) which specifically stimulates C-kinase (a protein kinase found in higher concentrations in neuronal tissue than in any other). The stimulation of this enzyme by phorbol esters (Castagna *et al.* 1982) has provided us with a tool for specifically activating the diacylglycerol branch of the signal pathway with, it is hoped, little effect on other cellular functions. However, some caution is necessary in drawing too many conclusions from experiments where cells have been treated with phorbol esters; notwithstanding the possible intracellular generation of inositide-derived messengers discussed above, the production of diacylglycerol is probably confined to the plasma membrane in the first instance. Current concepts of C-kinase favour the view that it is a soluble protein that is activated on binding to a diacylglycerol-containing membrane (Kraft and Anderson 1983; Hirota *et al.* 1985), and if phorbol esters can permeate to every intracellular membrane in the cell, then they may activate C-kinase in membranes which would never normally see that enzyme.

Diacylglycerol may also be involved in cytoskeletal-membrane connections (Burn *et al.* 1985), in stimulating intracellular phospholipases (Dawson *et al.* 1984), and it can also act as a source of arachidonic acid for prostaglandin production (Cabot and Gatt 1976; Dawson and Irvine 1978; Mauco *et al.* 1978; Bell *et al.* 1979), though the quantitative contribution that inositides actually make to arachidonate release is not yet certain (see Irvine 1982b). Nevertheless, given sufficient levels of diacylglycerol lipase in a tissue (and it was in neuronal tissue that this enzyme was first fully characterized, Cabot and Gatt 1976) some arachidonate may be released, and here again a great potential exists for feedback and feed-forward regulation, especially if one considers the wide range of eicosanoids that can be derived from arachidonate and the degree to which the proportions of the various eicosanoids can differ between cells (for a review of eicosanoids in nervous tissue, see Wolfe 1982). Below we will discuss interactions between the inositide pathway and the adenylate cyclase system, and while doing so it should be remembered that PGE_1 probably uses cAMP as an intracellular second messenger (Haslam *et al.* 1978) and $PGF_{2\alpha}$ stimulates inositide metabolism (MacPhee *et al.* 1984) and that these two prostaglandins are known to synergize in stimulating cell division (Otto *et al.* 1982). So a given cell clearly has a great potential for modifying its metabolism and the metabolism of its neighbours, by altering the pattern of eicosanoids that it produces from free arachidonate. Eicosanoids have been suggested as playing a particularly important role as local hormones responsible for modulating presynaptic activity (Trevisani *et al.* 1982; Inoue *et al.* 1984).

The other principle messenger produced from inositides is inositol-1,4,5-trisphosphate [$Ins(1,4,5)P_3$], whose major function as we understand it at present, is to mobilize calcium (Berridge 1983; Streb *et al.* 1983; Berridge and Irvine 1984) and it can therefore simplistically be equated with calcium. In some circumstances that equation may be too simplistic; in using $Ins(1,4,5)P_3$ to control another messenger (calcium), the cell introduces another step which can be controlled. Thus, changes in the loading or the location of calcium pools (the location aspect perhaps being a distribution between those pools which are accessible to receptor-stimulated mobilization and those that are not, see Putney 1982) could markedly alter the effectiveness of a given level of intracellular $Ins(1,4,5)P_3$. Above we discussed briefly the possibility of stimulated inositide turnover being located also within the cell, and alternative calcium pools for $Ins(1,4,5)P_3$ to mobilize may exist in, for example, the nucleus, which have not been discovered yet.

There are a number of lines of evidence pointing to a role for inositol lipids in the nervous system. The introduction of lithium as a mechanism for amplifying inositol lipid responses (Berridge *et al.* 1982) has helped to identify and localize those transmitters which act through the inositol lipids (Table 13.1). It is also apparent that neural tissue is a rich source of the enzymes involved in the bifurcating signal pathway based on DG and $Ins(1,4,5)P_3$ as second messengers. A particularly challenging problem for the future is to find out how these signal pathways function in identified neurones. Transmitters which act through the inositol lipids may either influence excitability and so contribute to the decision about whether or not to fire an action potential, or they may influence what happens when the action potential arrives at the synapse. Classically, the two 'halves' of the $Ptd Ins P_2$ pathway interact synergistically to produce full cell activation (Nishizuka 1984). The contribution of each half can be studied independently of the other by using pharmacological agents which bypass each second messenger. Phorbol esters can be used to mimic the action of

TABLE 13.1. *Identification and localization of transmitters capable of stimulating inositol lipid breakdown in neural tissue*

Transmitter	Tissue	Reference
Substance P	Rat hypothalamus	Watson and Downes (1983)
Neurotensin	Rat brain slices	Goedert *et al.* (1984)
Acetylcholine	Hippocampus	Janowsky *et al.* (1984)
	Rat cerebral cortex	Brown *et al.* (1984)
	Striatum	Gonzales and Crews (1984)
Norepinephrine	Hippocampus	Janowsky *et al.* (1984)
	Rat cerebral cortex	Brown *et al.* (1984); Schoepp *et al.* (1984)
5-hydroxytryptamine	Hippocampus	Janowsky *et al.* (1984)
	Cerebral cortex	Conn and Sanders-Bush (1984)
Histamine	Cerebellum, hippocampus	Daum *et al.* (1984)
	Cerebral cortex	Brown *et al.* (1984)
Vasopressin-like peptide	Rat superior cervical and coeliac ganglia	Hanley *et al.* (1984)

DG whereas calcium ionophores will mobilize calcium like IP_3. A labile intracellular store of calcium, which can be released by caffeine, has been described in mammalian neurones (Nerring and McBurney 1984). Release of calcium from this internal pool by $Ins(1,4,5)P_3$ could function to raise the resting level of calcium and so contribute to facilitation. In a hybrid neuroblastoma/glioma cell line (NE 108-15) bradykinin induces a transient decrease in $Ptd Ins P_2$ which might be responsible for the increased discharge of action potentials (Yano *et al.* 1984). Evidence is already available to show that the DG/C-kinase pathway can exert some subtle effects on the characteristics of each action potential. In *Aplysia* bag cells, phorbol esters increase the amplitude of the calcium-dependent action potentials (De Riemer *et al.* 1985). When tested on hippocampal pyramidal neurones, phorbol esters were able to enhance the firing of action potentials by blocking the potassium channels responsible for the slow after hyperpolarizations (Baraban *et al.* 1985b). There are clear indications, therefore, for an inositol lipid involvement in the action of certain transmitters which modulate firing patterns.

The inositol lipid pathway may also play a role in modulating the release of transmitters at synaptic endings. Release of acetylcholine from parasympathetic nerve endings during electrical stimulation can be potentiated by phorbol esters (Tanaka *et al.* 1984). Activation of the C-kinase pathway may also regulate exocytosis in neurosecretory cells (Pozzan *et al.* 1984). At present however, no transmitter has been identified which functions through the inositol lipids to modulate synaptic transmission. All of this evidence relating to the role of C-kinase has been derived from studies using phorbol esters to mimic the action of diacylglycerol and there must be reservations about such an approach (see above). Increasingly, instances are being found where phorbol esters, presumably stimulating kinase C can feed back negatively on inositide metabolism (for example, Mellors *et al.* 1985). Perhaps related to this phenomenon is the observation that phorbol esters markedly reduce the ability of acetylcholine to stimulate inositol lipid breakdown in rat hippocampal slices (Labarca *et al.* 1984) and a cloned neurosecretory cell line (Vicentini *et al.* 1985). Here again is potential for neurones to change their properties considerably if they alter their metabolism from one type (synergism between kinase C and calcium) to the other (antagonism between the two; see also Baraban *et al.* 1985a). The difference between the two types may be chronic (i.e. a matter of differentiation) or acute (perhaps controlled in turn by the levels of the same second messengers) or both types could even occur over a different time scale in the same cell (Drummond 1985), and an understanding of the mechanisms of interaction between the two principle messengers formed from inositides, diacyglycerol, and Ins$(1,4,5)P_3$, is an area where considerable work needs to be done.

It is clear from the immediately preceding paragraphs that even considering the two main inositide messengers and their inter-relationship is complex, and the recent identification of a third compound derived directly from Ptd Ins P_2 metabolism, inositol-1,3,4,-trisphosphate [Ins$(1,3,4)P_3$], opens up still further possibilities for modulatory mechanisms (Irvine *et al.* 1984b, 1985b). At present the function of this compound (if any) or even its route of synthesis are not established. In tissues where considerable amounts of Ins$(1,3,4)P_3$ are produced such as parotid or liver, there appears to be a time lag before its synthesis starts, but once started this synthesis is very rapid (Irvine *et al.* 1985b; Burgess *et al.* 1985). Other tissues produce much less of the (1,3,4) isomer as compared with Ins$(1,4,5)P_3$, for example the blowfly salivary gland (M. J. Berridge and R. F. Irvine unpublished), with, in that particular gland, no detectable time lag. *Limulus* photoreceptors however, produce mostly Ins$(1,3,4)P_3$ (Irvine *et al.* 1985a), and extremely quickly (J. E. Brown and R. F. Irvine, unpublished). In short, although Ins$(1,3,4)P_3$ seems to be produced rapidly by all cells when inositide turnover is stimulated, there is considerable variation in the relative amount of Ins$(1,3,4)P_3$ that accumulates, when it is produced and how quickly. Until some function is assigned to this compound we cannot make any predictions about these differences or the interactions of Ins$(1,3,4)P_3$ with other messengers, but whatever its function, it must inevitably introduce a further level of complexity into the picture discussed here.

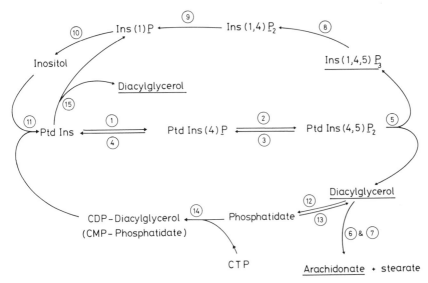

FIG. 13.1 Enzymes of inositide metabolism. Compounds with known second messenger functions are underlined. The enzymes are numbered as follows. 1. Ptd Ins kinase; 2. Ptd Ins(4)P kinase; 3. Ptd Ins(4,5)P_2 phosphomonoesterase; 4. Ptd Ins(4)P phosphomonoesterase; 5. Ptd Ins(4,5)P_2 phosphodiesterase; 6. Diacylglycerol lipase; 7. Monoacylglycerol lipase; 8. Ins(1,4,5)P_3 phosphatase; 9. Ins-(1,4)P_2 phosphatase; 10. Ins(1)P phosphatase; 11. CDP-diacylglycerol inositol phosphatidate transferase (Ptd Ins synthetase); 12. Diacyglycerol kinase; 13. Phosphatidate phosphohydrolase; 14. CTP phosphatidate cytidyl transferase; 15. Ptd Ins phosphodiesterase.

In conclusion, at least five potential second messengers can be produced from inositides (and with the diversity of known eicosanoids that number could be increased several-fold) and this in itself highlights the potential of the inositide system for modulation. Possible molecular mechanisms for the control of the intracellular concentrations of these messengers must now be considered.

PTD INS P_2 PHOSPHODIESTERASE

Ptd Ins P_2 is, in the first instance, probably the parent of all the messengers discussed above. Ptd Ins is in mass terms the ultimate source of them (Michell *et al.* 1981) and the recent suggestion that some Ptd Ins can be hydrolysed directly (Wilson *et al.* 1985) by the same enzyme but under different conditions (Low and Weglicki 1983; Irvine *et al.* 1984a; Wilson *et al.* 1984) is interesting in view of the fact that diacylglycerol will be generated from Ptd Ins but not Ins(1,4,5)P_3 (and thus no Ca^{2+}). Here lies another possible modification of the relative levels of second messengers. However, if Ptd Ins P_2 is the immediate source of diacylglycerol and Ins P_3, the control of its level in the cell and of Ptd Ins P_2 phosphodiesterase which hydrolyses it, is crucial.

Ptd Ins P_2 phosphodiesterase is controlled by as yet incompletely understood means. The original observation that it will not hydrolyse its substrate under physiological conditions (Downes and Michell 1982) leads to the hypothesis that it is controlled by the structure of its substrate (Irvine *et al.* 1984a, 1985c), because only with a non-bilayer substrate can it be induced to work. However, the various indirect pieces of evidence linking inositide catabolism with guanine-nucleotide-binding proteins (reviewed by Berridge and Irvine 1984) have received more direct confirmation (Haslam and Davidson 1984; Cockcroft and Gomperts 1985). In fact the two hypotheses of Ptd Ins P_2 phosphodiesterase control can be married by the suggestion that the enzyme is normally incapable of hydrolysing its substrate in a bilayer (Irvine *et al.* 1984a), but becomes capable of doing so when combined with the correct G-protein. In this context it is relevant to draw attention to the clear difference in calcium dependency (with Mg^{2+} present) between the human red blood cell membrane-bound enzyme and the soluble rat brain enzyme on the one hand, and the enzyme in the rabbit red blood cell and sea urchin eggs (both membrane-bound) on the other. The former pair are inactive until calcium is near millimolar (Downes and Michell 1982; Irvine *et al.* 1984b) whereas the latter begin to be active when calcium is 1–2 micromolar (Quist 1985; Whittaker and Irvine 1984). Could the difference between these be the presence or absence of different kinds of G-proteins?

If G-proteins are involved in the control of Ptd Ins P_2 phosphodiesterase (or any other enzyme of the inositide pathway) then a route of interaction with cyclic nucleotides is immediately obvious. Our present understanding of these proteins is that Ns and Ni at least have two subunits (the β and y) in common, and the α subunits which are unique to each protein can compete for these subunits (see Gilman 1984). It hardly needs stating that a potential competition for the β and y subunits by the α subunits of the putative inositide G-protein(s) exists, and that some interactions between adenyl-cyclase-stimulating agonists and inositide-stimulating agonists could well be explained in these terms.

A more direct interaction of Ptd Ins P_2 phosphodiesterase with cyclic nucleotides is suggested by the dramatic inhibition of inositide turnover caused by cyclic AMP in platelets (Nishizuka 1984). It cannot be ruled out that this effect is indirect in that cyclic AMP limits substrate availability by an effect on inositide kinases and phosphomonoesterases (Berridge and Irvine 1984), but the simplest explanation is an effect either on the phosphodiesterase or on the G-protein coupling mechanism. Finally, even if substrate structure is not the acute control on Ptd Ins P_2 phospho-diesterase, there is no doubt that the enzyme's activity is modified by the substrate, and chronic control of the rate of inositide turnover by changing the phospholipid composition of the membrane, for instance the sphingomyelin content (Dawson *et al.* 1985) is a possibility.

Another potential long-term influence on inositide metabolism may be polyamines (Smith and Wells 1984). These can alter calcium homeostasis (Koenig *et al.* 1983) and it has recently been suggested that diacylglycerol may mediate increases in the rate-limiting enzyme of their synthetic pathway, ornithine decarboxylase (Otani *et al.* 1984). An effect of cationic polymers to modify inositide-metabolizing enzymes has been known for some time (Irvine *et al.* 1979), and how all these observations could link together to form another regulatory system is an interesting area for exploration.

INOSITIDE KINASES

Altering the generation of second messengers by limiting the availability of Ptd Ins P_2 is clearly a potent way of modifying cell metabolism, and the enzymes making Ptd Ins P_2 must be under a number of control mechanisms. The most obvious control on a kinase is feed-back regulation by the enzyme's product, and this has been observed for Ptd Ins kinase (Michell *et al.* 1967) and Ptd Ins P kinase (Van Rooijen *et al.* 1985), though in both instances neither the presentation of the substrate nor the ionic environment were entirely physiological. We have however, recently confirmed an inhibition of Ptd Ins P kinase by Ptd Ins P_2 in a bilayer and in 'physiological' saline (P. D. Smith and R. F. Irvine, unpublished).

It is essential to the hypothesis which states that some Ptd Ins disappearance in stimulated tissues is due to phosphorylation to replace hydrolysed Ptd Ins P_2, (Michell *et al.* 1981; Berridge 1983), for the inositide kinases to be stimulated shortly after cell activation. Evidence for an activation by phorbol esters (presumably working through kinase C) has been obtained in several tissues (Aloyo *et al.* 1983; Taylor *et al.* 1984; De Chaffoy de Courcelles *et al.* 1984; Halenda and Feinstein 1984), but these are tissues where the calcium and diacylglycerol halves of the inositide pathway act together synergistically to produce cell activation; where they do not (see above) the picture may be different and herein lies another possible route of differentiation by which a neurone can prolong or forshorten an inositide response.

The effect of cyclic nucleotides on inositide kinases is confused at present. A cyclic AMP-stimulated Ptd Ins P kinase was originally reported by Torda (1972), and Enyedi *et al.* (1984) have also reported an increase in inositide labelling *in vitro* by addition of the cyclic AMP-dependant protein kinase. In neither example was the effect very large, and the physiological significance is not yet defined. Nevertheless, recent evidence (Sugden *et al.*

1985) has shown that cyclic AMP and inositides do not always work contrary to one another as they do in platelets, and the interactions between these pathways must vary between tissues, (and therefore could vary between neurones). The modulation of Ptd Ins kinase by ACTH (Jolles *et al.* 1980; Zwiers *et al.* 1982) is at present difficult to fit into our meagre understanding of the regulation of inositide kinases, but must represent another facet of nervous tissue's ability to modulate and modify the inositide pathways. The heterogeneous sub-cellular distribution of inositide kinases (see above) may eventually explain some of the present confusions and contradictions.

PTD INS *P* AND PTD INS *P*₂ PHOSPHOMONOESTERASES

The regulation of these enzymes, if they are regulated by anything other than substrate availability, is not known (Irvine 1982a; Irvine *et al.* 1985c), despite the possibility that many of the instances of apparent regulation of inositide kinases could in fact be caused by alterations in phosphomonoesterase activity. Ptd Ins *P* phosphomonoesterase and Ptd Ins *P*₂ phosphomonoesterase are both classically soluble and Mg^{2+}-dependent (Dawson and Thompson 1964; Irvine 1982a), but recent reports of distinct membrane-bound, Mg^{2+}-independent activities (e.g. Mack and Palmer 1984; Smith and Wells 1984) may open an entirely new view on the regulation of inositide levels by these enzymes.

PTD INS SYNTHETASE

There are a number of reports of stimulation of Ptd Ins synthesis *de novo* following cell stimulation (for example, Farese 1982; Farese *et al.* 1984), though it is not entirely clear at which point the synthesis is regulated, that is, whether Ptd Ins synthetase is stimulated or whether there is a greater availability of its indirect substrate phosphatidic acid by increased acylation of glycerophosphate. Some of the agonists that stimulate Ptd Ins synthesis are not classical inositide agonists in that they do not stimulate an inositide phosphodiesterase (Farese 1982; Farese *et al.* 1984), but stimulation of synthesis by carbachol in the exocrine pancreas has been reported (Chapman *et al.* 1983) and we do not know whether the mechanism of stimulation by the two groups of agonists is the same. Present evidence favours an increase in enzyme synthesis *de novo* being responsible (Farese 1982), and here is another potential longterm modulation of the inositide pathways.

In summarizing the above section and the previous two, the enzymes which control the total level of inositides in the cell, their levels relative one to another, and also their subcellular location, are still at present imperfectly understood; but given the important generation of second messengers from inosities, and their multifunctional nature, it is in these enzymes that a great potential exists for modifying the size and nature of the physiological response that an inositide-stimulating agonist would induce in a cell.

More direct modifications could be caused by regulating the catabolism of the second messengers formed from inosities, and this also offers the possibility of changing one message while not changing another.

DIACYLGLYCEROL

Initially, the production of diacylglycerol is controlled by Ptd Ins P_2 phosphodiesterase, and the amount of Ptd Ins P_2 (discussed above), though recent evidence from platelets suggests that a Ca^{2+}-stimulated phosphodiesteratic hydrolysis of Ptd Ins occurs (Wilson *et al*. 1985) which therefore opens new possibilities for control of this particular second messenger. Ptd Ins hydrolysis may represent a means of further raising diacylglycerol without increasing calcium [via Ins(1,4,5)P_3]; it is possible that the production of Ins(1,3,4)P_3, which circumstantial evidence suggests does not mobilize calcium, serves a similar purpose (Irvine *et al*. 1985b).

Diacylclycerol is removed by two routes. Probably the major route of inactivation is by phosphorylation to phosphatidate, and diacylglycerol kinase is still not well studied. It is (like many enzymes discussed here) part soluble and part membrane-bound (Lapetina and Hawthorne 1971; Kanoh and Akesson 1978); the distribution of such enzymes may of course *in vivo* be entirely different (Irvine 1982a). Diacylglycerol kinase has also been reported as being associated with microtubules (Daleo *et al*. 1976), an observation that has not to our knowledge been explored further, but an interesting one in the light of inositol's ability to overcome colchicine effects (Murray *et al*. 1951).

A more longterm regulation of diacylglycerol has been reported by Drummond and Raeburn (1984) who depleted GH$_3$ cells of inositol by chronic incubation with lithium with the result that phosphatidate (and hence diacylglycerol presumably formed by phosphatidate phosphohydrolase) accumulated because of the reduced inositide synthesis. This is more a pharmacological than a physiological phenomenon, but of more direct relevance to the long term modulation of diacylglycerol levels is the regulation of phosphatidate phosphohydrolase in liver by ethanol, cyclic nucleotides and corticosteroids (see Butterworth *et al*. 1984 for references); no such regulation has been sought in nervous tissue (teleological arguments lead Brindley and his co-workers to the logical conclusion that in liver phosphatidate phosphohydrolase is primarily regulating triacylglycerol synthesis), but if any such phenomena do occur they could modulate chronically levels of diacylglycerol in resting and stimulated neurones.

The deactivation of diacylglycerol by diacylglycerol lipase is as yet of uncertain significance (Irvine 1982b). In tissues rich in this enzyme (or enzymes, see Chau and Tai 1981) such as nervous tissue (Cabot and Gatt 1976) or platelets (Bell *et al*. 1979), a considerable proportion of diacylglycerol may be 'inactivated' in this way (for example Mauco *et al*. 1984), and the fact that this 'inactivation' can in turn generate other second messengers from the liberated arachidonate, as discussed above, is an added complication to diacylglycerol metabolism.

INOSITOL (1,4,5) TRISPHOSPHATE

The acute control of the synthesis of this compound lies in Ptd Ins P_2 phosphodiesterase, but after the initial pulse of production (probably 15–30 s) the rate of synthesis will be more dependent on the rate of re-synthesis of Ptd Ins P_2 as discussed above. In fact, if the inositide kinases are stimulated, then *in vivo* it may be that Ptd Ins P_2 levels hardly fall at all, and the decrease of Ptd Ins P_2 observed under experimental conditions may be artefacts of overstimulation. The recent work of Drummond and Raeburn (1984) emphasizes to what degree cells will maintain their Ptd Ins P_2 levels even under conditions of considerable Ptd Ins depletion.

Ins$(1,4,5)P_3$ is catabolized by a specific and very active enzyme that is probably bound to the plasma membrane (Downes *et al*. 1982; Storey *et al*. 1984; Seyfred *et al*. 1984), though other sub-cellular locations have not been excluded; in blowfly salivary glands dispersed by sonication the activity is primarily soluble (R. F. Irvine and M. J. Berridge, unpublished). The enzyme is very specific in that only the 5 phosphate is removed (Downes *et al*. 1982) and that Ins$(1,3,4)P_3$ is not hydrolysed (Irvine *et al*. 1984b), and its high activity is clearly illustrated by the very tight control kept on Ins$(1,4,5)P_3$ levels in stimulated cells (for example see Berridge 1983; Irvine *et al*. 1985b). It is not yet known if this enzyme is regulated or is purely substrate-limited; there is no doubt that a defect in its activity could have a profound effect on cell function (Berridge 1984b), and modulation of its activity would be a potent force on cellular calcium homeostasis.

INS $(1,3,4)P_3$

As little is known at present of the route of synthesis or degradation of this compound (or indeed of its function), it is not possible to speculate on the control of its levels. If the route of synthesis is by phosphodiesteratic hydrolysis of a Ptd Ins$(3,4)P_2$, then the time course of Ins$(1,3,4)P_3$'s production, and the low level of the parent lipid prior to stimulation (Irvine *et al*. 1985b) suggests that Ptd Ins (4)P-3-kinase (if such exists) is tightly controlled, and greatly activated on cell stimulation; furthermore it must be closely associated with the receptor site. If an $Ins(1,4,5)P_3$ isomerase exists, (and several attempts by us to detect it have failed so far) then it too must be tightly controlled before cell stimulation.

CYCLIC NUCLEOTIDES

The various demonstrable and possible interactions between the cyclic nucleotide and inositide signalling pathways have been mentioned a number of times in the preceding pages. Clearly, they vary between cells from a negative interaction (Nishizuka 1984) to little or no interaction (Williamson *et al*. 1981), to a positive interaction (Sugden *et al*. 1985), and there are many points, some discussed above, at which they can interact. If within the brain and central nervous system all these possibilities exist then the range of interactions between signal pathways is virtually limitless.

CONCLUSIONS

The preceding sentence can be applied to the overall picture of inositide signalling and its various regulatory features. The principal message of the Introduction was that despite the variety of neurotransmitters, especially in different combinations, the intracellular signalling systems are much fewer in number. Thus, the degree of subtlety that higher neurone function (including conscious thought) must employ over and above the sheer three-dimensional complexity of the 'wiring', is limited by the degree of subtlety to which the intracellular signalling systems can be modified. There are many potential levels of regulation discussed above, which could act over different time scales, and these may in turn reflect different aspects of neurone function. Some of the changes in inositide metabolism may be controlled by a stable differentiation of cell metabolism, whereas others are short-term regulatory effects; could these play roles in the formation of, respectively, long- and short-term memory? Leaving aside such specific speculations, there is no doubt that the inositide signalling system with its variety of different messengers and multiplicity of control points, is ideally suited to play an important part in higher neuronal function.

References

Aloyo, V. J., Zwiers, H., and Gispen, W. H. (1983). Phosphorylation of B-50 protein by calcium-activated, phospholipid-dependent protein kinase and B-50 protein kinase. *J. Neurochem*. **41**, 649–53.

Baker, R. R. and Chang, H.-Y. (1981). A comparison of lysophisphatidyl-choline acyl transferase activities in neuronal nuclei and microsomes isolated from immature rabbit cerebral cortex. *Biochim. Biophys. Acta*, **666**, 223–9.

Baraban, J. M., Gould, R. J., Peroutka, S. J., and Snyder, S. H. (1985a). Phorbol ester effects on neutrotransmission: Interaction with neurotransmitters and calcium in smooth muscle. *Proc. Nat. Acad. Sci. USA* **82**, 604–7.

——, Snyder, S. H., and Alger, B. E. (1985b). Protein kinase C regulates ionic conductance in hippocampal pyramidal neurons: Electrophysiological effects of phorbol esters. *Proc. Nat. Acad. Sci. USA*, **82**, 2538–42.

Bell, R. L., Kennedy, D. A., Stanford, N., and Majerus, P. W. (1979). Diglyceride lipase: a pathway for arachidonate release from human platelets. *Proc. Nat. Acad. Sci. USA*, **76**, 3238–41.

Berridge, M. J. (1981). Phosphatidylinositol hydrolysis: a multifunctional trans-ducing mechanism. *Mol. Cellular endocrinol.* **24**, 115–40.

—— (1983). Rapid accumulation of inositol trisphosphate reveals agonists hydrolyse polyphosphoinositides instead of phosphatidylinositol. *Biochem. J.* **212**, 849–58.

—— (1984a). Inositol trisphosphate and diacylglycerol as second messengers. *Biochem. J.* **220**, 345–60.

—— (1984b). Oncogenes, inositol lipids and cellular proliferation. *Biotechnology*, **2**, 541–6.

——, Downes, C. P., and Hanley, M. R. (1982). Lithium amplifies agonist-dependent phosphatidylinositol responses in brain and salivary glands. *Biochem. J.* **206**, 587–95.

——, and Irvine, R. F. (1984). Inositol trisphosphate, a novel second messenger in cellular signal transduction. *Nature (Lond.)* **312**, 315–21.

Brown, E., Kendall, D. A., and Nahorski, S. R. (1984). Inositol phospholipid hydrolysis in rat cerebral cortical slices: I. Receptor characterization. *J. Neurochem.* **42**, 1379–87.

Burgess, G., McKinney, J. S., Irvine, R. F., and Putney, J. W. (1985). Inositol 1,4,5-trisphosphate and inositol 1,3,4-trisphosphate formation in Ca^{2+}-mobilizing hormone-activated cells. *Biochem. J.* **232**, 237–43.

Burn, P., Putman, A., Meyer, R. K., and Burger, M. M. (1985). Diacylglycerol in large actin complexes and in the cytoskeleton of activated platelets. *Nature (Lond.)* **314**, 469–72.

Butterworth, S. C., Martin, A., and Brindley, D. N. (1984). Can phosphorylation of phosphatidate phosphohydrolase by a cyclic AMP-dependent mechanism regulate its activity and subcellular distribution and control hepatic glycerolipid synthesis? *Biochem. J.* **222**, 487–93.

Cabot, M. C. and Gatt, S. (1976). Hydrolysis of neutral glycerides by lipases of rat brain microsomes. *Biochem. Biophys. Acta*, **431**, 105–15.

Castagna, M., Takai, Y., Kaibuchi, K., Sano, K., Kikkawa, U., and Nishizuka, Y. (1982). Direct activation of calcium-activated phospholipid-dependent protein kinase by tumor-promoting phorbol esters. *J. Biol. Chem.* **257**, 7847–51.

Chapman, B. A., Wilson, J. S., Colley, P. W., Picola, R. C., and Somes, J. B. (1983). Increased phospholipid synthesis in the stimulated rat and human pancreas. *Biochem. Biophys. Res. Comm.* **115**, 771–6.

Chau, L. -Y. and Tai, H. -H. (1981). Release of arachidonate from diglyceride in human platelets requires the sequential action of a diglyceride lipase and a monoglyceride lipase. *Biochem. Biophys. Res. Comm.* **100**, 1688–95.

Choquette, D., Hakim, G., Filotep, A. G., Plishker, G. A., Bostwick, J. R., and Penniston, J. T. (1984). Regulation of plasma membrane Ca^{2+} ATPases by lipids of the phosphatidylinositol cycle. *Biochem. Biophys. Res. Comm.* **125**, 908–15.

Cockcroft, S. and Gomperts, B. D. (1985). Role of guanine nucleotide binding protein in the activation of polyphosphoinositide phosphodiesterase. *Nature (Lond.)* **314**, 534–6.

Collins, C. A. and Wells, W. W. (1983). Identification of phosphatidylinositol kinase in rat liver lysosomal membranes. *J. Biol. Chem.* **258**, 2130–4.

Conn, P. J. and Sanders-Bush, T. (1984). Selective 5HT-2 antagonists inhibit serotonin stimulated phosphatidylinositol metabolism in cerebral cortex. *Neuropharmacol.* **23**, 993–6.

Coolican, S. A. and Hathaway, D. R. (1984). Effect of L-phosphatidylinositol on vascular smooth muscle Ca^{2+}-dependent protease. *J. Biol. Chem.* **259**, 11627–30.

Daleo, G. R., Piras, M. M., and Piras, M. R. (1976). Diglyceride kinase activity of microtubules. Characterization and comparison with the protein kinase and ATPase activity associated with vinblastine-isolated tubulin of chick embryonic muscle. *Eur. J. Biochem.* **68**, 339–46.

Daum, P. R., Downes, C. P., and Young, J. M. (1984). Histamine stimulation of inositol 1-phosphate accumulation in lithium-treated slices from regions of guinea pig brain. *J. Neurochem.* **43**, 24–32.

Dawson, R. M. C., Hemington, N. L., and Irvine, R. F. (1985). The inhibition of diacylglycerol-stimulated intracellular phospholipases by phospholipids with a phosphocholine-containing polar group. A possible physiological role for sphingomyelin. *Biochem. J.* **230**, 61–8.

——, and Irvine, R. F. (1978). Possible role of lysosomal phospholipases in inducing tissue prostaglandin synthesis. *Adv. Prostaglandin Thromboxane Res.* **3**, 47–54.

——, ——, Bray, J., and Quinn, P. J. (1984). Long-chain unsaturated diacylglycerols cause a perturbation in the structure of phospholipid bilayers rendering them susceptible to phospholipase attack. *Biochem. Biophys. Res. Comm.* **125**, 836–42.

——, and Thompson, W. (1964). The triphosphoinositide phosphomonoesterase of brain tissue. *Biochem. J.* **91**, 244–50.

De Chaffoy de Courcelles, D., Roevens, P., and Van Belle, H. (1984). 12-0-Tetradecanoylphorbol 13-acetate stimulates inositol lipid phosphorylation in intact human platelets. *FEBS Lett.* **173**, 389–93.

De Riemer, S. A., Strung, J. A., Albert, K. A., Greengard, P., and Kaczmarek, L. K. (1985). Enhancement of calcium current in *Aplysia* neurones by phorbol ester and protein kinase C. *Nature (Lond.)*, **313**, 313–6.

Downes, C. P. and Michell, R. H. (1982). The control by Ca^{2+} of the polyphosphoinositide phosphodiesterase and the Ca^{2+}-pump ATPase in human erythrocytes. *Biochem. J.* **202**, 53–8.

——, Mussat, M. C., and Michell, R. H. (1982). The inositol trisphosphate phosphomonoesterase of the human erythrocyte membrane. *Biochem. J.* **203**, 169–77.

Drummond, A. H. (1985). Bidirectional control of cytosolic free calcium by thyrotropin-releasing hormone in pituitary cells. *Nature*, **315**, 752–5.

——, and Raeburn, C. A. (1984). The interaction of lithium with thyrotropin-releasing hormone-stimulated lipid metabolism in pituitary tumour cells. *Biochem. J.* **224**, 129–36.

Enyedi, A., Fargo, A., Sarkdi, B., and Gardos, G. (1984). Cyclic AMP-dependent protein kinae and Ca^{2+}-calmodulin stimulate the formation of poly-phosphoinositides in a sarcoplasmic reticulum preparation of rabbit heart. *FEBS Lett.* **176**, 235–8.

Farese, R. V. (1982). The role of inositide phospholipids in the action of steroidogenic hormones. *Cell Calcium* **3**, 441–50.

——, Barnes, D. E., Davis, J. S., Standaert, M. L., and Pollet, R. J. (1984). Effects of insulin and protein synthesis inhibitors on phospholipid metabolism, diacylglycerol levels, and pyruvate dehydrogenase activity in BC3H-1 cultured monocytes. *J. Biol. Chem.* **259**, 7094–100.

Gilman, A. G. (1984). G proteins and dual control of adenylate cyclase. *Cell*, **36**, 577–9.

Goedert, M., Pinnock, R. D., Downes, C. P., Mantyh, P. W., and Emson, P. C. (1984). Neurotensin stimulates inositol phospholipid hydrolysis in rat brain slices. *Brain Res.* **323**, 139–97.

Gonzales, R. A. and Crews, F. T. (1984). Characterization of the cholinergic stimulation of phosphoinositide hydrolysis in rat brain slices. *J. Neurosci.* **4**, 3120–7.

Halenda, S. P. and Feinstein, M. B. (1984). Phorbol myristate acetate stimulates formation of phosphatidyl inositol 4-phosphate and phosphatidyl inositol 4,5-bisphosphate in human platelets. *Biochem. Biophys. Res. Comm.* **124**, 507–13.

Hanley, M. R., Benton, H. P., Lightman, S. L., Todd, K., Bone, E. A., Fretten, P., Palmer, S., Kirk, C. J., and Michell, R. H. (1984). A vasopressin-like peptide in the mammalian sympathetic nervous system. *Nature (Lond.)*, **309**, 258–61.

Haslam, R. J. and Davidson, M. (1984). Guanaine nucleotides decrease the free $[Ca^{2+}]$ required for secretion of serotonin from permeabilized blood platelets. Evidence of a role for a GTP-binding protein in platelet activation. *FEBS Lett.* **174**, 90–4.

——, ——, Davies, T., Lynham, J. A., and McClenaghan, M. D. (1978). Regulation of blood platelet function by cyclic nucleotides. *Adv. Cyclic Nucleotide Res.* **9**, 533–53.

Hawthorne, J. N. (1983). Polyphosphoinositide metabolism in excitable membranes. *Biosci. Rep.* **3**, 887–904.

Hirota, K., Hirota, T., Aguilera, G., and Catt, K. J. (1985). Hormone-induced redistribution of calcium-activated phospholipid-dependent protein kinase in pituitary gonadotrophs. *J. Biol. Chem.* **260**, 3243–6.

Hökfelt, T., Everitt, B., Holets, V. R., Meister, B., Melander, T., Schalling, M., Staines, W., and Lundberg, J. M. (1986). Coexistence of peptides and other active molecules in neurons: diversity of chemical signalling potential. In *Fast and Slow Chemical Signalling in the Nervous System* (eds L. L. Iversen and E. Goodman), pp. 205–31. Oxford University Press, Oxford.

Inoue, T., Ito, Y., and Takeda, T. (1984). Prostaglandin-induced inhibition of acetylcholine release from neuronal elements of dog tracheal tissue. *J. Physiol.* **349**, 553–70.

Irvine, R. F. (1982a). The enzymology of stimulated inositol lipid turnover. *Cell Calcium*, **3**, 295–309.

—— (1982b). How is the level of free arachidonic acid controlled in mammalian cells? *Biochem. J.* **204**, 3–16.

——, Anderson, R. E., Rubin, L. J., and Brown, J. E. (1985a). Inositol 1,3,4-trisphosphate concentration is changed by illumination of *Limulus* ventral photoreceptors. *Biophys. J.* **47**, 38a.

——, Änggård, E. A., Letcher, A. J., and Downes, C. P. (1985b). Metabolism of inositol 1,4,5, trisphosphate and inositol 1,3,4-trisphosphate in rat parotid glands. *Biochem. J.* **229**, 505–11.

——, and Dawson, R. M. C. (1980). The mechanism and function of phosphatidylinositol turnover. *Biochem. Soc. Trans.* **8**, 376–7.

——, —— and Freinkel, N. (1982). Stimulated phosphatidylinositol turnover. A brief appraisal. In *Contemporary Metabolism* 2, (ed. N. Freinkel), pp. 301–42. Plenum, New York and London.

——, Hemington, N., and Dawson, R. M. C. (1979). The calcium-dependent phosphatidylinositol-phosphodiesterase of rat brain. *Eur. J. Biochem.* **99**, 525–30.

——, Letcher, A. J., and Dawson, R. M. C. (1984a). Phosphatidyl-inositol-4,5-bisphosphate phosphodiesterase and phosphomonoesterase activities of rat brain. *Biochem. J.* **218**, 177–85.

——, ——, Lander, D. J., and Dawson, R. M. C. (1985c). The enzymology of phosphoinositide catabolism, with particular reference to phosphatidylinositol, 4,5-bisphosphate phosphodiesterase. In *Inositol and Inositides* (eds J. E. Bleasdale, J. Eichberg, and G. Hauser), pp. 123–35. Humana Press, Clifton, U.S.A.

——, ——, —— and Downes, C. P. (1984b). Inositol trisphosphates in carbachol-stimulated rat parotid glands. *Biochem. J.* **223**, 237–43.

Iversen, L. L. (1984). Amino acids and peptides: fast and slow chemical signals in the nervous system? *Proc. Roy. Soc. Lond. Ser. B* **221**, 245–60.

Janowsky, A., Labarca, R., and Paul, S. M. (1984). Characterization of neurotransmitter receptor-mediated phosphatidylinositol hydrolysis in the rat hippocampus. *Life Sci.* **35**, 1953–61.

Jergil, B. and Sundler, R. (1983). Phosphorylation of phosphatidylinositol in rat liver golgi. *J. Biol. Chem.* **258**, 7968–73.

Jolles, J., Zwiers, H., Van Dongen, C. J., Schotman, P., Wirtz, K. W. A., and Gipsen, W. H. (1980). Modulation of brain polyphosphoinositide metabolism by ACTH-sensitive protein phosphorylation. *Nature (Lond.)* **286**, 623–5.

Kanoh, H. and Akesson, B. (1978). Properties of microsomal and soluble diacylglycerol kinase in rat liver. *Eur. J. Biochem.* **85**, 225–32.

Koenig, H., Goldstone, A., and Lu, A. (1983). Polyamines regulate calcium fluxes in a rapid plasma membrane response. *Nature (Lond.)* **305**, 530–4.

Kraft, A. S. and Anderson, W. B. (1983). Phorbol esters increase the amount of Ca^{2+}, phospholipid dependent protein kinase associated with plasma membrane. *Nature (Lond.)* **301**, 621–3.

Labarca, R., Janowsky, A., Patel, J., and Paul, S. M. (1984). Phorbol esters inhibit agonist-induced (^{3}H) inositol-1-phosphate accumulation in rat hippocampal slices. *Biochem. Biophys. Res. Comm.* **123**, 703–9.

Lapetina, E. G. and Hawthorne, J. N. (1971). The diglyceride kinase of rat cerebral cortex. *Biochem. J.* **122**, 171–9.

Lassing, I. and Lindberg, U. (1985). Specific interaction between phosphatidylinositol 4,5-bisphosphate and profilactin. *Nature (Lond.)* **314**, 472–4.

Low, M. G. and Weglicki, W. B. (1983). Resolution of myocardial phospholipase C into several forms with distinct properties. *Biochem. J.* **215**, 325–34.

MacPhee, C. H., Drummond, A. H., Otto, A. M., and Jimenez de Asua, L. (1984). Prostaglandin $F_{2\alpha}$ stimulates phosphatidylinositol turnover and increases the cellular content of 1,2-diacylglycerol in confluent resting Swiss 3T3 cells. *J. Cellular Physiol.* **119**, 35–40.

Mack, S. E. and Palmer, F. B. St.C. (1984). Evidence for a specific phosphatidylinositol 4-phosphate phosphatase in human erythrocyte membranes. *J. Lipid Res.* **25**, 75–85.

Mauco, G., Chap, H., Simon, M. F., and Douste-Blazy, L. (1978). Phosphatidic acid and lysophosphatidic acid production in phospholipase C- and thrombin-treated platelets. Possible involvement of a platelet lipase. *Biochimie*, **60**, 653–61.

——, Dangelmaier, C. A. and Smith, J. B. (1984). Inositol lipids, phosphatidate and diacylglycerol share stearoylarachidonoyl glycerol as a common backbone in thrombin-stimulated human platelets. *Biochem. J.* **224**, 933–40.

Mellors, A., Stalmach, M. E., and Cohen, A. (1985). Co-mitogenic tumour promoters suppress the phosphatidylinositol response in lymphocytes during early mitogenesis. *Biochim. Biophys Acta*, **833**, 181–8.

Michell, R. H., Harwood, J. L., Coleman, R., and Hawthorne, J. N. (1967). Characteristics of rat liver phosphatidylinositol kinase. *Biochim. Biophys. Acta*, **144**, 648–9.

——, Kirk, C. J., Jones, L. M., Downes, C. P., and Creba, J. A. (1981). The stimulation of inositol lipid metabolism that accompanies calcium mobilization in stimulated cells: defined characteristics and unanswered questions. *Philosoph. Trans. Roy. Soc. B* **296**, 123–37.

Murray, M. R., de Lam, H. E., and Chargaff, E. (1951). Specific inhibition by meso-inositol of the colchicine effect on rat fibroblasts. *Exp. Cell Res.* **2**, 165–77.

Nerring, J. R. and McBurney, R. N. (1984). Role for microsomal Ca storage in mammalian neurones? *Nature (Lond.)* **309**, 158–60.

Nishizuka, Y. (1984). The role of protein kinase C in cell surface signal transduction and tumour promotion. *Nature (Lond.)* **308**, 693–8.

Otani, S. Matsui, I., Kuramoto, A., and Morisawa, K. (1984). Induction of ornithine decarboxylase in guinea-pig lymphocytes and its relation to phospholipid metabolism. *Biochim. Biophys. Acta*, **800**, 96–101.

Otto, A. M., Nilsen-Hamilton, M., Boss, B. D., Ulrich, M. -O., and Jimenez de Asua, L. (1982). Prostaglandins E_1 and E_2 interact with prostaglandin $F_{2\alpha}$ to regulate initiation of DNA replication and cell division in Swiss 3T3 cells. *Proc. Nat. Acad. Sci. USA*, **79**, 4992–6.

Pozzan, T., Gatti, G., Dozio, N., Vicentini, L. M., and Meldolesi, M. (1984).

Ca^{2+}-dependent and -independent release of neurotransmitters from PC12 cells: a role for protein kinase C activation? *J. Cell Biol.* **99**, 628–38.

Putney, J. W. (1982). Inositol lipids and cell stimulation in mammalian salivary gland. *Cell Calcium*, **3**, 369–83.

Quist, E. E. (1985). Ca^{2+} stimulated phospholipid phosphoesterase activities in rabbit erythrocyte membranes. *Arch. Biochem. Biophys.* **236**, 140–9.

Schmitt, F. O. (1984). Molecular regulators of brain function: a new view. *Neurosci.* **13**, 991–1001.

Schoepp, D. D., Knepper, S. M., and Rutledge, C. O. (1984). Norepinephrine stimulation of phosphoinositide hydrolysis in rat cerebral cortex is associated with the alpha$_1$-adrenoceptor. *J. Neurochem.* **43**, 1758–61.

Seyfred, M. A., Farrell, L. E., and Wells, W. W. (1984). Characterisation of D-myo-inositol 1,4,5-trisphosphate phosphatase in rat liver plasma membranes. *J. Biol. Chem.* **259**, 13204–8.

Smith, C. D. and Wells, W. W. (1983). Phosphorylation of rat liver nuclear envelopes. II. Characterization of *in vitro* lipid phosphorylation. *J. Biol. Chem.* **258**, 9368–73.

——, and —— (1984). Characterization of a phosphatidylinositol 4-phosphate-specific phosphomonoesterase in rat liver nuclear envelopes. *Arch. Biochem. Biophys.* **235**, 529–37.

Storey, D. J., Shears, S. B., Kirk, C. J., and Michell, R. H. (1984). Stepwise enzymatic dephosphorylation of inositol 1,4,5-trisphosphate to inositol in liver. *Nature (Lond.)* **312**, 374–6.

Streb, H., Irvine, R. F., Berridge, M. J., and Schulz, I. (1983). Release of Ca^{2+} from a nonmitochondrial intracellular store in pancreatic acinar cells by inositol-1,4,5-trisphosphate. *Nature (Lond.)* **306**, 67–9.

Sugden, D., Vanecek, J., Klein, D. C., Thomas, T. P., and Anderson, B. (1985). Activation of protein kinase C potentiates isoprenaline-induced cyclic AMP accumulation in rat pinealocytes. *Nature (Lond.)* **314**, 359–61.

Takai, Y., Kishimoto, A., Kikkawa, U., Mori, T., and Nishizuka, Y. (1979). Unsaturated diacylglycerol as a possible messenger for the activation of calcium-activated, phospholipid-dependent protein kinase system. *Biochem. Biophys. Res. Comm.* **91**, 1218–24.

Tanaka, C., Taniyama, K., and Kusunoki, M. (1984). A phorbol ester and A23187 act synergistically to release acetylcholine from the guinea-pig ileum. *FEBS Lett.* **175**, 165–9.

Taylor, M. V., Metcalfe, J. C., Hesketh, T. R., Smith, G. A., and Moore, J. P. (1984). Mitogens increase phosphorylation of phosphoinositides in thymocytes. *Nature (Lond.)* **312**, 462–5.

Torda, C. (1972). Cyclic AMP-dependent diphosphoinositide kinase. *Biochim. Biophys. Acta*, **286**, 389–95.

Trevisani, A., Biondi, C., Belluzzi, O., Borasio, P. E., Capuzzo, A., Ferretti, M. E., and Perri, V. (1982). Evidence for increased release of prostaglandins of E-type in response to orthodromic stimulation in the guinea-pig superior cervical ganglion. *Brain Res.* **236**, 375–81.

Van Rooijen, L. A. A., Rossowska, M., and Bazan, N. G. (1985). Inhibition of phosphatidylinositol-4-phosphate kinase by its product phosphatidylinositol 4-5 bisphosphate. *Biochem. Biophys. Res. Comm.* **126**, 150–5.

Vicentini, L. M., Di Virgilio, F., Ambrosini, A., Pozzan, T., and Meldolesi, J. (1985). Tumour promoting phorbol 12-myristrate, 13-acetate inhibits phosphoinositide hydrolysis and cytosolic Ca^{2+} rise induced by the activation of muscarinic receptors in PC12 cells. *Biochem. Biophys. Res. Comm.* **127**, 310–7.

Viveros, O. H., Diliberto, E. J., and Daniels, A. J. (1983). Biochemical and functional evidence for the cosecretion of multiple messengers from single and multiple compartments. *Fed. Proc. Am. Soc. Exp. Biol.* **42**, 2923–8.

Watson, S. P. and Downes, C. P. (1983). Substance P induced hydrolysis of inositol phospholipids in guinea-pig ileum and rat hypothalamus. *Eur. J. Pharmacol.* **93**, 245–53.

Whittaker, M. and Irvine, R. F. (1984). Inositol, 1,4,5-trisphosphate micro-injection activates sea urchin eggs. *Nature* **312**, 636–9.

Williamson, J. R., Cooper, R. H., and Hoek, J. B. (1981). Role of calcium in the hormonal regulation of liver metabolism. *Biochim. Biophys. Acta*, **639**, 243–95.

Wilson, D. B., Bross, T. E., Hofmann, S. L. and Majerus, P. W. (1984). Hydrolysis of polyphosphoinositides by purified sheep seminal vesicle phospholipase C enzymes. *J. Biol. Chem.* **259**, 11718–24.

——, Neufeld, E. J., and Majerus, P. W. (1985). Phosphoinositide interconversion in thrombin-stimulated human platelets. *J. Biol. Chem.* **260**, 1046–51.

Wolfe, L. S. (1982). Eicosanoids: prostaglandins, thromboxanes, leukotrienes and other derivatives of carbon-20 unsaturated fatty acids. *J. Neurochem.* **38**, 1–14.

Yano, K., Higashida, H., Inoue, R., and Nozawa, Y. (1984). Bradykinin-induced rapid breakdown of phosphatidylinositol 4,5-bisphosphate in neuro-blastoma × glioma hybrid NG108–15 cells. *J. Biol. Chem.* **259**, 10201–7.

Zwiers, H., Jolles, J., Aloyo, V. J., Ostreicher, A. B., and Gispen, W. H. (1982). ACTH and synaptic membrane phosphorylation in rat brain. *Prog. Brain Res.* **56**, 405–17.

14

Coexistence of peptides and other active molecules in neurons: diversity of chemical signalling potential

TOMAS HÖKFELT, BARRY EVERITT, VICKY R. HOLETS, BJÖRN MEISTER, TOR MELANDER, MARTIN SCHALLING, WILLIAM STAINES, and JAN M. LUNDBERG

INTRODUCTION

The occurrence of chemical mediation of transmission of nerve impulses between neurons and between neurons and effector cells was recognized in the beginning of this century (see Stjärne *et al.* 1981). This led to an interest in identifying potential chemicals which could be involved in this process. For many decades the number of candidates was small and included compounds such as catecholamines and acetylcholine. In fact, with regard to the central nervous system only acetylcholine was a generally accepted transmitter for a long period (see Eccles 1964). The pioneering work of Vogt (1954) showed that noradrenaline has an uneven distribution in the brain, suggesting the possibility that this monoamine also could have a transmitter role. This view was greatly supported by the histofluorescence studies of Hillarp, Falck, Carlsson, Dahlström, Fuxe and their collaborators, who for the first time could demonstrate the cellular localization of a transmitter candidate in the microscope (Carlsson *et al.* 1962; Dahlström and Fuxe 1964, 1965; Fuxe 1965a,b). It fairly soon became clear that the noradrenaline precursor dopamine may also well have a role of its own in the central nervous system, possibly as a transmitter in basal ganglia (see Carlsson *et al.* 1958). In the 1960s a new group of compounds, the amino acids, came into focus, and evidence was presented that γ-aminobutyric acid (GABA) and glycine may represent important inhibitory transmitters in the brain and that glutamate may act as an excitatory transmitter (see Fonnum 1978). The 1970s have seen the explosive development in the area of peptides which today by far outnumber the classical transmitters in the brain and periphery (see Snyder 1980). The rapid progress in biochemistry, turning out new peptides and new members of peptide families continuously has left us in a situation, where our understanding of the possible physiology of these peptides is fragmentary. A transmitter role has, however, been suggested for some of these peptides, for example for substance P (Otsuka and Takahashi 1977; Pernow 1983). With this background it is no surprise that our view on chemical transmission has become more complex and differentiated, and it is apparent that the various types of compounds differ in terms of several of their characteristics, such as time course and mechanisms of action at the receptor site. For example, amino acids seem to be involved in fast transmission, whereas neuropeptides and monoamines often exert effects with slow onset and a long duration (see Bloom 1979; Iversen 1984; Schmitt 1984). Another view that has emerged is that neurons may contain not only one of these compounds but may produce, store and perhaps release several messengers (see Hökfelt *et al.* 1980a,b, 1982, 1984a,b,c; Cuello 1982; Chan-Palay and Palay 1984). This work is up till now very much based on histochemical studies, mainly immunohistochemistry (see Coons 1958), which permits visualization of several antigens in one cell. We would therefore first like to briefly discuss some aspects of this method, which has been the basis for our work.

ASPECTS OF METHODOLOGY

Demonstration of multiple antigens in a neuron can be achieved using various approaches. First, the 'adjacent section method' can be employed. Consecutive sections, as thin as possible, are incubated with different primary antisera. This procedure is critically dependent on the relation between the size of the immunoreactive cell bodies and the thickness of the sections. Provided that the sections are thin enough, the same cell body can often be identified in two or even more consecutive sections. When using cryostat sections, which normally are cut at about 10–15 μm, it may sometimes be difficult to carry out proper identification. Going down to 5 μm thickness improves the situation, but it may then be difficult to get a satisfactory immunostaining with some antisera. Ideally, plastic or epoxy resin embedded material should be used, allowing sections to be cut at 1 μm or even less. Immunostaining of such sections is often still problematic, however. Secondly, elution and restaining methods can be used. As first shown by Nakane (1968) it is possible to elute one staining pattern and reincubate the same section with a new antiserum and analyse for identity of labelled cells (Fig. 14.1a,b). Using the peroxidase method, dyes with various colours can be used, and in this way double labelling can be seen directly in the same section. If fluorescence methods are used, the

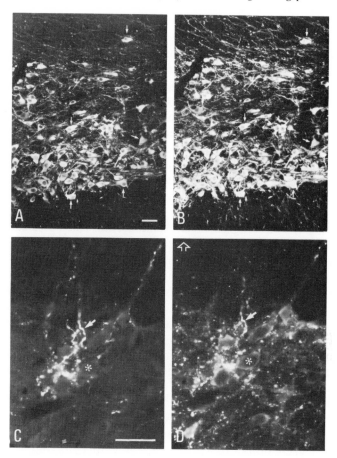

FIG. 14.1 (A–D). Immunofluorescence micrographs demonstrating two different possibilities to visualize two antigens in one cell body. (A,B) Section from the rostral pars compacta of the substantia nigra, first incubated with antiserum to cholecystokinin (CCK) and FITC conjugated antibodies (A). After elution with acid potassium permanganate the same section has been reincubated with tyrosine hydroxylase antiserum and FITC conjugated antibodies (B). Note that virtually all neurons (small arrows denote examples) contain both CCK- and tyrosine hydroxylase-like immunoreactivities. (C,D) The sympathetic lateral column is shown in a section incubated simultaneously with antiserum to 5-HT raised in guinea-pig (C) and to substance P raised in rat (monoclonal antibody) (D) followed by incubation with a mixture of secondary antibodies consisting of rhodamine (red) labelled goat-anti-guinea-pig antibodies and FITC (green) labelled goat anti-rat antibodies. By switching between proper filter combinations, 5-HT (C) and substance P (D)-positive nerve endings are demonstrated separately. Note that some nerve endings (thick arrows) contain both compounds, but that there are many substances P-positive/5-HT-negative nerve endings (c.f. C and D). Asterisks point to the same cell body, and open arrow points laterally. Bars indicate 50 μm.

first staining pattern has to be photographed before elution (Fig. 14.1a,b). A widely used procedure is the one of Tramu *et al.* (1978), which is based on acid potassium permanganate as eluent. The disadvantages with this approach are that the elution procedure may partially or completely destroy antigens. Thus, in our experience it is impossible to use certain antisera for restaining. Accordingly, negative results obtained after restaining, i.e. indicating lack of coexistence, should be interpreted with caution, since it may merely mean that the second antigen has been destroyed. Clearly, this type of study is preferably carried out with 'powerful' antisera, which give a strong immunostaining, even when the antigen has been partially destroyed. It should also be emphasized that proper control incubations have to be carried out to ensure complete elution of the first antibody. Finally, if antisera raised in different species are available, two antigens in the same section can be visualized using secondary antibodies labelled with different dyes, for example, red rhodamine and green fluorescein-isothiocyanate (FITC) (Fig. 14.1c,d). This approach has not been extensively used up till now, mainly because antisera are usually raised in rabbits and proper combination possibilities have been lacking. With the introduction of more and more monoclonal antibodies as well as antisera raised in goats, guinea-pigs, and other species, this convenient approach will be applied more often. Thus, switching between filter combinations for red and green fluorescence, one can directly decide whether coexistence is present or not. Furthermore, under favourable conditions, i.e. if the fibre networks are not too dense, one can also demonstrate coexistence in single fibres. It should, however, be pointed out that careful control experiments have to be carried out to avoid false positives, for example, due to the fact that the second antisera react with each other. It will be necessary to carry out such control experiments for each combination of primary and secondary antisera.

The immunohistochemical technique represents an extremely powerful method with apparent advantages and a virtually unlimited potential. It has, however, also certain drawbacks, especially concerning specificity and sensitivity. With regard to specificity, it has become obvious that the peptide field represents a particularly difficult situation, where families of peptide exist which may give rise to considerable identity problems when using immunological techniques for identification. The antisera raised against a certain peptide may crossreact with other members of the family and, of course, also with other peptides containing related amino acid sequences. In most immunohistochemical and radioimmunological studies one therefore prefers to use expressions such as 'substance P-like immunoreactivity', 'substance P-immunoreactive', or 'substance P-positive'. The sensitivity problem is of equal importance from many points of view. It is apparent that antisera are 'differentially sensitive' and that factors such as fixative and immunohistochemical method may markedly influence the sensitivity. It seems therefore highly unlikely that the methods used today represent the final answer and that with further improvements, we will be able to detect further cell bodies and fibres containing a certain antigen. This aspect should be born in mind especially when discussing problems related to coexistence, for example, whether or not a certain peptide occurs only in a subpopulation of neurons. Thus, negative immunohistochemical results should be interpreted with caution.

DIFFERENT TYPES OF COEXISTENCE: OVERVIEW

An increasing number of neuron systems has been described characterized by coexistence of potential synaptic messengers. It is important to note that different types of combinations of compounds can be recognized (Table 14.1). Much of the work has focused on the occurrence of a classical transmitter, such as noradrenaline, together with one or more peptides in the same neuron. However, there are also many examples of neurons containing more than one peptide, in which so far no classical transmitter has been demonstrated. Such neurons are, for example, found in the hypothalamus representing neurosecretory cells which release their contents into the blood stream, either into the hypophysial portal capillary system in the median eminence or into the systemic circulation in the posterior lobe of the pituitary gland. When discussing peptide-peptide coexistence situations, it should be remembered that for a long time there has been no good histochemical markers for several of the classical transmitters such as acetylcholine, glutamate, glycine, and histamine. Only recently have techniques been developed to visualize, directly or indirectly, some of them and only further studies can demonstrate whether there exist neurons releasing only peptides and no classical transmitters or whether all neurons have a classical transmitter.

Increasing evidence suggests that ATP may act as transmitter in several peripheral systems (see Burnstock 1985). For example, it is known that electrical stimulation of sympathetic nerves to the vas deferens causes a biphasic response, whereby the initial twitch contraction may be related to ATP release and the second, slower contraction to noradrenaline (see

TABLE 14.1. *Different types of coexistence situations*

Classical transmitter + peptide(s)
Classical transmitter + classical transmitter
Classical transmitter + peptide + ATP
Peptide + peptide
Peptide + leukotriene(?)

Burnstock 1985). These nerves also contain the peptide neuropeptide Y (NPY), which in these tissues has been shown to inhibit release of noradrenaline (Allen *et al*. 1982; Lundberg *et al*. 1982b; Ohhashi and Jacobowitz 1983; Stjärne and Lundberg 1984). Thus, these neurons may contain three different types of messengers, noradrenaline, ATP, and a peptide.

A further example of coexistence is neurons that may contain more than one classical transmitter. Thus, evidence has been presented that some of the pontine 5-hydroxytryptamine (5-HT) neurons may also contain GABA (Belin *et al*. 1983). Some neurons in the arcuate nucleus may contain both GABA and dopamine (Everitt *et al*. 1984b). Magnocellular neurons in the posterior hypothalamus contain both GABA and histamine (Takeda *et al*. 1984).

Finally, in the search for novel types of messenger molecules, we have recently focused on a new group of compounds belonging to the arachidonic acid metabolites, the leukotrienes (see Samuelsson 1983). Using different types of biochemical analytical techniques and bioassay, evidence has been presented that under certain conditions leukotriene C_4 (LTC_4), LTD_4 and LTE_4 can be formed in the brain (Dembinska-Kiec *et al*. 1984; Lindgren *et al*. 1984; Moskowitz *et al*. 1984). The exact site of formation has not been fully characterized, but there is evidence that formation does not occur in blood vessels (Moskowitz *et al*. 1984) and nervous tissue may therefore be responsible. Additional immuno-histochemical evidence suggests that one system that may produce an LTC_4-like compound is the neurosecretory luteinizing hormone releasing hormone (LHRH)-neurons involved in control of LH from the anterior pituitary gland (Hulting *et al*. 1984, 1985).

DIFFERENTIAL COEXISTENCE SITUATIONS IN THE CNS

Numerous coexistence situations have been encountered in the brain and spinal cord. The first cases mostly included catecholamines and 5-HT as classical transmitter, but more recently there are also multiple examples involving GABA (Table 14.2). A general phenomenon observed in relation to coexistence between classical transmitter and peptide is that only a subpopulation of certain systems exhibit a certain combination of messengers. The first example in the brain where this was observed were 5-HT neurons in the lower medulla oblongata (Dahlström and Fuxe 1964) containing substance P-like immunoreactivity (LI) (Chan-Palay *et al*. 1978; Hökfelt *et al*. 1978), and also a thyrotropin releasing hormone (TRH)-like peptide (Hökfelt *et al*. 1980b; Johansson *et al*. 1981; Gilbert *et al*. 1982). In contrast, there is so far no evidence that the rostral 5-HT

TABLE 14.2. *Coexistence of classical transmitters and peptides in the central nervous system (selected cases)*

Classical transmitter	Peptide	Brain region (species)	References
Dopamine	Neurotensin	Ventral mesencephalon area (rat)	Hökfelt *et al.* 1984a
	CCK	Ventral mesencephalon area (rat, man)	Hökfelt *et al.* 1980c,d
Norepinephrine	Enkephalin	Locus coeruleus (cat)	Charnay *et al.* 1982
	NPY	Medulla oblongata (man, rat)	Hökfelt *et al.* 1983; Everitt *et al.* 1984a
		Locus coeruleus (rat)	Everitt *et al.* 1984a
Epinephrine	Neurotensin	Medulla oblongata (rat)	Hökfelt *et al.* 1984a
	NPY	Medulla oblongata (rat)	Everitt *et al.* 1984a
5-HT	Substance P	Medulla oblongata (rat, cat)	Chan-Palay *et al.* 1978; Hökfelt *et al.* 1978; Johansson *et al.* 1981; Lovick and Hunt 1983
	TRH	Medulla oblongata (rat)	Johansson *et al.* 1981
	Enkephalin	Medulla oblongata, pons (cat)	Glazer *et al.* 1981; Hunt and Lovick 1982
ACh	VIP	Cortex	Eckenstein and Baughman 1984
	Substance P	Pons (rat)	Vincent *et al.* 1983
	Galanin	Basal forebrain (rat)	Melander *et al.* 1985
GABA	Somatostatin	Thalamus (cat)	Oertel *et al.* 1983
	Somatostatin	Hippocampus (rat)	Jirikowski *et al.* 1984
	Somatostatin	Cortex (rat, cat, monkey)	Schmechel *et al.* 1984
	CCK	Cortex (cat, monkey)	Hendry *et al.* 1984; Somogyi *et al.* 1984
	NPY		
	Enkephalin	Ventral pallidum (rat)	Zahm *et al.* 1985
	Enkephalin	Retina (chicken)	Watt *et al.* 1984
	Motilin	Cerebellum (rat)	Chan-Palay *et al.* 1981
	Opioid peptide	Basal ganglia (rat)	Oertel and Mugnaini 1984

cell groups in the pontine and mesencephalic raphe areas contain SP- or TRH-like immunoreactivites (Hökfelt *et al.* 1978). Thus, only a subpopulation of the central 5-HT neurons seem to contain these two particular peptides. Furthermore, as shown in Table 14.3, the proportions of 5-HT neurons containing SP- and TRH-like immunoreactivities in the medullary raphe nuclei seem to vary in relation to the subregion in this area. For example, the neurons in the medullary arcuate nucleus have a high proportion of 5-HT-positive, peptide-negative neurons, whereas the nucleus raphe pallidus seems to have a high proportion of neurons containing all three compounds (Johansson *et al.* 1981).

The differential colocalization of peptides and transmitters can also be exemplified on the basis of catecholamine neurons. In the central nervous

TABLE 14.3. *Quantitative evaluation of 5-hydroxytryptamine (5-HT), thyrotropin releasing hormone (TRH) and substance P (SP) immunoreactive cell bodies in the rostral (a) and caudal (b) medulla oblongata of|a colchicine treated rat.|Comparison of the relative distribution within the nuclei.*

	Region	5-HT (%)	TRH (%)	SP (%)	No. of cells counted (= 100 %)
(a)	NRM	53.7	24.4	21.9	1256
	ARC rm	76.6	21.9	1.6	192
	PPP	56.4	17.3	26.3	1099
	ARC rl	78.8	18.7	2.4	860
	SPP	62.2	18.2	19.6	1285
	Others	63.3	18.4	18.4	109
	Total (a)	62.2	19.9	17.9	4801
(b)	NRP	42.6	32.1	25.4	1525
	ARC cm	58.1	31.0	11.4	465
	NRO	52.2	22.9	24.9	3016
	PO	49.6	14.8	35.7	704
	ARC cl	60.4	27.5	12.2	1037
	Others	64.4	14.4	21.3	160
	Total (b)	51.7	25.1	23.2	6907
	Total (a + b)	56.0	23.0	21.0	11,708

This table is based on five series of sections spaced at different intervals. The total number of cells counted refers to the total number of immunoreactive cells. The borderline between rostral and caudal medulla oblongata has been put approximately between the caudal ending of nucleus raphe magnus and the rostral beginning of nuclei raphe obscurus and pallidus. ARC = arcuate region (rm = rostral medial; rl = rostral lateral; cm = caudal medial; *cl* = caudal lateral). NRM = nucleus raphe magnus. NRO = nucleus raphe obscurus. NRP = nucleus raphe pallidus. PO = 'paraolivar' region. PPP = 'parapyramidal' region. SPP = 'suprapyramidal' region. From Johansson *et al.* (1981).

system a number of catecholamine neurons have been shown to contain a neuropeptide Y (NPY)-like peptide (Everitt *et al.* 1984a) and to a minor extent a neurotensin (NT)-like peptide (Hökfelt *et al.* 1984a). In Table 14.4 these findings are summarized, demonstrating that, for example, almost all neurons in the noradrenergic A1 and adrenergic Cl cell groups in the ventrolateral medulla oblongata contain NPY-like immunoreactivity (nomenclature according to Dahlström and Fuxe 1964; Hökfelt *et al.* 1985). in the locus coeruleus 25–30 per cent of all neurons contain this peptide, whereas no dopamine neurons have been shown to be NPY-immunoreactive (A8–A17). It could be pointed out, however that for example the arcuate nucleus contains very high numbers of NPY-positive cells but they are completely separated from the dopamine neurons of the A12 group (Everitt *et al.* 1984a). Neurotensin (NT) shows, in comparison, only a minor degree of coexistence. Thus, some cells of the small sized adrenaline cells of the so called dorsal strip (C2 dorsal) in the nucleus tractus solitarii and a small proportion of the A10 dopamine cells in the ventral tegmental area are NT-immunoreactive (Hökfelt *et al.* 1984a). A substantial portion of the dopamine cells of the A12 group in the arcuate nucleus do, however, exhibit NT-LI.

The above mentioned systems have been chosen to exemplify the apparent heterogeneity of certain types of transmitter-defined neurons with regard to presence or absence of certain peptides. Thus, it seems as if 5-HT and catecholamine neurons can be subdivided on the basis of content of certain peptides. As discussed above, we should, however, judge these data with caution since the negative findings may represent insufficient sensitivity of our technique. In this context it may be pointed out that the occurrence of peptides in cell bodies is almost always demonstrated after colchicine treatment, i.e. after arrest of axonal transport and accumulation of peptides synthesized in the cell body (see Dahlström 1971). It may therefore be anticipated that the more peptide a neuron synthesizes, the more peptide should accumulate in the cell body and the more easily it should be detectable with immunohistochemistry. Thus, one could perhaps make the statement that the neurons which most easily escape detection are the ones which produce the smallest amount of peptide and where the peptides may be of least importance.

The locus coeruleus discussed above represents an unique nucleus since it almost completely consists of noradrenergic cell bodies. In most other regions there is a considerable heterogeneity of various types of transmitter specific neurons, and it has been possible to demonstrate that these neurons in turn are heterogenous in terms of coexistence situations. Such an example is the arcuate nucleus in the basal hypothalamus involved in the gating and regulation of neuroendocrine information to the anterior and intermediate lobe of the pituitary gland. Thus, this nucleus contains

TABLE 14.4. *Summary of differential coexistence patterns of neuropeptide Y (NPY)- and neurotensin (NT)-like immunoreactivity in catecholamine neuron groups.*

Area	Catecholamine	CA/NPY coexistence	CA/NT coexistence
(1) Ventrolateral medulla oblongata	NA (A1 group)	Yes, major	No
	A (C1 group)	Yes, complete	No
(2) Dorsal medulla oblongata (dorsal vagal complex)	NA (A2 group)	Yes, minor	No
	A (C2 group)	Yes, complete	No
	A (C2, dorsal group)	(?)	Yes, partial
(3) Area postrema	NA	No	No
(4) Pons, locus coeruleus	NA (A6 group)	Yes, partial	No
(5) Pons	NA (A4 group)	Yes, major	No
	NA (A5 group)	No	No
	NA (A7 group)	No	No
(6) Mesencephalon, ventral tegmental area and substantia nigra	DA (A8–10 groups)	No	Yes, minor (A10)
(7) Hypothalamus, periventricular groups including arcuate nucleus	DA (A12 and A14 groups)	No	Yes, partial (A12)
(8) Zona incerta	DA (A11 and A13 groups)	No	No
(9) Olfactory bulb	DA (A15)	No	No

For fine neuroanatomical details, particularly with regard to the nucleus of the solitary tract (A2/C2 groups), see Everitt *et al.* (1984a,b) and Hökfelt *et al.* (1984a,b,c), from where this table has been taken. A = adrenaline; DA = dopamine; NA = noradrenaline.

somatostatin, NPY, NT, growth hormone releasing factor (GRF), galanin, a variety of opioid peptide-immunoreactive neurons as well as neurons containing classical transmitters such as dopamine, GABA and in all probability acetylcholine (for references see Everitt *et al.* 1985). Analysis of possible coexistence situations have revealed that several combinations of compounds can be observed (Meister *et al.* 1985; Everitt *et al.* 1985). Thus, dopamine coexists with GABA, NT and/or with GRF. On the other hand, NPY and somatostatin-positive neurons seem essentially to represent separate populations and so far no classical transmitters have been observed in these neurons (Fig. 14.2). The neuroendocrine system offers good possibilities to evaluate the significance of these coexistences.

The functional significance of subgrouping is still very unclear. Holets *et al.* (1986) have contributed some interesting information to this

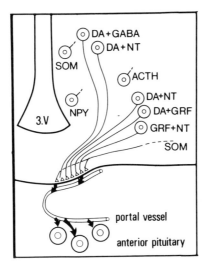

FIG. 14.2. Schematic illustration of various peptide and transmitter containing neurons in the arcuate nucleus of the basal hypothalamus of the rat. Some neuron populations contain more than one putative messenger compound, for example dopamine (DA) + γ-aminobutyric acid (GABA), DA + neurotensin (NT), DA + growth hormone releasing factor (GRF), or GRF + NT. These neuron populations project to the external layer of the median eminence and may directly or indirectly, via the portal vessels, influence hormone secretion from the anterior pituitary gland. Other neurons seem to contain only one compound, for example somatostatin (SOM), neuropeptide Y (NPY) or adrenocorticotropin (ACTH). The latter neurons, of course, also contain other cleavage products of the proopiomelanocortin precursor. These SOM, NPY and ACTH neurons do not project to the external layer of the median eminence but in central direction or give rise to local fiber networks. Note that somatostatin nerve endings in the external layer of the median eminence do not originate in the arcuate nucleus but from outside basal hypothalamus. 3.V = third ventricle.

problem. In a detailed analysis of the locus coeruleus it was observed that about 25–30 per cent of all noradrenaline neurons contained NPY-LI. Furthermore, the projections of these neurons were analyzed using retrograde tracing with a fluorescent dye, Fast Blue, combined with immunohistochemistry (see Skirboll *et al*. 1984). It was observed that noradrenergic neurons in the locus coeruleus containing NPY preferentially projected to the hypothalamus giving rise to bilateral projections, whereas only a smaller proportion of these neurons projected to the cerebral cortex and the spinal cord (Holet *et al*. 1986). In contrast to the hypothalamic projections, the latter noradrenaline/NPY neurons give rise almost exclusively to ipsilateral projections. These findings indicate that subdivision of transmitter specific groups of neurons by a peptide may be related to the target areas of the projections.

It is interesting in this context that transplantation of mesencephalic dopamine neurons to the striatum have indicated that dopamine neurons lacking CCK-LI shows a preferential ingrowth into striatum (Schultzberg *et al*. 1984) which normally is not innervated by dopamine/CCK neurons (Hökfelt *et al*. 1980c,d). Dopamine neurons containing CCK-LI, on the other hand, seem to give rise to fibres mainly within the transplant extending for only a very short distance into the head of the caudate nucleus (Schultzberg *et al*. 1984). Whether or not the peptides in these cases indeed have anything to do with the regulation of ingrowth appears very uncertain but would perhaps be worth while to explore.

SPECIES VARIATIONS OF COEXISTENCE IN PATTERNS

To what extent coexistence combinations have been maintained during phylogeny has only been studied to a limited extent. One could, however, mention some studies on dopamine/cholecystokinin-immunoreactive neurons in the ventral mesencephalon. Thus, dopamine neurons in this area contain CCK-like immunoreactivity in mouse, rat, cat, monkey, and probably man (see Hökfelt *et al*. 1980c,d, 1985). However, certain differences can be observed (Table 14.5). Thus, in the cat the zona compacta of the substantia nigra contained the highest proportions of dopamine/CCK coexistence. In the monkey coexistence has so far only been observed in the midline area, i.e. the area corresponding to the A10 dopamine group. In the rat marked regional distributions have been observed. In the rostral parts, most dopamine cells in the zona compacta contain CCK-like immunoreactivity, but the peptide is present in fewer cells in the caudal parts of this zone. All cells in the pars lateralis contain both dopamine and CCK-like immunoreactivity. The ventral tegmental area contains dopamine/CCK coexistence at all levels. In the mouse coexistence

TABLE 14.5. *Dopamine-CCK coexistence in different regions of three species**

	Rat	Cat	Monkey	Mouse
Substantia nigra (A9 group)				
Pars compacta			0?	0?
Rostral	++++			
Mid	+++	++++		
Caudal	++++			
Pars lateralis	+++++			
Pars reticulata	0	0		
Ventral tegmental area (A10 group)			++	++
Anterior	+++	++		
Posterior	++++	+		

*An estimation of the approximate percentage of DA cell bodies containing CCK-like immunoreactivity (LI) is given: CCK-LI is present in all (+++++), in many (++++), in moderate numbers (+++), in low numbers (++), in single (+) or in none (0) of the DA cell bodies. From Hökfelt *et al.* (1985).

of this type has been observed mainly in the ventral tegmental area (A10) (Sundblad, Staines, Hökfelt, Jonsson, Frey and Rehfeld, in preparation). Further studies of other types of coexistence situations as well as studies including more diverse species are in progress.

DIFFERENTIAL RELEASE OF COEXISTING MESSENGER MOLECULES

A key question is to what extent coexisting messengers can produce selective and differential responses. Several possibilities to achieve this can be considered. One extreme would be that the neuron always releases all types of messenger molecules at the same time and that, for example, selectivity and specificity is dependent on type and distribution of the receptors. Another alternative is that the neuron has the capacity to release the messengers differentially. The former possibility will be discussed below. The latter mechanism has been approached by studying whether the messengers are stored in the same or in different subcellular organelles. It is well known that nerve endings in general contain at least two types of vesicle, the synaptic vesicles with a diameter of about 50 nm and a larger type of vesicle (diameter about 100 nm), often characterized by an electron dense-core and termed large dense-core vesicles. Subcellular fractionation analysis of the cat salivary gland (Lundberg *et al.* 1981) and of the rat vas deferens (Fried *et al.* 1985) have revealed that classical transmitters and peptides may be partially stored in different compartments. Thus, acetylcholine and noradrenaline can be found both in a light and a heavy fraction which, upon ultrastructural analysis, contains small synaptic vesicles and large dense-core vesicles, respectively. In contrast vasoactive intestinal polypeptide (VIP) (coexisting with acetylcholine) and NPY (coexisting with noradrenaline) principally only appear in the heavy fraction, suggesting storage exclusively in large dense-core vesicles. Thus, if mechanisms exist allowing selective activation of the two types of vesicles upon arrival of nerve impulses, it should be possible to obtain differential release. We have suggested that such activation could be frequency coded—at low activity small vesicles are activated causing release only of the classical transmitter, at higher frequencies in addition large vesicles are activated resulting in exocytotic release of both peptide and classical transmitter (see Lundberg and Hökfelt 1983).

INTERACTIONS BETWEEN PEPTIDES AND CLASSICAL TRANSMITTERS

Possible interactions between a classical transmitter and a peptide, possibly released from the same nerve endings, have been explored only to a limited extent. Perhaps the best evidence for interaction has been obtained in studies on the peripheral nervous system, especially the salivary gland of the cat and vas deferens of the rat, and on autonomic ganglia of the bull-frog (Jan and Jan 1983). Some results from studies on central nervous tissue are also available.

Interaction at the post-synaptic level

Immunohistochemical studies on the cat salivary gland have revealed that both the sympathetic and parasympathetic nerves contain a biologically active peptide. Thus, NPY is present in a population of the sympathetic noradrenergic fibres, mainly innervating blood vessels (Lundberg *et al.* 1982b) and VIP occurs in the parasympathetic cholinergic fibres (Lundberg *et al.* 1979). This classical experimental model offers favourable conditions for analysis of the role of transmitters and peptides in the control of secretion and blood flow through the gland as shown by Lundberg, Änggård and collaborators (see Lundberg *et al.* 1982a). Briefly, these experiments have shown that acetylcholine induces both secretion and increase in blood flow, effects which are both atropine-sensitive. VIP alone increases blood flow but has no apparent effect on secretion. However, VIP potentiates acetylcholine-induced secretion and when acetylcholine and VIP are infused together additive effects on blood flow are observed (see Lundberg *et al.* 1982a). With regard to sympathetic nerves, both noradrenaline and NPY cause vasoconstriction, whereby NPY exhibits a slowly developing, long lasting effect (Lundberg and Tatemoto 1982). Taken together these findings indicate that both in sympathetic and parasympathetic nerves peptides and classical transmitters co-operate at the post-synaptic level to cause a physiological response.

Evidence for co-operation at the post-synaptic level has also been obtained in studies in central neurons. Thus, Barbeau and Bedard (1981) have studied the effect of intravenously administered TRH on the stretch reflex in chronically spinalized or 5-HT neurotoxin-treated rats. A marked activation of the stretch reflex was observed, similar to the effect seen after administration of the 5-HT precursor 5-hydroxytryptophan (5-HTP) (see also Andén *et al*. 1964). Interestingly, the effect of TRH could be blocked by previous administration of a 5-HT antagonist. These findings suggested to Barbeau and Bedard (1981) that TRH may act at a site closely associated with the 5-HT receptor. In view of the similarity to the response caused by 5-HTP, it may be suggested that TRH and 5-HT cooperate at synapses in the ventral horn of the spinal cord, possibly directly or indirectly influencing motoneurons and the muscles involved in the stretch reflex.

In this context it may be relevant to discuss results from some behavioural studies. Thus, sexual behaviour in the male rat was studied after application of 5-HT, substance P or TRH (Hansen *et al*. 1983). Several parameters such as number of mounts and intromissions, latencies (mount, intromission and ejaculation) as well as postejaculatory interval were analyzed. 5-HT (50 μg), TRH (10 μg), or substance P (10 μg) or combinations of these drugs were administered intrathecally via a catheter introduced according to the technique of Yaksh and Rudy (1976). No significant effects of 5-HT were observed in these experiments when given alone. Furthermore, only very small effects were observed with TRH on mount and intromission latencies, indicating an increase in these parameters. Similarly, no effects were observed with substance P alone (L. Svensson and S. Hansen, unpublished data). However, when given together 5-HT and TRH caused a marked increase in both mount and intromission latency (from about 0.2 min to more than 7 min). None of the other parameters studied were affected by the combined administration of 5-HT and TRH (Hansen *et al*. 1983). Thus, these results are in agreement with the results obtained in the spinal model described above, suggesting an interaction of 5-HT and TRH at a post-synaptic level.

Interaction at the presynaptic level

A second type of interaction has been observed in the rat vas deferens where noradrenaline and NPY coexist in sympathetic nerves (Lundberg *et al*. 1982b). Here NPY in a dose-dependent manner inhibits the electrically induced contraction of vas deferens, which seems to be due to inhibition of noradrenaline release (Allen *et al*. 1982; Lundberg *et al*. 1982b; Ohhashi and Jacobowitz 1983; Stjärne and Lundberg 1984).

Evidence for interaction of peptide and classical transmitter in the release process has also been obtained from studies on central nervous system. Thus, in the cat a large proportion of the nigral dopamine neurons contain a CCK-like peptide and, correspondingly, dense networks of CCK- and tyrosine hydroxylase-immunoreactive fibres can be seen in the nucleus caudatus-putamen (Hökfelt *et al*. 1985). Thus, a fairly rich source of tissue, in all probability with nerve endings containing both dopamine and CCK-LI, is available for *in vitro* release studies (Markstein and Hökfelt 1984). Thus, the effects of CCK-8 were studied on basal and electrically evoked tritium outflow from slices of the nucleus caudatus-putamen after preincubation with tritiated dopamine. It was shown that tritium outflow was Ca^{2+} dependent and abolished by tetrodotoxin. The sulphated, but not the unsulphated form of CCK-8 inhibited both basal and electrically evoked tritium outflow in concentrations down to 10^{-14}M.

A different type of interaction in the release process has been described for substance P at the spinal cord level. Thus, Mitchell and Fleetwood-Walker (1981) analysed the effect of peptides on potassium induced release of 5-HT in spinal cord slices. Addition of either substance P or TRH did not influence potassium induced tritium outflow from slices. However, if cold 5-HT was added to the bath in a concentration known to activate the inhibitory 5-HT presynaptic receptor, substance P, but not TRH, counteracted the inhibition of tritium outflow caused by cold 5-HT. These experiments were interpreted to indicate that substance P blocks the 5-HT presynaptic receptor, and thus 5-HT induced inhibition of release.

Receptor-receptor interactions

Evidence has recently been presented that peptides can regulate the binding characteristics of monoamine receptors (see Agnati *et al*. 1984). Such interactions have been described, e.g. between dopamine and CCK peptides, and between 5-HT and CCK peptides. However, this interaction seems to occur in areas not at the moment directly concerned with coexistence of these compounds, for example, the striatum and cortex in the rat. On the other hand, interactions between α_2-agonists and NPY in the medulla oblongata as well as substance P and 5-HT in the spinal cord may reflect phenomena related to coexistence. Thus, substances P (10 μM) can significantly increase the number of 5-HT binding sites and reduce the affinity for 5-HT (Agnati *et al*. 1980) and these actions could partly be counteracted by a substance P antagonist (Agnati *et al*. 1983a). NPY in similar concentrations selectively influenced the characteristics of α_2

adrenergic binding sites, but did not influence α_1- and β-ligand binding (Agnati *et al.* 1983b), increasing the number of *p*-aminoclonidine binding sites. Interestingly, such an effect of NPY was lacking in membrane preparations from spontaneously hypertensive rats (Agnati *et al.* 1983a). This type of interaction could be a basis for co-operativity observed between peptides and classical transmitters.

PEPTIDE-PEPTIDE INTERACTION IN THE PERIPHERY AND SPINAL CORD

Recently a novel peptide has been discovered on the basis of recombinant DNA and molecular biological techniques. Thus, analysis of the nucleotide sequence of the calcitonin gene provided evidence for the existence of a calcitonin gene-related peptide (CGRP) (Amara *et al.* 1982). Using antibodies to this peptide, it could be demonstrated that CGRP is expressed in nervous tissues, and a characteristic and unique distribution pattern was observed for the peptide with immunohistochemistry including presence in primary sensory neurons (Rosenfeld *et al.* 1983). Several groups have observed that these CGRP primary sensory neurons are identical to previously described substance P systems (Gibson *et al.* 1984; Lee *et al.* 1985; Wiesenfeld-Hallin *et al.* 1984). Several biological effects have been described for CGRP, for example, it exerts a vasodilatory action when administered peripherally (Rosenfeld *et al.* 1983; Brain *et al.* 1985). In view of the coexistence of substance P and CGRP in primary sensory neurons, we have attempted to analyze possible sites of interaction both in the periphery as well as in the spinal cord. These studies have revealed interesting types of interactions (Wiesenfeld-Hallin *et al.* 1984; Lundberg *et al.* 1985). In the periphery substance P and CGRP show a differential pattern of activities (Fig. 14.3). Both compounds decrease blood pressure, but with regard to several other parameters no similarities can be observed. Thus, substance P strongly increases extravasation and also increases insufflation pressure in the lungs, whereas CGRP does not affect these functions. On the other hand, CGRP markedly increases heart rate, ventricular rate and ventricular tension, which do not seem to be affected by substance P. The complexity of this system is further substantiated by the fact that these neurons in addition contain several other members of the tachykinin family (see Hua *et al.* 1985), neurokinin A (neurokinin-α, substance K, neuromedin L; Kimura *et al.* 1983; Kangawa *et al.* 1983; Nawa *et al.* 1983; Minamino *et al.* 1984) and neuropeptide K (Tatemoto *et al.* 1985). Neurokinin A exerts in some cases effects similar to substance P and in others to CGRP and in addition some opposite to CGRP. Thus, neurokinin A increases extravasation and insufflation pressure as does substance P, but decreases heart rate, ventricular rate and ventricular

	Extravasation (Evan's Blue)	Blood pressure	Insufflation pressure	Heart rate	Ventricular rate	Ventricular tension
Saline	0	0	0	0	0	0
Substance P	↑↑↑	↓↓	↑↑	0	0	0
Neurokinin A (α, Substance K, Neuromedin L)	↑↑	↓↓	↑↑↑	↓↓	↓↓	↓↓
Calcitonin Gene Related Peptide	0	↑↑↑	0	↑↑↑	↑↑↑	↑↑
Capsaicin	↑↑	↓↓	↑↑↑	↓↓	↓↓	↑
Neurokinin B (β, Neuromedin K)	↑↑	↓↓	↑↑	0	0	0

FIG. 14.3. Schematic illustration of the effects of various tachykinins as well as of the drug capsaicin on various parameters such as extravasation, blood pressure, insufflation pressure, heart rate, ventricular rate and tension. Substance P, neurokinin A and calcitonin gene related peptide (CGRP) are present in the same neurons. Neurokinin B has so far not been shown in these systems. Capsaicin is a drug which causes release of peptides from sensory neurons of the C type. Note differential effects of the various tachykinins on the various parameters. For further explanation see text and Lundberg *et al.* (1985). The data summarized in this figure have been extracted from that paper.

tension, which are effects opposite to the ones seen with CGRP. Since little is known about actual release of these compounds or interaction at the receptor level, these results are at the moment complex and difficult to explain. Focusing on substance P/CGRP interactions, it may be speculated that different tissues and components of tissues have different receptor populations. Thus, for example heart tissue may have CGRP receptors, whereas substance P receptors are lacking. Under such circumstances the neuron would not have to be able to differentially release CGRP and substance P, but specificity of action is achieved by absence or presence of the receptors for the respective peptide on the post-synaptic cells.

Substance P/CGRP interaction has also been analysed at the level of the spinal cord. Thus, these compounds were administered onto the lumbar spinal cord via an intrathecal catheter according to Yaksh and Rudy (1976). Local administration of substance P onto the spinal cord has previously been shown to cause a characteristic behaviour characterized by caudally directed biting and scratching (Hylden and Wilcox 1981; Piercey *et al.* 1981; Seybold *et al.* 1982). In our study the effects of substance P (1, 10, and 20 μg) could be confirmed, exhibiting a dose dependent behaviour lasting for a few minutes. Administration of CGRP alone in doses up to

20 μg do not induce any observable effects. However, if substance P (10 μg) and CGRP (20μg) were injected together, a marked increase in the duration of the response to more than 30 min was observed. The mechanism(s) underlying this dramatic effect may represent a further type of interaction of two compounds released from the same nerve endings and we are at present attempting to analyse such mechanisms.

EVIDENCE FOR NOVEL TYPES OF MESSENGERS: LEUKOTRIENES

Evidence has recently been presented that various members of the leukotriene family can be produced in the brain, possibly by nervous tissue (Dembinska-Kiec *et al.* 1984; Lindgren *et al.* 1984; Moskowitz *et al.* 1984). Thus, using identification by high performance liquid chromatography (HPLC), radioimmunoassay and bioassay, formation of LTC_4, LTD_4, and LTE_4 was demonstrated in slices of brain tissue incubated with arachidonic acid and the ionophore A23187 (Lindgren *et al.* 1984). LTC_4 synthesis was inhibited by nordihydroguaiaretic acid in a concentration of 30 μM. It was also demonstrated that LTC_4 biosynthesis shows regional patterns with the lowest amounts formed in cerebellum and highest in hypothalamus and median eminence, being almost ten times higher in the hypothalamus than in the cerebellum. Attempts were also made to determine the cellular localization by immuno-histochemistry. Using an antiserum raised against LTC_4 conjugated to bovine serum albumin (BSA) (Aeringhaus *et al.* 1982), numerous immunoreactive cell bodies were observed in many parts of the brain (Lindgren *et al.* 1984). However, in control experiments, where the anti-serum was absorbed with LTC_4-BSA conjugate, it was not possible to abolish the staining patterns, except for one region—the median eminence. Here a dense fibre network was seen in the lateral part of the median emin-ence. This staining was absent after incubation with control serum, suggesting the occurrence of LTC_4-LI in these fibres. This analysis was extended with comparative studies demonstrating that the LTC_4-positive fibres were identical to LHRH nerve endings, showing coexistence of the two compounds (Hulting *et al.* 1985). These results indicate that an LTC_4-like compound could be involved in control of LH release from the anterior pituitary gland. Consequently, the effect of leuko-trienes on LH release from rat anterior pituitary cells *in vitro* has been analysed. Thus, LTC_4 in the picomolar range released LH but not growth hormone (Hulting *et al.* 1984, 1985). In contrast, the LTC_4 precursor LTB_4 had no effect on the release of these two hormones. The stimula-tory effect of LTC_4 was seen after 0.5 h incubation, but not after 3 h, suggesting a fairly rapid but transient effect. In contrast, LHRH released LH with a slow onset but with a longer duration. These findings suggest that LH release from gonadotrops in the anterior pituitary may be under dual control, a fast acting factor related to an LTC_4-like compound in addition to the conventional LH releaser, LHRH.

These findings support earlier evidence for a messenger role of LTC_4. Thus, Palmer *et al.* (1980, 1981) have demonstrated that LTC_4, when applied iontophoretically in the cerebellum, causes a long lasting activation of Purkinje cells. The effect had a slow onset and a long duration. This activation was achieved at micromolar concentrations of LTC_4 in the pipette, indicating an extraordinary high potency. Finally, it has recently been shown that high affinity binding sites for LTC_4 are present in homogenates of rat central nervous tissue with a K_D of about 30 nM (Schalling *et al.* 1985). Thus, leukotrienes or a similar compound(s) may represent a novel type of messenger molecules in the nervous system. Clearly, the results described above are very preliminary and need further substantiation.

Acknowledgements

This research was supported by grants from the Swedish Medical Research Council (O4X–2887), Knut and Alice Wallenbergs Stiftelse, Petrus och Augusta Hedlunds Stiftelse, and Magnus Bergvalls Stiftelse. We thank Ms W. Hiort, Ms S. Nilsson, and Ms A. Peters for excellent technical assistance, and Ms E. Björklund for skilful help in preparing the manuscript.

References

Aehringhaus, U., Wölbling, R. H., König, W., Patrono, C., Peskar, B. M., and Peskar, B. A. (1982). Release of leukotriene C_4 from human polymorphonuclear leucocytes as determined by radioimmunoassay. *FEBS Lett.* **146**, 111–4.

Agnati, L. F., Fuxe, K., Benfenati, F., Battistini, N., Härfstrand, A., Hökfelt, T., Cavicchioli, L., Tatemoto, K., and Mutt, V. (1983a). Failure of neuropeptide Y in vitro to increase the number of α_2-adrenergic binding sites in membranes of medulla oblongata of the spontaneous hypertensive rat. *Act Physiol. Scand..* **119**, 309–12.

——, ——, ——, ——, ——, ——, Tatemoto, K., and Mutt, V. (1983b). Neuropeptide Y in vitro selectively increases the number of α_2-adrenergic binding sites in membranes of the medulla oblongata of the rat. *Acta Physiol. Scand.* **118**, 293–5.

——, ——, ——, ——, Zini, I., Camurri, M., and Hökfelt, T. (1984). Postsynaptic effects of neuropeptide comodulators at central monoamine synapses. In *Neurology and Neurobiology, Vol. 8B: Catecholamines, Part B: Neuropharmacology and Central Nervous System—Theoretical Aspects* (eds

E. Usdin, A. Carlsson, A. Dahlström, and J. Engel) pp. 191–8. Alan R. Liss, Inc., New York.

——, ——, Zini, I., Lenzi, P., and Hökfelt, T. (1980). Aspects on receptor regulation and isoreceptor identification. *Med. Biol.* **58**, 182–7.

Allen, J. M., Tatemoto, K., Polak, J. M., Hughes, J., and Bloom, S. R. (1982). Two novel related peptides, neuropeptide Y (NPY) and peptide YY (PYY) inhibit the contraction of the electrically stimulated mouse vas deferens, *Neuropeptides*, **3**, 71–7.

Amara, S. G., Jones, V., Rosenfeld, M. G., Ong, E. S., and Evans, R. M. (1982). Alternative RNA-processing in calcitonin gene expression generates mRNAs encoding different polypeptide products. *Nature*, **298**, 240–4.

Andén, N.-E., Jukes, M., and Lundberg, A. (1964). Spinal reflexes and monoamine liberation. *Nature (Lond.)* **202**, 1222–3.

Barbeau, H., and Bédard, P. (1981). Similar motor effects of 5-HT and TRH in rats following chronical spinal transection and 5,7-dihydroxytryptamine injection, *Neuropharmacol.* **20**, 477–81.

Belin, M. F., Nanopoulos, D., Disier, M., Aguera, M., Steinbusch, H., Verhofstad, A., Maitre, M., and Pujol, J. F. (1983). Immunohistochemical evidence for the presence of γ-aminobutryric acid and serotonin in one nerve cell. A study on the raphe nuclei of the rat using antibodies to glutamate decarboxylase and serotonin. *Brain Res.* **275**, 329–39.

Bloom, F. E. (1979). Contrasting principles of synaptic physiology: peptidergic and non-peptidergic neurons. In *Central Regulation of the Endocrine System* (eds K. Fuxe, T. Hökfelt, and R. Luft) Nobel Foundation Symposium 42, pp. 173–187. Plenum Press, New York.

Brain, S. D., Williams, T. J., Tippins, J. R., Morris, H. R., and MacIntyre, I. (1985). Calcitonin gene-related peptide is a potent vasodilator. *Nature*, **313**, 54–6.

Burnstock, G. (1985). Purinergic mechanisms broaden their sphere of influence. *TINS* **8** (1985) 5–6.

Carlsson, A., Falck, B., and Hillarp, N.-Å. (1962). Demonstration of catecholamines with a histochemical fluorescence method. *Acta Physiol. Scand.* **56**, Suppl. **56**, 1–28.

——, Lindqvist, M., Magnusson, T., and Waldeck, B. (1958). On the presence of 3-hydroxytryptamine in brain. *Science* **127**, 471.

Chan-Palay, V., Jonsson, G., and Palay, S. L. (1978). Serotonin and substance P coexist in neurons of the rat's central nervous system. *Proc. Nat. Acad. Sci. USA*, **75**, 1582–6.

——, Nilaver, G., Palay, S. L., Beinfeld, M. C., Zimmerman, E. A., Wu, J.-Y., and O'Donohue, T. L. (1981). Chemical heterogeneity in cerebellar Purkinje cells: evidence and coexistence of glutamic acid decarboxylase-like and motilin-like immunoreactivities. *Proc. Nat. Acad. Sci. USA*, **78**, 7787–91.

——, and Palay, S. L. (eds) (1984). *Coexistence of Neuroactive Substances in Neurons*. John Wiley & Sons, New York.

Charnay, Y., Léger, L., Dray, F., Beród, A., Jouvet, M., Pujol, J. F., and Dubois, P. M. (1982). Evidence for the presence of enkephalin in catecholaminergic neurons of cat locus coeruleus. *Neurosci. Lett.* **30**, 147–51.

Coons, A. H. (1958). Fluorescent antibody methods. In *General Cytochemical Methods* (ed. J. F. Danielli) pp. 399–422. Academic Press, New York.

Cuello, A. C. (ed.) (1982). *Co-transmission*. MacMillan, London and Basingstoke.

Dahlström, A. (1971). Effects of vinblastine and colchicine on monoamine containing neurons of the rat with special regard to the axoplasmic transport of amine granules. Acta Neuropathol. Suppl. **5**, 226–37.

——, and Fuxe, K. (1964). Evidence of the existence of monoamine-containing neurons in the central nervous system. I. Demonstration of monoamines in the cell bodies of brain stem neurons. *Acta Physiol. Scand.* **62**, Suppl. 232, 1–55.

——, and —— (1965). Evidence for the existence of monoamine containing neurons in the central nervous system. II. Experimentally induced changes in the intraneuronal levels of bulbospinal neurons system. *Acta Physiol. Scand.* **64**, Suppl. 247, 5–36.

Dembinska-Kieć, A., Simmet, T., and Peskar, B. A. (1984). Formation of leukotriene C_4-like material by rat brain tissue. *Eur. J. Pharmacol.* **99**, 57–62.

Eccles, J. C. (ed.) (1964). *The Physiology of Synapses*. Springer-Verlag, Berlin.

Eckenstein, F., and Baughman, R. W. (1984). Two types of cholinergic innervation in cortex, one co-localized with vasoactive intestinal polypeptide. *Nature*, **309**, 153–5.

Everitt, B. J., Hökfelt, T., Terenius, L., Tatemoto, K., Mutt, V., and Goldstein, M. (1984a). Differential co-existence of neuropeptide Y (NPY)-like immunoreactivity with catecholamines in the central nervous system of the rat, *Neurosci*. **11**, 443–62.

Everitt, B. *et al.* (1986). The hypothalamic arcuate nucleus-median eminence complex: Immunohistochemistry of transmitters, peptides and DARPP-32 with special reference to coexistence in dopamine neurons. *Brain Res. Rev.*, submitted.

——, ——, Wu, J.-Y., and Goldstein, M. (1984b). Coexistence of tyrosine hydroxylase-like and gamma-aminobutyric acid-like immunoreactivities in neurons of the arcuate nucleus. *Neuroendocrinol*. **39**, 189–91.

Fonnum, F. (ed.) (1978). *Amino Acids as Chemical Transmitters*, NATO Advanced Study Institutes Series. Series A: Life Sciences. Plenum Press, New York.

Fried, G., Terenius, L., Hökfelt, T., and Goldstein, M. (1985). Evidence for differential localization of noradrenaline and neuropeptide Y (NPY) in neuronal storage vesicles isolated from vas deferens. *J. Neurosci*. **5**, 450–8.

Fuxe, K. (1965a). Evidence for the existence of monoamine neurons in the central nervous system. III. The monoamine nerve terminal. *Z. Zellforsch. Mikrosk. Anat*. **65**, 573–96.

—— (1965b). Evidence for the existence of monoamine neurons in the central nervous system. IV. The distribution of monoamine nerve terminals in the central nervous system. *Acta Physiol. Scand*. **64**, Suppl. 247, 39–85.

Gibson, S. J., Polak, J. M., Bloom, S. R., Sabate, I. M., Mulderry, P. M., Ghatei, M. A., McGregor, G. P., Morrison, J. F. B., Kelly, J. S., Evans, R. M., and Rosenfeld, M. G. (1984). Calcitonin gene-related peptide immunoreactivity in the spinal cord of man and of eight other species. *J. Neurosci*. **4**, 3101–11.

Gilbert, R. F. T., Emson, P. C., Hunt, S. P., Bennett, G. W., Marsden, C. A., Sandberg, B. E. B., Steinbusch, H., and Verhofstad, A. A. J. (1982). The effects of monoamine neurotoxins on peptides in the rat spinal cord, *Neurosci*. **7**, 69–88.

Glazer, E. J., Steinbusch, H., Verhofstad, A., and Basbaum, A. I. (1981). Serotonin neurons in nucleus raphe dorsalis and paragigantocellularis of the cat contain enkephalin. *J. Physiol. (Paris)*, **77**, 241–5.

Hansen, S., Svensson, L., Hökfelt, T., and Everitt, B. J. (1983). 5-hydroxytryptamine-thyrotropin releasing hormone interactions in the spinal cord: effects on parameters of sexual behaviour in the male rat, *Neurosci. Lett.* **42**, 299–304.

Hendry, S. H. C., Jones, E. G., DeFelipe, J., Schmechel, D., Brandon, C., and Emson, P. C. (1984). Neuropeptide-containing neurons of the cerebral cortex are also GABAergic. *Proc. Nat. Acad. Sci. USA*, **81**, 6526–30.

Hökfelt, T., Everitt, B. J., Theodorsson-Norheim, E., and Goldstein, M. (1984a). Occurrence of neurotensin-like immunoreactivity in subpopulations of hypothalamic, mesencephalic, and medullary catecholamine neurons. *J. Comp. Neurol.* **222**, 543–59.

——, Johansson, O., and Goldstein, M. (1984b). Chemical anatomy of the brain. *Science*, **225**, 1326–34.

——, ——, Ljungdahl, Å., Lundberg, J. M., and Schultzberg, M. (1980a). Peptidergic neurons. *Nature (Lond.)* **284**, 515–21.

——, Ljungdahl, A., Steinbusch, H., Verhofstad, A., Nilsson, G., Brodin, E., Pernow, B., and Goldstein, M. (1978). Immunohistochemical evidence of substance P-like immunoreactivity in some 5-hydroxytryptamine-containing neurons in the rat central nervous system, *Neurosci.* **3**, 517–38.

——, Lundberg, J. M., Lagercrantz, H., Tatemoto, K., Mutt, V., Lundberg, J. M., Terenius, L., Everitt, B. J., Fuxe, K., Agnati, L. F., and Goldstein, M. (1983). Occurrence of neuropeptide Y (NPY)-like immunoreactivity in catecholamine neurons in the human medulla oblongata, *Neurosci. Lett.* **36**, 217–22.

——, ——, Schultzberg, M., Johansson, O., Ljungdahl, Å., and Rehfeld, J. (1980b). Coexistence of peptides and putative transmitters in neurons. In *Neural Peptides and Neuronal Communication* (eds E. Costa and M. Trabucchi) pp. 1–23. Raven Press, New York.

——, ——, Skirboll, L., Johansson, O., Schultzberg, M., and Vincent, S. R. (1982). Coexistence of classical transmitters and peptides in neurons, In *Co-transmission* (ed. A. C. Cuello) pp. 77–126. MacMillan, London and Basingstoke.

——, Mårtensson, R., Björklund, A., Kleinau, S., and Goldstein, M. (1985a). Distributional maps of tyrosine hydroxylase immunoreactive neurons in the rat brain. In *Handbook of Chemical Neuroanatomy* (eds A. Björklund and T. Hökfelt) Vol. 2, Classical Transmitters in the CNS, Part I, pp. 277–379. Elsevier, Amsterdam.

——, Rehfeld, J. F., Skirboll, L., Ivemark, B., Goldstein, M., and Markey, K. (1980c). Evidence for coexistence of dopamine and CCK in mesolimbic neurones. *Nature (Lond.)* **285**, 476–8.

——, Skirboll, L., Everitt, B., Meister, B., Brownstein, M., Jacobs, T., Faden, A., Kuga, S., Goldstein, M., Markstein, R., Dockray, G., and Rehfeld, J. (1985b) Distribution of cholecystokinin-like immunoreactivity in the nervous system with special reference to coexistence with classical neurotransmitters and other neuropeptides. In *Neuronal Cholecystokinin* (eds J. J. Vanderhaeghen and J. Crawley), *Ann. N.Y. Acad. Sci.* **448**, 255–74.

——, ——, Rehfeld, J. F., Goldstein, M., Markey, K., and Dann, O. (1980c). A subpopulation of mesencephalic dopamine neurons projecting to limbic areas contains a cholecystokinin-like peptide: evidence from immunohistochemistry combined with retrograde tracing, *Neurosci.* **5**, 2093–124.

Holets, V., Hökfelt, T., Terenius, L., Rökaeus, Å., and Goldstein, M. (1985). Occurrence and projection of locus coeruleus noradrenaline neurons containing NPY- and galanin-like immunoreactivities. To be submitted.

Hua, X., Theodorsson-Norheim, E., Brodin, E., Lundberg, J. M., and Hökfelt, T. (1985). Co-existence of multiple tachykinins (neurokinin A, neuropeptide K and substance P) in capsaicin-sensitive sensory neurons in the guinea-pig. *Regulatory Peptides* (in press).

Hulting, A.-L., Lindgren, J.-Å., Hökfelt, T., Eneroth, P., Werner, S., Patrono, C., and Samuelsson, B. (1985). Leukotriene C_4 as a mediator of LH release from rat anterior pituitary cells. *Proc. Nat. Acad. Sci. USA*, **82**, 3834–8.

——, ——, ——, Heidvall, K., Eneroth, P., Werner, S., Patrono, C., and Samuelsson, B. (1984). Leukotriene C_4 stimulates LH secretion from rat pituitary cells *in vitro*. *Eur. J. Pharmacol*. **106**, 459–60.

Hunt, S. P. and Lovick, T. A. (1982). The distribution of serotonin, metenkephalin and β-lipotropin-like immunoreactivity in neuronal perikarya of the cat brain stem, *Neurosci. Lett*. **30**, 139–45.

Hylden, J. L. K. and Wilcox, G. L., (1981). Intrathecal substance P elicits a caudally-directed biting and scratching behavior in mice. *Brain Res*. **217**, 212–5.

Iversen, L. L. (1984). Amino acids and peptides: fast and slow chemical signals in the nervous system. *Proc. Roy. Soc. Lond*. **B221**, 245–60.

Jan, Y. N. and Jan, L. Y. (1983). A LHRH-like peptidergic transmitter capable of 'action at a distance' in autonomic ganglia. *TINS*, **6**, 320–5.

Jirikowski, G., Reisert, I., Pilgrim, Ch., and Oertel, W. H. (1984). Coexistence of glutamate decarboxylase and somatostatin immunoreactivity in cultured hippocampal neurons of the rat. *Neurosci. Lett*. **46**, 35–9.

Johansson, O., Hökfelt, T., Pernow, B., Jeffcoate, S. L., White, N., Steinbusch, H. W. M., Verhofstad, A. A. J., Emson, P. C., and Spindel, E. (1981). Immunohistochemical support for three putative transmitters in one neuron: coexistence of 5-hydroxytryptamine-, substance P-, and thyrotropin releasing hormone-like immunoreactivity in medullary neurons projecting to the spinal cord, *Neurosci*. **6**, 1857–81.

Kangawa, K., Minamino, N., Fukuda, A., and Matuso, H. (1983). Neuromedin K: A novel mammalian tachykinin identified in porcine spinal cord. *Biochem. Biophys. Res. Comm*. **114**, 533–40.

Kimura, S., Okada, M., Sugita, Y., Kanazawa, I., and Munekata, E. (1983). Novel neuropeptides, neurokinin α and β isolated from porcine spinal cord. *Proc. Jpn. Acad*. **59B**, 101–4.

Lee, Y., Kawai, Y., Shiosaka, S., Takami, K., Kiyama, H., Hillyard, C. J., Girgis, S., MacIntyre, I., Emson, P. C., and Tohyama, M. (1985). Coexistence of calcitonin gene-related peptide and substance P-like peptide in single cells of the trigeminal ganglion of the rat: immunohistochemical analysis. *Brain Res*. **330**, 194–6.

Lindgren, J. Å., Hökfelt, T., Dahlén, S. E., Patrono, C., and Samuelsson, B. (1984). Leukotrienes in the rat central nervous system. *Proc. Nat. Acad. Sci. USA*, **81**, 6212–6.

Lovick, T. A. and Hunt, S. P. (1983) Substance P-immunoreactive and serotonin-containing neurones in the ventral brainstem of the cat, *Neurosci. Lett*. **36**, 223–8.

Lundberg, J. M., Franco-Cereceda, A., Hua, X., Hökfelt, T., and Fischer, J. A. (1985). Co-existence of substance P and calcitonin gene-related peptide-like immunoreactivities in sensory nerves in relation to cardiovascular and bronchoconstrictor effects of capsaicin, *Eur. J. Pharmacol.* **108**, 315–9.

——, Fried, G., Fahrenkrug, J., Holmstedt, B., Hökfelt, T., Lagercrantz, H., Lundgren, G., and Änggård, A. (1981). Subcellular fractionation of cat submandibular gland: comparative studies on the distribution of acetylcholine and vasoactive intestinal polypeptide (VIP). *Neurosci.* **6**, 1001–10.

——, Hedlung, B., Änggård, A., Fahrenkrug, J., Hökfelt, T., Tatemoto, K., and Bartfai, T. (1982a). Costorage of peptides and classical transmitters in neurons. In *Systemic Role of Regulatory Peptides* (eds S. R. Bloom, J. M. Polak, and E. Lindenlaub) pp. 93–119. Schattauer, Stuttgart and New York.

——, and Hökfelt, T. (1983). Coexistence of peptides and classical neurotransmitters. *TINS*, **6**, 325–33.

——, ——, Schultzberg, M., Uvnäs-Wallensten, K., Köhler, C., and Said S. (1979). Occurrence of vasoactive intestinal polypeptide (VIP)-like immunoreactivity in certain cholinergic neurons of the cat: evidence from combined immunohistochemistry and acetylcholine esterase staining, *Neurosci.* **4**, 1539–59.

——, and Tatemoto, K. (1982). Pancreatic polypeptide family (APP, BPP, NPY and PYY) in relation to sympathetic vasoconstriction resistant to α-adrenoceptor blockade. *Acta Physiol. Scand.* **116**, 393–402.

——, Terenius, L., Hökfelt, T., Martling, C. R., Tatemoto, K., Mutt, V., Polak, J., Bloom, S., and Goldstein, M. (1982b). Neuropeptide Y (NPY)-like immunoreactivity in peripheral noradrenergic neurons and effects of NPY on sympathetic function, *Acta Physiol. Scand.* **116**, 477–80.

Markstein, R. and Hökfelt, T. (1984). Effect of cholecystokinin-octa-peptide on dopamine release from slices of cat caudate nucleus, *J. Neurosci.* **4**, 570–5.

Meister, B., Hökfelt, T., Vale, W. W., and Goldstein, M. (1985). Growth hormone releasing factor (GRF) and dopamine coexist in hypothalamic arcuate neurons. *Acta Physiol. Scand.* **124**, 133–6.

Melander, T., Staines, W. A., Hökfelt, T., Rökaeus, Å., Eckenstein, F., Salvaterra, P. M., and Wainer, B. H. (1985). Galanin-like immunoreactivity in cholinergic neurons of the septum-basal forebrain complex projecting to the hippocampus of the rat. *Brain Res.* (in press).

Minamino, N., Kangawa, K., Fukuda, A., and Matsuo, H. (1984). A novel mammalian tachykinin identified in porcine spinal cord. *Neuropeptides*, **4**, 157–66.

Mitchell, R. and Fleetwood-Walker, S. (1981). Substance P, but not TRH, modulates the 5-HT autoreceptor in ventral lumbar spinal cord, *Eur. J. Pharmacol.* **76**, 119–20.

Moskowitz, M. A., Kiwak, K. J., Hekimian, K., and Levine, L. (1984). Synthesis of compounds with properties of leukotrienes C_4 and D_4 in gerbil brains after ischemia and reperfusion. *Science*, **224**, (1968). 886–9.

Nakane, P. K. (1968). Simultaneous localization of multiple tissue antigens using the peroxidase-labelled antibody method: a study in pituitary glands of the rat. *J. Histochem. Cytochem.* **16**, 557–60.

Nawa, H., Hirose, T., Takashima, H., Inayama, S., and Nakanishi, S. (1983).

Nucleotide sequences of cloned cDNAs for two types of bovine brain substance P precursor. *Nature*, **306**, 32–6.

Oertel, W. H., Graybiel, A. M., Mugnaini, E., Elde, R. P., Schmechel, D. E., and Kopin, E. J. (1983). Coexistence of glutamic acid decarboxylase- and somatostatin-like immunoreactivity in neurons of the feline nucleus reticularis thalami. *J. Neurosci.* **3**, 1322–32.

——, and Mugnaini, E. (1984). Immunocytochemical studies of GABAergic neurons in rat basal ganglia and their relations to other neuronal systems. *Neurosci. Lett.* **47**, 233–8.

Ohhashi, T. and Jacobowitz, D. M. (1983). The effects of pancreatic polypeptides and neuropeptide Y on the rat vas deferens. *Peptides*, **4**, 381–6.

Otsuka, M. and Takahashi, T. (1977). Putative peptide neurotransmitters. *Ann. Rev. Pharmacol. Toxicol.* **17**, 425–39.

Palmer, M. R., Mathews, W. R., Hoffer, B. J., and Murphy, R. C. (1981). Electrophysiological response of cerebellar Purkinje neurons to leukotriene D_4 and B_4. *J. Pharmacol. Exp. Ther.* **219**, 91–96.

——, ——, Murphy, R. C., and Hoffer, B. J. (1980). Leukotriene C elicits a prolonged excitation of cerebellar Purkinje neurons. *Neurosci. Lett.* **18**, 173–80.

Pernow, B. (1983). Substance P. *Pharmacol. Rev.* **35**, 85–141.

Piercey, M. F., Dobry, P. J. K., Schroeder, L. A., and Einspahr, F. J. Behavioral evidence that substance P may be a spinal cord sensory neurotransmitter. *Brain Res.* **210**, 407–12.

Rosenfeld, M. G., Mermod, J. -J., Amara, S. G., Swanson, L. W., Sawchenko, P. E., Rivier, J., Vale, W. W., and Evans, R. M. (1983). Production of a novel neuropeptide encoded by the calcitonin gene via tissue-specific RNA processing. *Nature*, **304**, 129–35.

Samuelsson, B. (1983). Leukotrienes: mediators of immediate hypersensitivity reactions and inflammation. *Science*, **220**, 568–75.

Schalling, M., Neil, A., Terenius, L., Hökfelt, T., Lindgren, J. -Å., and Samuelsson, B. (1985). Leukotriene C_4 binding sites in the rat central nervous system. *Eur. J. Pharmacol.*, in press.

Schmechel, D. E., Vickrey, B. G., Fitzpatrick, D., and Elde, R. P. (1984). GABAergic neurons of mammalian cerebral cortex: widespread subclass defined by somatostatin content. *Neurosci. Lett.* **47**, 227–32.

Schmitt, F. O. (1984). Molecular regulators of brain function: a new view. *Neurosci.* **13**, 991–1001.

Schultzberg, M., Dunett, S. B., Björklund, A., Stenevi, U., Hökfelt, T., Dockray, G. J., and Goldstein, M. (1984). Dopamine and cholecystokinin immunoreactive neurons in mesencephalic grafts reinnervating the neostriatum: evidence for selective growth regulation. *Neurosci.* **12**, 17–32.

Seybold, V. S., Hylden, J. L. K., and Wilcox, G. L. (1982). Intrathecal substance P and somatostatin in rats: behaviors indicative of sensation. *Peptides*, **3**, 49–54.

Skirboll, L., Hökfelt, T., Norell, G., Phillipson, O., Kuypers, J. G. J. M., Bentivoglio, M., Catsman-Berrevoets, C. E., Visser, T. J., Steinbusch, H., Verhofstad, A., Cuello, A. C., Goldstein, M., and Brownstein, M. (1984). A method for specific transmitter identification of retrogradely labeled neurons: immunofluorescence combined with fluorescence tracing. *Brain Res. Rev.* **8**, 99–127.

Snyder, S. (1980). Brain peptides as neurotransmitters. *Science*, **209**, 976–83.

Somogyi, P., Hodgson, A. J., Smith, A. D., Nunzi, M. G., Gorio, A., and Wu, J. -Y. (1984). Different populations of GABAergic neurons in the visual cortex and hippocampus of cat contain somatostatin- or cholecystokinin-immunoreactive material. *Neurosci.* **14**, 2590–603.

Stjärne, L., Hedqvist, P., Lagercrantz, H., and Wennmalm, Å. (eds) (1981). *Chemical Neurotransmission*. Academic Press, London.

——, and Lundberg, J. M. (1984). Neuropeptide Y (NPY) depresses the secretion of ³H-noradrenaline and the contractile response evoked by field stimulation in rat vas deferens. *Acta Physiol. Scand.* **120**, 477–9.

Takeda, N., Inagaki, S., Shiosaka, S., Taguchi, Y., Oertel, W. H., Tohyama, M., Watanabe, T., and Wada, H. (1984). Immunohistochemical evidence for the coexistence of histidine decarboxylase-like and glutamate decarboxylase-like immunoreactivities in nerve cells of the magnocellular nucleus of the posterior hypothalamus of rats. *Proc. Nat. Acad. Sci. USA*, **81**, 7647–50.

Tatemoto, K., Lundberg, J. M., Jörnvall, M., and Mutt, V. (1986). Neuropeptide K: isolation, structure and biological activities of a novel brain tachykinin. *Nature* submitted.

Tramu, G., Pillez, A., and Leonardelli, J. (1978). An efficient method of antibody elution for the successive or simultaneous location of two antigens by immunocytochemistry, *J. Histochem. Cytochem.* **26**, 322–4.

Vincent, S. R., Satoh, K., Armstrong, D. M., and Fibiger, H. C. (1983). Substance P in the ascending cholinergic reticular system, *Nature*, **306**, 688–91.

Vogt, M. (1954). The concentration of sympathin in different parts of the central nervous system and normal conditions and after the administration of drugs. *J. Physiol. (Lond.)* **123**, 451–81.

Watt, C. B., Su, Y. T., and Lam, D. M. -K. (1984). Interactions between enkephalin and GABA in avian retina. *Nature*, **311**, 761–3.

Wiesenfeld-Hallin, Z., Hökfelt, T., Lundberg, J. M., Forssmann, W. G., Reinecke, M., Tschopp, F. A., and Fischer, J. A. (1984). Immunoreactive calcitonin gene-related peptide and substance P coexist in sensory neurons to the spinal cord and interact in spinal behavioural responses of the rat. *Neurosci. Lett.* **52**, 199–204.

Yaksh, T. L. and Rudy, T. A. (1976). Chronic catheterization of the spinal subarachnoid space. *Physiol. Behav.* **17**, 1031–6.

Zahm, D. S., Zaborszky, L., Alones, V. E., and Heimer, L. (1985). Evidence for the coexistence of glutamate decarboxylase and metenkephalin immunoreactivities in axon terminals of rat ventral pallidum. *Brain Res.* **325**, 317–21.

15

Neuropeptides and monoamines as slow signals: discussion

S. FREEDMAN AND K. WATLING

It is clear that superimposed upon the 'fast' signalling process within the nervous system, there exists a secondary control system which can modify responses to the primary signal. This system has the characteristic of being operated by a wide diversity of chemical transmitters and can occur over a much longer time scale than that described in the previous section. The speakers during this session were able to demonstrate how this regulation can occur at a variety of levels including ion channel regulation, complex cotransmission, differential release processes, selective activation of secondary messengers and intracellular protein phosphorylation.

Some of the major mechanisms associated with this slow signalling process are the small currents such as Im, I_{Ca}, and the Ca^{2+}-activated K^+ currents which Dr D. A. Brown described as important modulators of neuronal firing rates. It was pointed out in discussion that many of the neurotransmitters which can modify these currents (including acetylcholine) are linked to inositol phospholipid turnover. The overall effect on the cell, however, may not be entirely explicable in terms of the calcium mobilizing effects of inositol-1,4,5-trisphosphate [$Ins(1,4,5)P_3$].

A similar point was made during Dr R. F. Irvine's presentation where much interest was focused upon the role of diacyl glycerol in the cellular response. During discussion Dr Brown indicated that this was an area that his group and many other workers were interested in. It had recently been reported that the inhibitory effects of acetylcholine in the hippocampus could be replicated by phorbol esters (S. H. Snyder and colleagues, Baltimore). This was an effect on the Ca^{2+}-activated K^+ current seen in this area rather than replication of the M current since the cells do not go into the full bursting behaviour and the full depolarisation that is normally associated with M current inhibition. The mechanism for M current inhibition is as yet unclear, although one possibly may include inhibition of adenylate cyclase, though as yet there is no evidence to support this hypothesis. The role of cyclic GMP was raised, since in some systems cyclic

GMP has been reported to imitate the actions of acetylcholine. The problem with this idea is that guanylate cyclase is activated by an entry of calcium and so in many cases a response in cyclic GMP will be secondary to calcium movement.

The role of protein phosphorylation as a means of regulating signal transduction was discussed extensively by Dr S. I. Walaas. Particular attention was paid to the interaction of dopamine with the protein DARPP-32. One point raised in discussion concerned the suggested association of this protein with D-1 dopamine receptors. The pharmacology of the response has not been characterized in great detail, although the antipsychotic agent fluphenazine does block dopamine-induced phosphorylation of DARPP-32. The interaction of selective antagonists such as sulpiride and SCH-23390 with this system would provide an interesting insight into DARPP-32 and dopamine receptor classification. One point raised concerned evidence in the mammalian brain for the phosphorylation of a channel protein by neurotransmitters. Dr Walaas was of the opinion that the evidence from intact tissue was very weak and that transmitter regulation of ion-channel phosphorylation remains to be demonstrated.

In addressing the question of fast and slow chemical signalling in the nervous system, Dr T. M. Jessell focused his attention on the array of synaptic transmitters that mediate the transfer of sensory information from dorsal root ganglion (DRG) neurones to dorsal horn neurones. The neuropeptides present within sub-populations of primary sensory afferents, notably substance P, appear to mediate predominantly slow synaptic responses at primary afferent synapses. Dr Jessell's work using DRG and dorsal horn neurones in cell culture has so far identified only two classes of compounds as potential fast sensory neurotransmitters, namely, excitatory amino acids and nucleotides. Thus, L-glutamate, quisqualate and kainate induce rapid depolarization responses in more than 95 per cent of dorsal horn neurones in culture. In contrast, only a minority of cells in culture, approximately 15–20 per cent, show a similar response to ATP. Clearly, therefore, there are many neurones sensitive only to amino acids and not to ATP. In terms of which presynaptic neurones might release neurotransmitters, immunocytochemical studies suggest that 30–40 per cent of the DRG neurones may contain neuropeptides. Taken in conjunction with the suggestion that over 90 per cent of these neurones may be releasing glutamate, these data imply a considerable degree of coexistence between glutamate and neuropeptides in these neurones. Approximately 10–14 per cent of dorsal horn neurones examined in culture respond to substance P with the effect being manifested most clearly in terms of changes in the threshold of the neurones to small depolarizing stimuli. Neuropeptides released from specific subtypes of

DRG neurones are likely, therefore, to modulate the excitability of sub-classes of dorsal horn neurones that receive synaptic input via fast sensory transmitters.

In his chapter, Dr R. F. Irvine drew attention to the diversity of potential intracellular second messengers derived from inositol phospholipids, and suggested that in addition to the short-term effects induced by diacyl glycerol and inositol trisphosphates, subtle long-term modifications of neuronal activity could be effected by changes in the relative proportions of these second messengers. Much attention was focussed on $Ins(1,4,5)P_3$ whose major function appears to be to mobilize cell calcium. In response to the question of whether the effects of $Ins(1,4,5)P_3$ could be mediated via calcium influxes as a result of a direct action on plasma membrane calcium permeability, Dr Irvine commented that this was very much an open question because although changes in calcium influx have been observed in response to $Ins(1,4,5)P_3$, the inositides themselves may modulate calcium ATPase activity. The recent discovery that some tissues, such as parotid or liver, are able to produce considerable amounts of a second inositol-trisphosphate, $Ins(1,3,4)P_3$, raises the possibility that it too might function as an intracellular second messenger. Although there is considerable variation both in the rate of production and in the relative amounts of $Ins(1,3,4)P_3$ produced as compared to $Ins(1,4,5)P_3$, little information is available concerning either its origins or function. Moreover, there is no evidence to support the existence of an isomerase enzyme able to interconvert these two inositol trisphosphates.

The question of whether a cell could respond selectively to either $Ins(1,4,5)P_3$ or diacyl glycerol formed as a result of inositide breakdown was also raised in discussion. Dr Irvine suggested two possibilities: firstly, in some cells $Ins(1,3,4)P_3$ could conceivably be produced in preference to $Ins(1,4,5)P_3$; alternatively, in some cells phosphatidylinositol may be hydrolysed directly yielding diacyl glycerol and inactive glyceryl monophosphate.

Regulation of neuronal function by coexisting messengers was the general theme of the chapter by Dr T. Hökfelt and co-workers. Consider-able interest was generated by his suggestion that neurones may have the capability of differentially releasing transmitters. Dr Hökfelt was complimented on the finding that neuropeptide Y was solely located in the large vesicle fraction from the sympathetic nerves to the vas deferens. The view was expressed that these large vesicles originate in the cell body where protein and peptide synthesis occur, whereas the small vesicles originate locally. Attention was drawn to the finding by Dr Walaas that synapsin I was only found in association with the small vesicles. Thus, it was suggested that control of the phosphorylation of synapsin I, perhaps by presynaptic modulators, could selectively operate release from the small vesicles, but not the large ones. This interesting possibility is currently under investigation.

The possible role of frequency control in the differential release of small vesicles compared with large vesicles was discussed. It was felt that long-term regulation of both fast transmitters and slow transmitters by this mechanism would require complex integration. The release of a slow transmitter such as LHRH could occur just as readily as a fast transmitter, but the difference would occur as the frequency of stimulation was raised. Thus, increasing the frequency would selectively increase the release of slow transmitter, but would not affect the fast transmitter.

Part III
CHEMICALLY ADDRESSED NEURAL COMMUNICATIONS

16

Chemical information processing in the brain: prospect from retrospect

FRANCIS O. SCHMITT

According to the Kuhnian view of the nature of scientific progress (Kuhn 1970), most scientific research is what he terms 'normal sciences'; it is essentially puzzle-solving. Frequently, through the application of new techniques, new knowledge is developed in the framework of already existing concepts and theories. In contrast, 'revolutionary science' is the invention or discovery of a *new* concept, a paradigm shift that leads to new ways of understanding basic mechanisms.

Judged by these criteria, the discovery that synaptic transmission is mediated by chemical, rather than by electrical, action as was previously supposed, was indeed revolutionary science. Loewi (1921) showed that impulses in the vagus nerve are transmitted at synaptic junctions not bioelectrically, but biochemically; the vagal control of the heartbeat was shown to be mediated by acetylcholine. The details and generality of this discovery are described by Loewi (1936). This discovery and the discovery of the comparable role of noradrenaline as the chemical mediator of sympathetic fibers are well described by Dale (1936) in which he also referred to the important work of Cannon and his colleagues. These two descriptions were contained in the lectures given by Loewi and by Dale when they shared the award of a Nobel Prize in 1936.

Like most revolutionary discoveries, this idea was viewed with much skepticism or outright opposition by leading neurophysiologists at the time. However, it was soon found that other substances besides acetylcholine and noradrenaline may act as mediators of post-synaptic or post-junctional excitation or inhibition. In the rapid expansion of transmitterology that followed, many substances were reported in the literature to act like transmitters, but many of these claims were not supported by unequivocal evidence. To avoid resulting confusion, six criteria were agreed upon, all of which had to be met before a candidate, or 'putative', transmitter could be accepted as a true neurotransmitter. Less than a dozen such classical transmitters have been discovered over the years.

The subsequent explosive development that began roughly in 1960 resulted in the formation of *neuroscience* as an emergent field and of a new community of scientists. During this period, neurophysiological concepts had become widely based on the chemical role of classical neurotransmitters. By application of immunohistofluorescence techniques the classical neurotransmitter characteristic of a given neuron could be determined; thus, large circuits of neurons could be chemically labelled or coded.

A working hypothesis that I proposed rather tentatively in 1980*, and more explicitly later (Schmitt 1985) was based on two closely interrelated ideas.

(1) Neurons may chemically intercommunicate by the mediation not only of the dozen-odd classical neurotransmitters but also by peptides, hormones, 'factors', other specific proteins, and by many other kinds of *informational substances* (ISs), a term that seemed more generally applicable than 'neuroactive substances' that was previously used (Schmitt 1979, 1982; Chan-Palay and Palay 1984).

(2) Alongside of, and in parallel with, synaptically linked, 'hardwired' neuronal circuitry, that forms the basis for conventional neurophysiology and neuroanatomy, and that operates through conventional synaptic junctions, there is a system that I call 'parasynaptic'. In parasynaptic neuronal systems ISs may be released at points, frequently relatively remote from target cells, which they reach by diffusion through the extracellular fluid. Such a system has all the specificity and selectivity characteristic of the conventional synaptic mode; in the parasynaptic case the receptors that provide the specificity and selectivity are on the surface of the cells where they can be contacted by, and bind to, the IS ligands diffusing in the extracellular fluid.

Experimental results described during the early 1980s seem in good agreement with this working hypothesis. A few examples may be mentioned.

The concept that the chemical coding of neuronal circuits based alone on the classical neurotransmitter they characteristically contain, and the notion of the chemical and physiological primacy of neurotransmitters, had to be revised because of the now widely confirmed discovery (Chan-Palay *et al.* 1978; Hökfelt *et al.* 1980; see also Chan-Palay and Palay 1984) that, in the neurons of particular brain tracts, certain peptides may coexist with

---

*Presented in February, 1980, in a series of lectures at the University of Texas, Medical Branch, Galveston, during the tenure of a Cecil and Ida Green Scholar appointment.

the classical neurotransmitter characteristic of that tract. In view of the many kinds of peptides (over 70 known at this writing) and the fact that many of these are thought to act as neurotransmitters, there is a substantial possibility that many combinations and permutations of peptides and classical neurotransmitters may coexist within individual neurons (Iversen 1984). Although there is yet little evidence for it, the possibility also exists that other types of ISs may also coexist within neurons.

The discovery in recent years of new peptides and of their significance in neurobiology is expanding very rapidly. The current interest in these phenomena even exceeds that shown by neuroscientists in the biogenic amines several decades ago. We must be prepared, therefore, to discover that such peptides may function in ways completely unknown at this time. Included in recent discoveries are new hormones, as well as proteinaceous substances called 'factors' that may be required for the functioning and survival of neurons.

Discoveries in this field were frequently made, not by scientific design, but by fortuitous, serendipitous, sometimes intuitive means. However, by the application of recombinant DNA technology, it is now possible, from cDNA clones, to determine nucleotide sequences, hence amino acid sequences, of proteins encoded by brain-specific mRNAs. Antisera to corresponding proteins are being used immunocytochemically to localize such newly discovered proteins in the brain (Milner and Sutcliffe 1983; Sutcliffe *et al*. 1983). By such direct chemical analytic methods it has, for example, been estimated that at least 30 000 distinct mRNA species, hence also proteins, are expressed in the mammalian brain. By contrast, it has, over the years, been possible to discover but several dozen brain-specific proteins by conventional neurochemical methods.

Because many ISs, such as hormones and factors, may occur in very low ($\sim$picomolar) concentrations, it may be impossible to detect them in bulk brain samples by conventional neurochemical methods. However, detection may now be readily made by application of DNA probe techniques.

It is thus apparent that we are emerging from the 'transmitter-centred' era of neurophysiology and neurochemistry, and are now embarking on an historic period of discovery of previously undreamed of varieties of ISs, a substantial fraction of which may be required to subserve brain function and its regulation. Interest would be particularly high in those that may play a role in higher functions of the human brain.

The greatly increased kinds and numbers of ISs, together with the parasynaptic mode by which many of these ISs are delivered to target neurons, as proposed in this working hypothesis, provide greatly heightened flexibility and plasticity that may be required for the development, normal functioning and survival of neurons.

Discovery, by the application of new techniques, of many kinds and numbers of neurobiologically important ISs would, in the Kuhnian scheme, probably be considered to be an expression of normal, not revolutionary, science.

However, the above working hypothesis suggests that, in addition to the above, such ISs might activate intracellular and intercellular biochemical reactions required for the control of vital processes of a kind that could be triggered *only* in the parasynaptic mode, *not* by neurotransmitters delivered in the conventional synaptic mode.

If this possibility should in the future be confirmed by substantial experimental evidence, then the proposed concept may eventually become a candidate for the Kuhnian appellation of 'revolutionary' science.

References

Chan-Palay, V., Jonsson, G., and Palay, S. L. (1978). Serotonin and substance P coexist in neurons of the rat central nervous system. *Proc. Nat. Acad. Sci. USA*, **75**, 1582–6.

——, and Palay, S. L. (eds) (1984). *Coexistence of Neuroactive Substances in Neurons*. John Wiley and Sons, New York.

Dale, H. H. (1936). Some recent extensions of the chemical transmission of the effects of nerve impulses. In *Nobel Lectures, Physiology or Medicine*, pp. 402–13. Elsevier, Amsterdam. (Published for the Nobel Foundation 1965.)

Hökfelt, T., Lundberg, J. M., Schultzberg, M., Johansson, O., Ljungdahl, A., and Rehfeld, J. (1980). Co-existence of peptides and putative transmitters in neurons. In *Neural Peptides and Neuronal Communication* (eds F. Costa and M. Trabucci), pp. 7–23. Raven Press, New York.

Iversen, L. L. (1984). Amino acids and peptides: fast and slow chemical signals in the nervous system? The Ferrier Lecture. *Proc. Roy. Soc. Lond.* **B-221**, 245–60.

Kuhn, T. S. (1970). *The Structure of Scientific Revolutions* (2nd edn). University of Chicago Press, Chicago, Ill.

Loewi, O. (1921). Über humorale Übertragbarkeit der Nernerven wirkung. *Pflüg. Arch. Gesamte Physiol.* **189**, 239–42.

—— (1936). The chemical transmission of nerve action. In *Nobel Lectures, Physiology or Medicine*, pp. 416–29. Elsevier, Amsterdam. (Published for the Nobel Foundation 1965.)

Milner, R. J., and Sutcliffe, G. (1983). Gene expression in rat brain. *Nucleic Acid Res.* **11**, 5497–520.

Schmitt, F. O. (1979). The role of structural, electrical, and chemical circuitry in brain function. In *The Neurosciences: Fourth Study Program* (eds F. O. Schmitt and F. G. Worden), pp. 5–20. MIT Press, Cambridge, Mass.

—— (1982). A protocol for molecular genetic neuroscience. In *Molecular Genetic Neuroscience* (eds F. O. Schmitt, S. J. Bird, and F. E. Bloom), pp. 1–9. Raven Press, New York.

—— (1985). Molecular regulators of brain function: a new view. *Neurosci.* **13**, 991–1001.

Sutcliffe, J. G., Milner, R. J., Shinnick, T. M., and Bloom, F. E. (1983). Identifying the protein products of brain-specific genes using antibodies to chemically synthesize peptides. *Cell*, **33**, 671–82.

17

Serotonin, octopamine, and proctolin: two amines and a peptide, and aspects of lobster behaviour

EDWARD A. KRAVITZ

Amine-containing neurons in the vertebrate central nervous system are an enigma. Clustered mostly in brainstem nucleii in very small numbers, these cells ramify widely throughout the cerebral and cerebellar cortices, the basal ganglia and brainstem, and the spinal cord, often completely ignoring the highly ordered topography and geography of the areas traversed by their processes (Grzanna and Molliver 1980; Moore 1981; Moore and Bloom 1978, 1979). Within target areas of the varicose processes of the cells, nerve terminals are seen either with or without synaptic specializations in apposition to processes of the resident neurons (controversy exists in the literature on this point and it remains possible that species and regional differences in the appearance of terminals may exist cf. Beaudet and Descarries 1978; Pickel *et al*. 1981). Stimulation of amine-containing neurons or application of amines while recording from suspect target cells also produces varying results. Target cells may show enhanced, diminished, or unchanged electrical activity in response to amines and more than likely all categories of response are correct and appropriate (Aghajanian 1981; Kitai 1981; Siggins 1978). Despite this apparent randomness, it is also clear from pharmacological and behavioural studies that amines play important and essential roles in a wide variety of physiological processes. For example, amines and amine-related drugs influence mood and affect, sleep and the state of arousal, learning, and memory, and less subtle but no less important processes like the control of movement (Hunter *et al*. 1977; Iversen 1977; Jouvet 1977; Kelly 1977). A classic clinical example linking an amine deficit to a disease is found in Parkinson's disease where a significant dopamine-neuron loss in a nigro-striatal pathway is both a diagnostic and probable causative factor (Hornykiewicz 1973). How can one resolve the obviously important role of amine-containing neurons in behaviour that must be precisely controlled, with the very small numbers of amine-containing neurons in the brain and with their enormous, apparently random terminal fields of innervation?

Some clues to the resolution of this problem may be forthcoming from studies with invertebrate animals where amines seem to be concerned with portions of the behavioural repertoire of the animals and where unique identified amine-containing neurons that may play important roles in the behaviours are amenable to experimental analysis. An elegant example that illustrates this experimental approach comes from studies in the leech nervous system where serotonin figures prominently in swimming behaviour (Kristan and Nusbaum 1982–3; Kristan and Weeks 1983). The role of serotonin in swimming behaviour in this animal may be linked to further roles for serotonin in arousal, motivation, and biting that appear to be a part of the feeding behaviour of leeches (Lent and Dickenson 1984). Amines also are important in learning in invertebrates. This is documented in *Aplysia* where the cellular processes involved have been well worked out (Kandel and Schwartz 1982), and in *Drosophila* where genetic and biochemical analyses have been combined with behavioural studies to show clearly the important roles of amines and amine action on cyclic nucleotide systems in learning (Aceves-Pina *et al.* 1983; Byers *et al.* 1981; Tempel *et al.* 1984). Also in *Drosophila* mutations show that the timing of cyclical events can be altered by single gene defects, and that here again changes of amine levels are linked to these alterations (Livingstone and Tempel 1983). A further example comes from our studies on the role of amines in postural regulation in lobsters.

Posturing is an important part of the behavioural repertoire of animals. Lobsters assume an elevated 'aggressive'-looking stance, in which they rise high on their walking legs with their claws open in front of them and their abdomens loosely folded beneath them, at the start of agonistic encounters, when a 'winner' emerges from the encounter, when startled, and, in males, during a part of the mating cycle (Atema and Cobb 1980; Scrivener 1971). They assume a lowered 'submissive'-looking posture, with their walking legs and claws pointed forward and their abdomens gently arching upwards, during agonistic encounters (the 'loser'), during part of the mating cycle (female only), and in a playing-dead response seen best in juvenile lobsters. Similar 'aggressive'- and 'submissive'-looking postures can be induced by amines. The injection of serotonin into freely moving lobsters triggers a sustained 'aggressive'-looking pose while the injection of octopamine leads to the 'submissive'-looking stance (Livingstone *et al.* 1980). This report describes our studies over the past six years examining the cellular basis of these responses (for a more detailed review of amines in the lobster nervous system, see Kravitz *et al.* 1985). During this time it has become clear that the pentapeptide proctolin is associated with amines and amine neurons, and that like the amines, proctolin generates long-lasting physiological changes in target tissues (Kravitz *et al*. 1985; Schwarz *et al*. 1980; Siwicki and Bishop 1985).

The stance triggered by serotonin injection results from contraction of most or all of the postural flexor muscles in the body, while the octopamine-posture is governed by contraction of the opposing extensor muscles. Amine injection triggers these patterns of contraction by actions both at peripheral neuromuscular junctions and in the central nervous system (the ventral nerve cord). In the periphery the amines prime exoskeletal muscles to respond more vigorously. While differences exist in the details of the actions of the two amines (see below), there is no opposing action of serotonin and octopamine on flexor and extensor muscles. The opposition comes from actions on the ventral nerve cord where the two amines trigger the readout of opposing motor programmes.

ACTIONS OF AMINES AND THE PEPTIDE PROCTOLIN ON EXOSKELETAL MUSCLES

Lobster exoskeletal muscles are innervated by excitatory axons using glutamate as a transmitter compound and inhibitory axons using GABA. Muscles contain no amines. Therefore, serotonin and octopamine reach muscles via the general circulation and low concentrations of amines have been found in lobster haemolymph (Livingstone *et al*. 1980). Most postural muscles also contain no measurable proctolin. If proctolin exerts a physiological control role in these cases, it again would be through proctolin circulating in the haemolymph (Schwarz *et al*. 1984). In some cases, however, a direct proctolinergic innervation of skeletal muscle has been found (Siwicki and Bishop 1985).

In the neuromuscular preparation that has been best studied, the opener muscle of the dactyl of the walking leg, important long-lasting actions of both amines and of proctolin have been observed (Breen and Atwood 1983; Fisher and Florey 1983; Glusman and Kravitz 1982; Kravitz *et al*. 1985). Serotonin is the most potent, with actions seen on all three parts of the neuromuscular apparatus (Kravitz *et al*. 1985; Grundfest and Reuben 1961). Serotonin enhances transmitter release from both excitatory and inhibitory nerve terminals, induces a contracture and enhances the contractility of exoskeletal muscle fibres (Fig. 17.1). At excitatory junctions the actions of serotonin are complex, involving two components that are distinguishable by their time-courses (Glusman and Kravitz 1982): a rapid component ($T_{1/2} = 1-2$ min) and a much slower one ($T_{1/2} = 30$ min or more). In these same preparations octopamine and proctolin have much weaker (octopamine) or no effects (proctolin) on nerve terminals. These two substance, however, do produce actions similar to serotonin on muscle fibres (contracture, enhanced contractility). The opposing muscle (the

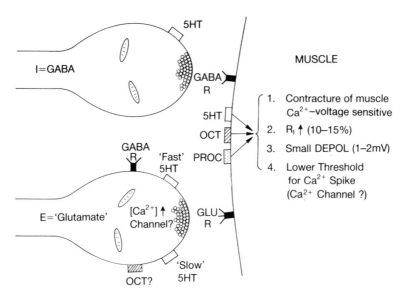

FIG. 17.1. Summary diagram of sites of action and effects of serotonin (5-HT), octopamine (Oct) and proctolin (PROC) on the lobster opener muscle neuromuscular preparation. (Reprinted in a slightly modified form from Kravitz *et al.* 1980, with permission.)

closer of the dactyl) shows a similar pattern of amine and peptide actions. In the several flexor and extensor muscle pairs we examined the generalization held that whatever effects serotonin, octopamine, and proctolin exert on one member of the pair, they also exert on the other (Harris-Warrick and Kravitz 1984). There is no opposition in the actions of amines on neuromuscular preparations. Rather they produce a priming action in which the tissues respond more vigorously to nerve stimulation for a sustained period of time ($T_{1/2} = 30$ min or more).

Detailed physiological studies show that the slower component in the enhancement of release caused by serotonin, results in part either from a change in the Ca^{2+} buffering machinery of nerve terminals or from a change in the Ca^{2+} sensitivity of the release process (Glusman and Kravitz 1982). Biochemical studies show that this serotonin-induced pre-synaptic change is linked to increases in cyclic AMP levels (M. F. Goy and E. A. Kravitz, unpublished observations). On the post-synaptic (muscle) side, physiological studies show that, in addition to generating a contracture, the amines and proctolin trigger the appearance of Ca^{2+}-action potentials. A preliminary voltage clamp analysis of the action of serotonin shows an enhanced Ca^{2+} current that may account for the appearance of the action potentials (Kravitz *et al.* 1980). The metabolic changes underlying the long-lasting post-synaptic physiological changes are not known, but probably do not involve alterations in cyclic nucleotide metabolism (Kravitz *et al.* 1985).

ACTIONS OF AMINES ON THE VENTRAL NERVE CORD

To explore the actions of amines on neurons of the ventral nerve cord, preparations were dissected that included one or more ganglia of the ventral nerve cord, the nerve roots going to postural flexor or extensor muscles (or both) and the muscles innervated by the nerve roots (see inset diagram Fig. 17.2). Actions of amines on the nerve cord were monitored by intracellular microelectrode recordings from muscle fibres or from

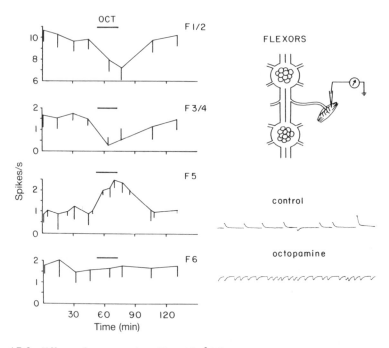

FIG. 17.2. Effect of octopamine (3×10^{-5} M) on motoneurons innervating the superficial flexor muscles. The inset diagram illustrates the experimental arrangement. On the lower right a sample intracellular recording is shown. The left side of the figure shows the average action potential frequency (one S.D. shown) in the five excitatory (F1, 2, 3, 4, 6) and one inhibitory (F5) motoneurons innervating the muscle. (Reprinted from Kravitz *et al.* 1983, with permission.)

	SLOW FLEXORS		SLOW EXTENSORS	
	E	I	E	I
SEROTONIN	↑	↓	↓	↑
OCTOPAMINE	↓	↑	↑	↓

FIG. 17.3. Actions of serotonin and octopamine on the firing of excitatory and inhibitory motoneurons innervating postural flexor and extensor muscles. (Reprinted from Kravitz *et al.* 1983, with permission.)

identified neurons in the central ganglia or by extracellular recordings from nerve trunks (Harris-Warrick and Kravitz 1984; Livingstone *et al.* 1980). The types of results seen are illustrated in Fig. 17.2 where the actions of octopamine on a nerve cord-flexor muscle preparation are shown. In the control situation, spontaneous excitatory and inhibitory synaptic potentials are recorded from muscle fibres (lower right side of Fig. 17.2). On octopamine application to the ganglion, however, dramatic reductions in the firing of the excitatory neurons innervating these muscles are seen, along with a significant enhancement in the firing of the inhibitory neuron. The time course and magnitude of these changes are shown on the left side of the Fig. 17.2. When the results of a large number of experiments are pooled it becomes clear that octopamine and serotonin trigger the readout of opposing motor programmes from the ventral nerve cord (Fig. 17.3). Octopamine application, which in intact animals leads to an extended posture, decreases the firing of excitors to the flexors and inhibitors to the extensors, and simultaneously increases the firing of excitors to extensors and inhibitors to flexors. Serotonin application, which leads to the flexed posture, causes an essentially opposite result: excitors to flexors and inhibitors to extensors increase in firing, excitors to extensors and inhibitors to flexors decrease in firing. It has long been known in invertebrate preparations that firing of single neurons can trigger the readout of complex motor patterns. Wiersma and Kennedy and their colleagues (Wiersma and Ikeda 1964; Evoy and Kennedy 1967) introduced the concept of command neurons to describe the single axons whose activation would produce these effects. We suspect, and are currently investigating the possibility, that serotonin and octopamine interact with the command neuron circuitry in such a way as to read out opposing flexor (serotonin) and extensor (octopamine) motor programmes.

Proctolin also is capable of turning on motor programmes, but swimmerets (Bradbury and Mulloney 1982) and the stomatogastric system (A. Selverston, unpublished observations) are the systems activated and not exoskeletal muscles.

CELLULAR LOCALIZATION OF SEROTONIN, OCTOPAMINE AND PROCTOLIN

In the studies described thus far, the three hormonal substances were either injected into intact animals or bath-applied to test preparations. We turned next to trying to find amine- and proctolin-containing neurons that might serve important roles in the postural control processes under investigation. We began by examining the general distribution of serotonin, octopamine and proctolin in the lobster nervous system, and then, using immunocytochemical procedures, attempted to determine the cellular localization of the three substances.

The distribution studies showed that low levels of amines and of proctolin were found throughout the lobster ventral nerve cord (Evans *et al.* 1976b; Livingstone *et al.* 1981; Schwarz *et al.* 1984). Much higher concentrations were found, however, at two locations along peripheral nerve trunks in the thoracic region of the ventral nerve cord (see diagram, Fig. 17.4). The two locations were: (1) close to a cluster of neurosecretory cells near a bifurcation of the roots into medial and lateral divisions; and (2) near the distal ends of the roots in a well known crustacean neurohemal organ, the pericardial organ (Evans *et al.* 1976a; Livingstone *et al.* 1981; Schwarz *et al.* 1984; Sullivan 1979; Sullivan *et al.* 1977). At both of these sites depolarization-induced release of amines and proctolin can be demonstrated. In addition, the synthesis of serotonin and octopamine from

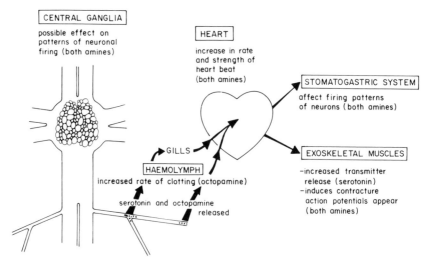

FIG. 17.4. Peripheral and central targets of amines. (Reprinted from Kravitz *et al.* 1980, with permission.)

precursor compounds can be shown and the nerve terminals involved in synthesis have been identified (Livingstone *et al.* 1981). Materials released from both of the peripheral release sites are carried in the haemolymph directly into the heart (via the gills from the more proximal site), and from the heart are distributed throughout the body. To our knowledge the release sites along second thoracic roots are the origin of the circulating amines.

In searching for amine neurons that might be important in the regulation of posture we anticipated that individual cells, or possibly, groups of cells, might have endings within central ganglia where influences on motor programmes could be felt, and also might have endings within the peripheral neurosecretory regions of thoracic second roots where amines could be released for actions on exoskeletal muscles and other peripheral targets (see Fig. 17.4). Using immunocytochemical methods and antibodies to serotonin, octopamine, and proctolin we began the search for such neurons. The serotonin results (Beltz and Kravitz 1983) will be presented here, but maps of the locations of proctolin-staining neurons recently have been completed and octopamine studies are in progress (Siwicki and Bishop 1985; B. Trimmer, unpublished observations). The peripheral root regions containing high levels of amines show a plexus of fibres and varicosities that stain intensely for serotonin whose origins can be traced to central ganglia (Fig. 17.5A). In the central ganglia a highly reproducible pattern of staining of individual cell bodies, neuropil regions and axonal tracts are seen (Fig. 17.5B and 17.6). From multiple sets of serotonin-stained ganglia, maps of the locations of cells and processes were constructed (Fig. 17.6 is a portion of such a map). Moreover, a particularly intriguing stained neuronal process was seen on either side of most of the thoracic ganglia (T1–T5; Fig. 17.5C). This branch had its origin in a medial serotonin-staining fibre bundle and after leaving the central trunk, divided into one branch that formed a set of varicosities within the neuropil region of the central ganglion and a second branch that travelled out the second root of that ganglion to form part of the peripheral plexus of neurosecretory endings. Since we were searching for cells that might supply endings to both central neuropil regions and peripheral neurosecretory complexes, we felt it important to learn the cellular origins of these dividing branches. Two pairs of large cells, in the fifth thoracic (T5) and in the first abdominal (A1) ganglia, were selected as possible candidates for the origins of these processes (Fig. 17.6). The selection was based on morphological criteria: these cells contributed processes to the medial serotonin-staining bundle and shared a similar morphology as far as their serotonin-stained branches could be traced; in certain preparations, the branches of the A1 cells could be followed out from second thoracic roots of the T5 ganglion towards the peripheral neurosecretory plexus. Using a

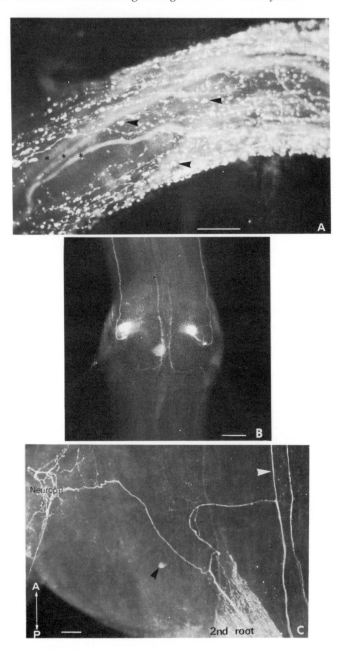

FIG. 17.5. Serotonin-like immunoreactivity in lobster central and peripheral tissues. (A) A photograph from a thoracic second root region. (B) Cell bodies staining for serotonin in the A1 abdominal ganglion. A large unpaired weakly

ANTERIOR

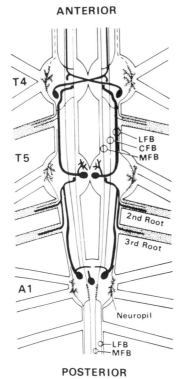

POSTERIOR

FIG. 17.6. Schematic diagram of the immunoreactive cell bodies, fibres, and neuropil of a part of the ventral nerve cord. This is a composite drawing of whole mount preparations of ten ventral nerve cords. Cell bodies are drawn as large, filled, round, or elongate circles. Heavy black lines represent immunoreactive fibres that have been traced to their cell bodies of origin. Fine lines indicate immunoreactive fibres that have not been connected with cell bodies. Each of the fine lines of the lateral fibre bundles (LFBs), central fibre bundles (CFBs), and midline fibre bundles (MFBs) represent several fibres. Dashed lines indicate fibres that have not been directly visualized in these immunohistochemical preparations, but which we believe exist because the patterns of staining are similar from ganglion to ganglion. Stippled regions represent fine processes and varicosities of neuropil and plexus regions. [Further details are available in Beltz and Kravitz (1983) from which this figure has been reprinted in a slightly modified form with permission.]

staining cell is also observed in this ganglion. (C) A view of a second thoracic ganglion showing a single fluorescent process that gives rise to arbors of central and peripheral varicosities. This process originates in one of the paired cells in the A1 ganglion (see Fig. 17.7). Experimental details are outlined in the text and fully described in Beltz and Kravitz 1983. Scale bar, 50 μm. (Reprinted with permission.)

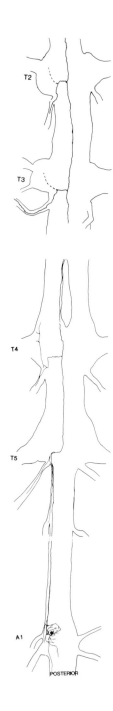

knowledge of the approximate location of the cell bodies of these cells and stimulating regions of the thoracic part of the ventral nerve cord into which the T5 and A1 cell axons had been traced, we attempted to identify these cells for physiological and anatomical studies. Confirmation of their identity depended upon the injection of Lucifer Yellow into suspected cells and, following fixation of the tissues, processing them for serotonin-immunostaining by a peroxidase technique. In all cases, the suspected dye-injected cells turned out to be serotonin-immunoreactive neurons (B. S. Beltz, unpublished observations). The identification was confirmed further recently by direct biochemical analyses of the serotonin contents of these cells by HPLC techniques (Siwicki *et al.* in preparation). To examine the morphological features of these cells, identified neurons were injected with horse radish peroxidase, tissues were incubated for several days, and then were fixed and processed for peroxidase staining. Such studies showed clearly that the A1 cells were the origin of the branch in question (Fig. 17.7). Like vertebrate aminergic neurons, these cells turn out to have enormous arbors of endings. Processes of these cells project in an anterior direction through all segments of the thoracic part of the nervous system as far as the suboesophageal ganglion, and give off an identical T-shaped branch in each ganglion. The T5 cells also contribute endings to the peripheral neurosecretory plexus, and in addition, have central sets of endings. Together, the T5 and A1 cells supply all the axons leaving the central nervous system to form the serotonergic portion of the thoracic root neurosecretory complex (B. S. Beltz, unpublished observations). Recently, using both immunocytochemical methods and direct chemical analysis, it has been found that the T5 and A1 serotonin-containing cells also contain proctolin (Kravitz *et al.* 1985; Siwicki *et al.* 1985). This raises the possibility that proctolin will be co-released with serotonin on activation of these cells.

THE T5 AND A1 SEROTONIN-CONTAINING NEURONS: THEIR POSSIBLE ROLE IN THE REGULATION OF POSTURE

In Fig. 17.8 a schematic drawing of a portion of one of the A1 cells is shown. The model, which is presently being tested, of how such a cell might

FIG. 17.7. Reconstruction of an A1 serotonin-immunoreactive cell (one of the paired cells). Horseradish peroxidase was injected into the cell body and allowed to diffuse for 2–3 days. The upper and lower parts of the figure (A1, T5, T4 ganglia; T3, T2 ganglia) are from two different injections, but the main features have been seen repeatedly in injections of A1 cells. Details are outlined in text. (Reprinted, with permission.)

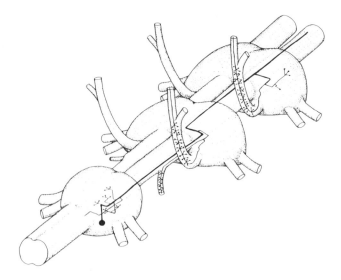

FIG. 17.8. Artists drawing of an A1 cell and a partial view of its endings in the A1, T5, and T4 ganglia of the ventral nerve cord. Note the T-shaped branch in the T4 ganglion. This is the branch illustrated in Fig. 17.5C and a comparable process from the same cell is found in the T3, T2, and T1 ganglia. (Reprinted, with permission.)

be involved in the regulation of posture is as follows. When a cell of this type is activated, serotonin (and proctolin?) is released from two separate sets of endings, one peripheral, the other central. Serotonin released into the haemolymph from the peripheral endings reaches exoskeletal muscles via the general circulation and primes the muscles to respond more vigorously. After exposure to serotonin, excitatory and inhibitory nerves to muscles are capable of releasing more transmitter, and muscle cells can contract more vigorously for a prolonged time period. Serotonin released from nerve endings in neuropil regions of the central nervous system provides the specificity for the flexed ('aggressive'?) posture. This results from either a controlling or a biasing action of the amine on a command circuitry directing the readout of a motor programme causing flexion. The firing patterns of many excitatory and inhibitory motoneurons are altered in a patterned way to generate the flexed posture. Excitatory and inhibitory motorneurons to opposing muscles increase their rates of firing. This may account for the initially surprising observation that the release of transmitter from both excitatory and inhibitory motor neuron terminals on exoskeletal muscle fibres was increased upon treatment with serotonin.

If the T5 and A1 serotonin-containing cells do function in this way, then they may be cells more concerned with a portion of the behavioural repertoire of these animals (the regulation of an 'aggressive'-looking stance) than with the point to point wiring that is the function of most of the neurons in the lobster nervous system. Such an idea is currently under active investigation.

Acknowledgements

This work was done in close collaboration with M. Livingstone, S. Glusman, B. Beltz, M. Goy, K. Siwicki, T. Schwarz, and R. Harris-Warrick, all of whom were students or post-doctoral fellows in the Kravitz laboratory. In addition, D. Cox, J. Gagliardi, E. Bodkin, M. Lee, M. LaFratta, and J. LaFratta all provided essential assistance. The work was supported by NIH grant NS 07488.

References

Aceves-Pina, E. O., Booker, R., Duerr, J. S., Livingstone, M. S., Quinn, W. G., Smith, R. F., Sziber, P. P., Tempel, B. L., and Tully, T. P. (1983). Learning and memory in *Drosophila*, studied with mutants. *Cold Spring Harbor Symp. Quant. Biol.* **XLVIII**, 831–9.

Aghajanian, G. K. (1981). The modulatory role of serotonin at multiple receptors in brain. In *Serotonin Neurotransmission and Behavior* (ed. B. L. Jacobs and A. Gelperin), pp. 156–85. MIT Press, Cambridge, MA U.S.A.

Atema, J. and Cobb, J. S. (1980). Social behavior. In *The Biology and Management of Lobsters* (ed. J. S. Cobb and B. F. Phillips) Vol. 1, pp. 409–50, Academic Press, N.Y., U.S.A.

Beaudet, O. J. and Descarries, L. (1978). The monoamine innervation of rat cerebral cortex—synaptic and non-synaptic axon terminals. *Neurosci.* **3**, 851–60.

Beltz, B. S. and Kravitz, E. A. (1983). Mapping of serotonin-like immunoreactivity in the lobster nervous system. *J. Neurosci.* **3**, 585–602.

Bradbury, A. G. and Mulloney, B. (1982). Proctolin activates and octopamine inhibits swimmeret beating. *Soc. Neurosci. Abs.* **8**, part 2, 736.

Breen, C. A. and Atwood, H. L. (1983). Octopamine—A neurohormone with presynaptic activity dependent effects at crayfish neuromuscular junctions. *Nature*, **303**, 716–8.

Byers, D., Davis, R. L., and Kiger, J. A., Jr (1981). Defect in cyclic AMP phosphodiesterase due to the dunce mutation of learning in *Drosophila melanogaster*. *Nature*, **289**, 79–81.

Evans, P. D., Kravitz, E. A., and Talamo, B. R. (1976a). Octopamine release at two points along lobster nerve trunks. *J. Physiol.* **262**, 71–89.

——, ——, ——, and Wallace, B. G. (1976b). The association of octopamine with specific neurons along lobster nerve trunks. *J. Physiol.* **262**, 51–70.

Evoy, W. H. and Kennedy, D. (1967). The central nervous organization underlying control of antagonistic muscles in the crayfish. I. Types of command fibers. *J. Exp. Zool.* **165**, 223–38.

Fischer, L. and Florey, E. (1983). Modulation of synaptic transmission and excitation-contraction coupling in the opener muscle of the crayfish, Astacvs leptodactylus, by 5-hydroxytryptamine and octopamine. *J. Exp. Biol.* **102**, 187–98.

Glusman, S. and Kravitz, E. A. (1982). The action of serotonin on excitatory nerve terminals in lobster nerve-muscle preparations. *J. Physiol.* **325**, 223–41.

Grundfest, H. and Reuben, J. P. (1961). Neuromuscular synaptic activity in lobster. In *Nervous Inhibition* (ed. E. Florey), pp. 92–104. Pergamon Press, Oxford.

Grzanna, R. and Molliver, M. E. (1980). The locus coeruleus in the rat: an immunohistochemical delineation. *Neurosci.* **5**, 21–40.

Harris-Warrick, R. M. and Kravitz, E. A. (1984). Cellular mechanisms for modulation of posture by octopamine and serotonin in the lobster. *J. Neurosci.* **4**, 1976–93.

Hornykiewicz, O. (1973). Dopamine in the basal ganglia. Its role and therapeutic implications (including the clinical use of L-DOPA). *Br. Med. Bull.* **29**, 172–8.

Hunter, B., Zornetzer, S. R., Jarvik, M. E., and McGaugh, J. L. (1977). Modulation of learning and memory: effects of drugs influencing neurotransmitters. In *Drugs, Neurotransmitters and Behavior* (eds L. L. Iversen, S. D. Iversen, and S. H. Snyder), Handbook of Psychopharmacology Vol. 8, pp. 531–77. Plenum Press, N.Y.

Iversen, S. D. (1977). Brain Dopamine systems and behavior. In *Drugs, Neurotransmitters and Behavior* (eds L. L. Iversen, S. D. Iversen, and S. H. Snyder), Handbook of Psychopharmacology Vol. 8, pp. 333–84. Plenum Press, N.Y.

Jouvet, M. (1977). Neuropharmacology of the sleep-waking cycle. In *Drugs, Neurotransmitters and Behavior* (eds L. L. Iversen, S. D. Iversen, and S. H. Snyder), Handbook of Psychopharmacology Vol. 8, pp. 233–93. Plenum Press, N.Y.

Kandel, E. R. and Schwartz, J. H. (1982). Molecular biology of learning: modulation of transmitter release. *Science*, **218**, 433–43.

Kelly, P. H. (1977). Drug induced motor behavior. In *Drugs, Neurotransmitters and Behavior* (eds. L. L. Iversen, S. D. Iversen, and S. H. Snyder), Handbook of Psychopharmacology Vol. 8, pp. 295–331. Plenum Press N.Y.

Kitai, S. T. (1981). Electrophysiology of the corpus striatum and brain stem integrating systems. In *Handbook of Physiology, Section 1, The Nervous System* (ed. V. B. Brooks), Vol. 2, part II, pp. 997–1015. Am. Physiol. Soc., Bethesda, MD U.S.A.

Kravitz, E. A., Beltz, B. S., Glusman, S., Goy, M., Harris-Warrick, R., Johnston, M., Livingstone, M., Schwarz, T., and Siwicki, K. K. (1985). The well-modulated lobster. The roles of serotonin, octopamine and proctolin in the lobster nervous system. In *Model Neural Networks and Behavior* (ed. A. I. Selverston), pp. 339–60. Plenum Publishers, N.Y.

——, Glusman, S., Harris-Warrick, R. M., Livingstone, M. S., Schwarz, T., and Goy, M. F. (1980). Amines and a peptide as neurohormones in lobsters: Actions on neuromuscular preparations and preliminary behavioral studies. *J. Exp. Biol.* **89**, 159–76.

Kristan, W. B. Jr and Nusbaum, M. P. (1982–3). The dual role of serotonin in leech swimming. *J. Physiol. Paris*, **78**, 743–7.

—— and Weeks, J. C. (1983). Neurons controlling the initiation, generation and modulation of leech swimming. *Soc. Exp. Biol. Symp.* **XXXVII**, 243–60.

Lent, C. M. and Dickenson, M. H. (1984). Serotonin integrates the feeding behavior of the medicinal leech. *J. Comp. Physiol. A*, **154**, 457–71.

Livingstone, M. S., Harris-Warrick, R. M., and Kravitz, E. A. (1980). Serotonin and octopamine produce opposite postures in lobsters. *Science*, **208**, 76–9.

——, Schaeffer, S. F., and Kravitz, E. A. (1981). Biochemistry and ultrastructure of serotonergic nerve endings in the lobster: serotonin and octopamine are contained in different nerve endings. *J. Neurobiol.* **12**, 27–54.

—— and Tempel, B. L. (1983). Genetic dissection of monoamine neurotransmitter synthesis in *Drosophila. Nature*, **303**, 67–70.

Moore, R. Y. (1981). The anatomy of central serotonin neuron systems in the rat brain. In *Serotonin Neurotransmission and Behavior* (eds B. L. Jacobs and A. Gelperin), pp. 35–71. MIT Press, Cambridge, MA, U.S.A.

—— and Bloom, F. E. (1978). Central catecholamine neuron systems: anatomy and physiology of the dopamine systems. *Ann. Rev. Neurosci.* **1**, 129–69.

—— and —— (1979). Central catecholamine neuron systems: anatomy and physiology of the norepinephrine and epinephrine systems. *Ann. Rev. Neurosci.* **2**, 113–68.

Pickel, V. M., Beckley, S. C., Joh, T. H., and Reis, D. J. (1981). Ultrastructural immunocytochemical localization of tyrosine hydroxylase in the neostriatum. *Brain Res.* **225**, 373–85.

Schwarz, T. L., Harris-Warrick, R. M., Glusman, S., and Kravitz, E. A. (1980). A peptide action in a lobster neuromuscular preparation. *J. Neurobiol.* **11**, 623–8.

——, Lee, G. M.-H., Siwicki, K. K., Standaert, D. G., and Kravitz, E. A. (1984). Proctolin in the lobster: the distribution, release and chemical characterization of a likely neurohormone. *J. Neurosci.* **4**, 1300–11.

Scrivener, J. C. E. (1971). Agonistic behavior of the American lobster, Homarus americanus (Milne-Edwards). *Fish. Res. Board Can. Tech. Rep.* **235**, 1–128.

Siggins, G. R. (1978). Electrophysiological role of dopamine in striatum: excitatory or inhibitory? In *Psychopharmacology: A Generation of Progress* (eds M. A. Lipton, A. DiMascio, and K. F. Killam), pp. 143–57. Raven Press, N.Y., U.S.A.

Siwicki, K. K., Beltz, B. S., Schwarz, T. L., and Kravitz, E. A. (1985). Proctolin in the lobster nervous system. *Peptides* (in press).

—— and Bishop, C. A. (1985). Mapping of proctolin-like immunoreactivity in the nervous systems of lobster and crayfish. *J. Comp. Anat.* (in press).

Sullivan, R. E. (1979). A proctolin-like peptide in crab pericardial organs. *J. Exp. Zool.* **210**, 543–52.

——, Friend, B. J., and Barker, D. L. (1977). Structure and function of spiny lobster ligamental nerve plexuses: evidence for synthesis, storage and secretion of biogenic amines. *J. Neurobiol.* **8**, 581–605.

Tempel, B. L., Livingstone, M. S., and Quinn, W. G. (1984). Mutations in the dopa decarboxylase gene affect learning in *Drosophila. Proc. Nat. Acad. Sci. USA*, **81**, 3577–81.

Wiersma, C. A. G., and Ikeda, K. (1964). Interneurons commanding swimmeret movements in the crayfish, *Procambarus clarki* (Girard). *Comp. Biochem. Physiol.* **12**, 509–25.

18

Aplysia neurosecretory cells: multiple populations of dense core vesicles

WAYNE SOSSIN, THANE KREINER, AND RICHARD H. SCHELLER

INTRODUCTION

Elucidating the roles of peptide messengers (neuropeptides) in the nervous system is an expanding branch of neuroscience. The ability of neuropeptides to modulate the actions of neurons by both synaptic and non-synaptic mechanisms (Krieger 1983) increases the complexity of unraveling nervous system interactions.

The gastropod mollusc *Aplysia californica* offers a number of advantages as a model system for studying neuropeptides. Our laboratory is concerned with the roles neuropeptides play in governing behaviour in *Aplysia*; it is our hope to follow the flow of information from neuropeptide genes to observable and quantifiable behaviour as patterns of the organism. An advantage of the *Aplysia* system for this task is the vast amount of accumulated information which defines neuronal circuits and the functions of these circuits in governing the behaviour of the animal (Kandel 1976, 1979).

The other main advantage of studying neuropeptides in *Aplysia* is the presence of large identifiable peptidergic neurons (Frazier *et al.* 1967). Many neurons devote as much as 50 per cent of their transcription and translation to the synthesis of neuropeptide precursors (Loh *et al.* 1977). Taking advantage of this situation, our laboratory has used differential screening procedures to isolate recombinant cDNA and genomic clones encoding *Aplysia* neuropeptide precursors. This technique involves first constructing a genomic library from sperm DNA or a cDNA library from poly A$^+$ RNA of the nervous system. Then mRNA is isolated from at least two cells or tissues, one where the particular neuropeptide of interest is expressed and another where it is not. A radioactive probe is made from the different RNAs and used to screen the library under conditions where only major products of the cells will be seen. Clones are selected which hybridize to one probe but not another. Using this strategy we have isolated clones for seven *Aplysia* neuropeptide precursors (Fig. 18.1). These precursors encode the egg laying hormone, ELH (Scheller *et al.* 1982), atrial peptides A and B (Scheller *et al.* 1983), the R14 peptide (Nambu *et al.* 1983), the L11 peptide (Taussig *et al.* 1984), the small cardioactive peptides, SCP_A and SCP_B, (Mahon *et al.* 1985), and the tetrapeptide Phe-Met-Arg-Phe-NH$_2$ (FMRFamide); (Schafer *et al.* 1985).

APLYSIA NEUROPEPTIDE PRECURSORS

All biologically active peptides are synthesized as parts of larger precursor molecules (Herbert and Uhler 1982). The active peptides are liberated from the precursors by endopeptidases which cleave at basic residues, a process which is conserved in a variety of species ranging from yeast to man. Other modifications of the amino acid chains include glycosylation, phosphorylation, acetylation, sulphation, and amidation (Loh *et al.* 1984). Neuropeptide precursors often give rise to multiple biologically active molecules, which has led to the term polyprotein. A major goal of peptide research is to understand the functional significance of the polyprotein precursor organization. The *Aplysia* neuropeptide precursors we have characterized illustrate the general features of polyproteins and will be discussed below.

The FMRFamide precursor

The most dramatic example of a polyprotein is illustrated by the FMRFamide precursor (Fig. 18.1a). The gene is present in a single copy per haploid genome and encodes at least 20 copies of the amidated tetrapeptide. The DNA sequence encoding the tetrapeptide varies; however, each peptide unit has an identical amino acid sequence, Phe-Met-Arg-Phe-NH$_2$. The spacer between the peptides is quite variable, having a generally acidic character which results in an approximately neutrally charged precursor. Preliminary characterization of several independent cDNA clones suggests that different mRNAs arise from alternate splicing of a single primary transcript (Schaefer *et al.* 1985).

The FMRFamide gene is expressed in about 100 cells throughout the *Aplysia* CNS including the giant cholinergic cell R2. R2 sends processes to the body wall where they impinge upon mucus secreting cells; however, the role of the peptide is not yet known.

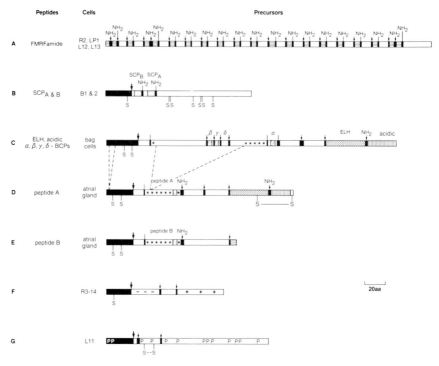

FIG. 18.1. Schematic representation of *Aplysia* neuropeptide precursors. The large arrows indicate positions of signal sequence cleavage, the smaller arrows internal proteolytic cleavages at dibasic residues, and the lines internal proteolytic cleavages at single basic residues. An NH$_2$ above the arrow or line indicates carboxyterminal amidation. Cysteine residues are indicated by an S written below the precursor and many of the known peptides are labelled. Homologous regions are indicated by equivalent symbols. In the R3–14 precursor shown in F, the asymmetric charge distribution is indicated by − and + symbols. In the L11 precursor, shown in G, the large number of proline residues are indicated by the symbol P.

The small cardioactive peptides A and B

The small cardioactive peptides A and B (SCP$_{A\&B}$) are 9 and 11 amino acids, respectively. These peptides share the same carboxyterminal seven amino acids and are both amidated. SCP$_{A\&B}$ are found in tandem on a single precursor directly after the signal sequence; however, unlike FMRFamide, the peptides are not identical (Fig. 18.1b; Mahon *et al.* 1985). The remainder of the precursor does not encode any known peptides and the six cysteine residues in this portion of the molecule suggest a highly cross-linked structure.

While the SCPs were originally isolated on the basis of their cardioexcitatory activities (Lloyd 1982), 90 per cent of the SCP immunoreactive material is localized to the buccal ganglion. Within the ganglion, the peptides are present in neurons involved in the mechanical aspects of feeding (Lloyd *et al.* 1984, 1985). The SCPs have modulatory actions on buccal muscle and oesophagus increasing the force of acetylcholine-induced contractions. These peptides also parallel the actions of serotonin in facilitating transmitter release from sensory neurons in the gill withdrawal reflex (Abrams *et al.* 1984).

The egg laying hormone gene family

Oviposition and the associated fixed action pattern are governed by a family of five genes expressed in the central nervous system and the atrial gland. These genes are 90 per cent homologous and give rise to a battery of peptides illustrated in Fig. 18.1c–e. The ELH precursor is expressed in the bag cell neurons which fire a long train of action potentials during each egg laying episode (Fig. 18.1c). The precursor encodes many potentially bioactive molecules including ELH and the α, β, γ bag cell peptides (BCPs). Again, as seen in the other precursors, the multiple peptides are related; however, in this case the homology is very limited and as a result some of the peptides have evolved independent targets. The molecular genetic studies suggest that the gene family has evolved in two general steps: first, duplication of a primordial peptide sequence to generate a polyprotein followed by duplication of large regions of the chromosome resulting in a set of five genes (Scheller *et al.* 1983).

During the bag cell discharge, ELH and other peptides contained on the precursor are released into the connective tissue sheath. This allows the molecules to interact with abdominal ganglion neurons; by diffusing into the haemolymph the peptides are also carried to distant targets. The combined actions of the peptide messengers on central neurons and the ovotestis elicit the egg laying behaviour.

The related genes encoding peptides A and B (Fig. 18.1d,e) are expressed in an exocrine gland situated at the distal portion of the hermaphroditic duct at the opening of the gonopore. These peptides most likely act as pheromones attracting animals to previously deposited egg masses; further behavioural studies are necessary to confirm these ideas (Susswein 1985).

The R14 and L11 precursors

Unlike the neuropeptides discussed above, where the peptide was identified before the gene, in R14 and L11 the precursor has been characterized before the peptide (Fig. 18.1f,g). These precursors have no internal homologies and given the large introns have probably evolved via recombination events which bring together previously unrelated sequences. Since the precise cues for precursor cleavage are not well defined, the first step in elucidating the roles for these peptides is to identify the cleavage products. By microinjecting labelled amino acids into the identified cells followed by isolation of the peptide products and microsequencing, we are able to determine the positions of processing sites. Having defined these sites, the set of potential biologically active molecules is defined for physiological analysis.

We estimate there may be as many as 50 precursors to biologically active peptides in *Aplysia*. The genes encoding these precursors may represent up to 0.5–1 per cent of the *Aplysia* haploid genome and the peptide products are likely to encode a substantial fraction of the central nervous systems specificity. There are many questions to be answered along the way to an understanding of the function of these peptides. Our previous work, described above, has identified a number of putative neuropeptides which are active in *Aplysia*. The task ahead is to follow the flow of information encoded in these peptides in an attempt to understand their roles in the *Aplysia* nervous system.

PEPTIDE PACKAGING

A basic issue in need of further investigation is the cellular pathway of the neuropeptide from translation to secretion. The precursors must be routed through the secretory pathway where many of the postranslational modifications take place. In addition, the neuropeptide must be selectively transported to the appropriate release site. Some of the still unanswered questions pertaining to neuropeptide processing are: Where are neuropeptides processed? How are they transported? What are the regulatory events that govern processing? An understanding of the formation and a definition of the contents of dense core vesicles (DCV) found in peptidergic neurons will be crucial to answering many of these questions.

Neuropeptides have been localized to dense core vesicles in a number of systems [opioid peptides (Klein *et al*. 1982), neuropeptide Y (Fried *et al*. 1985), VIP and substance P (Probert *et al*. 1981), R14 Peptide (Kaldany *et al*. 1986) and others]. Also there is evidence that neuropeptide processing occurs in the vesicles (reviewed in Gainer *et al*. 1985). As a beginning of our attempt to understand the role of DCV in neuropeptide function, we have undertaken a survey of dense core vesicle populations in a number of identified peptidergic neurons in *Aplysia*. By investigating the cell bodies we hope to be able to see the full range of DCV synthesized by the cell as opposed to examining the terminals which may only contain a subset of the total DCV.

In this study we will report on our findings on the size distributions of DCV in three neuronal populations, the bag cells, B1–B2, and R15. Further, we demonstrate that the SCP$_B$ peptide is indeed localized to the DCV in the case of buccal neurons B1 and B2. The data further suggest that there may be a great deal of complexity involved in vesicle formation. Each cell has a distinctive vesicle size distribution. In R15, there are two types of vesicles (granular and dense core), while in the bag cells there are at least two classes, both of which are dense cored and one is much larger than the other. In all three cell types the size distribution of the DCV significantly deviates from a normal distribution. Characteristic peaks and troughs in the vesicle histograms may point to physical constraints on vesicle sizes, or to the existence of a number of different classes of vesicles in each cell.

MATERIALS AND METHODS

Tissue preparation for EM

Abdominal and buccal ganglia from *Aplysia californica* ranging from 100–800 g were fixed for several hours in a trialdehyde fixative (4 per cent glutaraldehyde, 2.5 per cent paraformaldehyde, 1 per cent acrolein, 1 per cent DMSO, 100 mM $CaCl_2$ in PBS containing 0.3M sucrose, pH 7.4). Single neurons were dissected from the ganglia and fixed in the same fixative overnight; excess fixative was washed off with several changes of PBS, and the neurons were post-fixed in 2 per cent OsO_4 in PBS, 0.3M sucrose for 2 h, washed thoroughly in PBS and H_2O and stained *en bloc* with 2 per cent aqueous uranyl acetate for 2 h, dehydrated, and embedded in Epon-Araldite. Thin sections, 65 nm, were cut, mounted on formvar-coated copper slot grids, and contrasted with uranyl acetate and lead citrate for examination on the transmission electron microscope. All photographs used in the vesicle size analysis were made at ×14 000, and enlarged ×2.55.

Digitation of vesicle sizes

Electron micrographs were made of random sections of single cells through at least three different levels. Using a BIOQUANT program (R & M Biometrics) the diameters of single vesicles were measured on a Houston Instrument digitizing tablet. In some cases the areas and perimeters were also measured for the same vesicles. Control for resolution of the digitizing technique was effected by using circles of five sizes covering the range of vesicle sizes encountered and measuring these diameters repeatedly, but not successively.

Estimation of sphere diameters

The raw data was grouped into bin sizes of 10 nm. The limit of detection on the digitizer was 3 nm and therefore no additional information would be gained by a smaller bin size. Because of the large difference in sizes of the bag cell vesicles the data was originally grouped in 40 nm bins and then, for closer examination of the small DCV, regrouped into 10 nm bins. To transform the data from profile diameters to sphere diameters the method of Cruz-Orive was used (Cruz-Orive 1983). This method gives a distribution-free estimate of sphere diameters and takes into account both a minimum size cut-off and a capping angle. For these studies the minimum size was considered to be 30 nm to avoid confusing DCV with other cellular compartments. The capping angle was determined to be 35 degrees using the method suggested by Cruz-Orive (Cruz-Orive 1983). The algorithm was implemented on an IBM P.C. An example of the transformation is shown in Fig. 18.3A. As can be seen, no dramatic changes occur because of the transformation, although several of the bin sizes show significant frequency shifts. This is usually the case when there is a large shift upward in frequency from one bin to another. The shift is heightened in the transformed histogram due to the assumption that many of the profile diameters measured in the smaller bin size were oblique cuts of the immediately higher bin size.

Immunoelectron microscopy

Buccal ganglia were fixed overnight in increasing concentrations of paraformaldehyde to 8 per cent; all fixatives were buffered in PBS pH 7.4 and 0.3–0.7M sucrose, depending on the concentration of para-formaldehyde. Single B1 and B2 neurons were dissected from the ganglia, fixed an additional 8 h in 8 per cent paraformaldehyde, then washed in PBS repeatedly. The neurons were embedded in 10 per cent gelatin, and blocks containing the neurons cut from the gelatin, then fixed over-night in 8 per cent paraformaldehyde. The fix was again washed away and the gelatin embedded neurons infiltrated with 2.3M sucrose. These blocks were quickly frozen in freon at $-110°C$ and sectioned on a glass knife at the same temperature. Thin sections were collected on a loop of 2.3M sucrose, transferred to carbon-coated formvar-covered copper grids, cleared in 50 mM NH_4Cl in PBS for 30 min, blocked 10 min in 2 per cent BSA, and transferred to a 1:100 dilution of antisera for 1 h. Unreacted primary antibody was washed off, and the grids floated on an appropriate dilution (Slot and Geuze 1981) of 6-nm gold colloid-coated with protein A (Pharmacia) in PBS-Gly containing 33 per cent normal goat serum, complement free. The grids were then washed in PBS, followed by 1M NaCl for 30 min, and finally in water repeatedly. Controls for the specificity of the immune reaction included the use of preimmune sera and blocking the SCP_B specific antisera with the peptide prior to incubating with the tissue sections. Other procedures are as described by Bendayan (1984) and Roth *et al.* (1978).

RESULTS

B1–B2

B1 and B2 are neurons located on the dorsal surface of the buccal ganglia and send processes to the oesophagus. They are both known to synthesize two neuropeptides, SCP_A and SCP_B (Mahon *et al.* 1985). The neurons B1 and B2 are morphologically and biochemically very similar with the important distinction that B2 is cholinergic and B1 is not (Lloyd *et al.* 1985). Using antibodies to SCP_B followed by visualization with Protein A-coated colloidal gold (Roth *et al.* 1978; Bendayan 1984) we have localized SCP_B-like immunoreactivity to dense core vesicles in the soma of B1 and B2 (Fig. 18.2A,B).

The histogram of estimated vesicle sphere diameters for a total of 1220 B1/B2 DCV is shown in Fig. 18.3A. The vesicles have a mean size of 99 ± 26 nm. The distribution is significantly different from a normal distribution ($\psi^2, P < 0.001$). The shape of the histogram with a depression at the mean bin size (90–100 nm) led us to speculate on the possibility that the histogram reflected a bimodal distribution. To test this hypothesis a Q–Q plot was constructed from the histogram (Fowlkes 1979). The Q–Q plot transforms a gaussian distribution into a straight line by using x co-ordinates of standard deviation. The Q–Q plot of B1–B2 is seen in Fig. 18.3B. The plot shows a deviation from a straight line as seen by the different lines originating from best fits to the first or second part of the graph.

In an attempt to explore the distribution further, we looked at the distributions in individual cells. The results of three individual cells are shown in Fig. 18.4A–C. Upon close examination, similar peaks and troughs as seen in the original histogram are noted, but the frequencies of DCV at each peak and trough are quite variable in the different cells. Each single cell histogram consists of only 200 DCV; thus, the sampling error is larger and the histograms are not as smooth, since small changes in vesicle numbers make a corresponding larger change in frequency. One possible

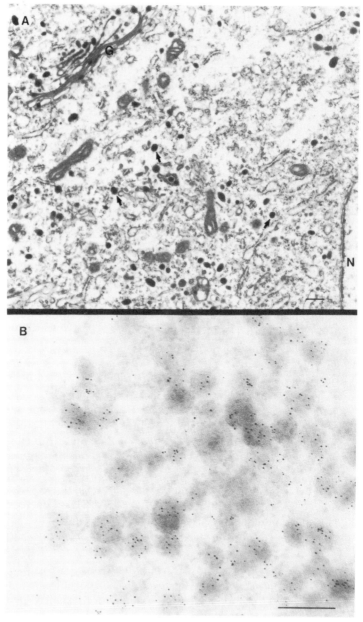

FIG. 18.2. Dense core granules in buccal ganglion neurons. (A) Electron micrograph of an osmium stained B1–B2 neuron. N, nucleus; G, golgi; arrows point to examples of DCV; bar = 300 nm (B) Cryoultramicrotome section of a B1–B2 neuron reacted with an antibody to SCP$_B$ followed by colloidal gold bound to protein A. The gold particles represent SCP$_B$ immunoreactivity and are localized to the DCV. Bar = 300 nm.

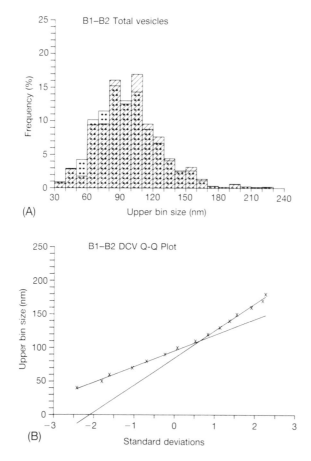

(A)

(B)

FIG. 18.3. Size distribution of B1–B2 dense core vesicles. (A) Vesicle distribution histogram. Starred areas represent the profile diameters, while the hatched areas represent the transformed sphere diameters as described in methods. $n = 1220$ vesicles. (B) Q–Q plot of sphere diameters from Fig. 18.2A. Lines are least square fits from the first six values and the last six values.

explanation for these data is that vesicles are formed in a number of discrete sizes, a possibility that will be discussed further.

Bag cells

The bag cells are located in two clusters of approximately 400 neurons each, situated on the rostral margin of the abdominal ganglion. Using the colloidal gold technique, ELH has been localized to the bag cell DCV (T. Kreiner and R. H. Scheller, unpublished). The granules are also seen at the

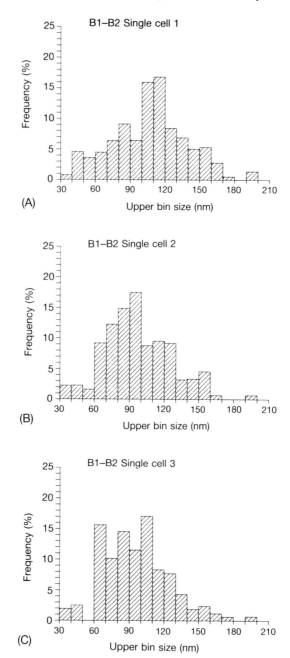

FIG. 18.4. Size distribution of dense core vesicles in three individual B1–B2 neurons. (A–C) Single B1–B2 cell DCV estimated sphere diameter histograms. $n = 200$ for each histogram.

putative release sites of ELH in the sheath as well as in the bag cell soma (Frazier *et al.* 1967).

Figure 18.5A is a typical electron micrograph of a bag cell soma. The histogram of estimated vesicle diameters (Fig. 18.5B) shows a major population with a mean of 140 nm as well as a smaller population of very large DCV with a mean of 400 nm. The large DCV account for approximately 5–15 per cent of the vesicles but because of their size account for at least 50 per cent of the volume of vesicles in the bag cell soma. The smaller population when examined more closely (smaller bin size, Fig. 18.5C) displays a marked bimodality. This can also be seen by the S-shaped curvature of the Q–Q plot (Fig. 18.5D). An S-shaped Q–Q plot suggests two populations corresponding to the flat parts of the S with a mixture of the populations in the middle. The large deviation seen in the Q–Q plot at the start and end of the distribution may reflect separate vesicle populations, but there are too few small vesicles (40–100 nm) to make any definite conclusion. Also at the upper end of the graph one is beginning to see the less prevalent large bag cell DCV, so that this part of the curve should probably be disregarded. The estimated vesicle diameter was fitted to two gaussians using a least squared iterative algorithm with starting estimates from an adjusted Q–Q plot (Fowlkes 1979). The fit is a good one (S.E.M. = 4.5) and shows two populations of roughly equal size: 114 ± 16 nm and 159 ± 20 nm.

R15

R15 is thought to be peptidergic (Kupferman and Weiss 1976) even though neither the neuropeptide nor the gene encoding the peptide products have yet been reported. The evidence consists of the white colour of the cell (diagnostic of secretory neurons), and the major low molecular weight peptides produced by R15. Furthermore, protease sensitive extracts of R15 cause an increased in body weight of up to 10 per cent due to water uptake (Kupferman and Weiss 1976). This along with the observation that R15 receives input from the osphradium, an organ sensitive to salt concentration, implies that the R15 neuron is involved in regulating water and salt balance.

The vesicles in R15 differ significantly from the other cells in this study by appearing in two different morphological classes (granular and dense core, see Fig. 18.6A). Both classes are present in roughly equal numbers but the granular population is slightly larger (mean of 123 ± 39 nm v. 112 ± 31 nm; Student's *t* test, $P < 0.01$). The two estimated sphere distributions are roughly similar in shape, with a flatter distribution than that seen in B1–B2 or the bag cells (Fig. 18.6B,C). The Q–Q plots are similar to B1–B2 with a deviation from a straight line but no classic S-shape to signal a clear bimodal distribution as seen for the bag cells (Fig.

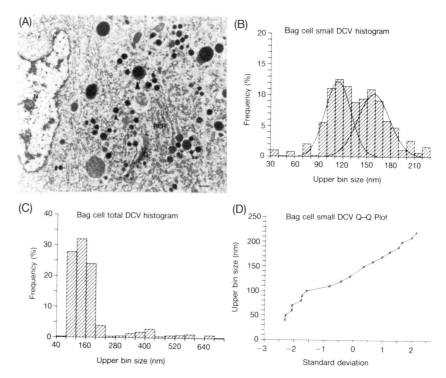

FIG. 18.5. Dense core vesicles in the bag cells. (A) Electron micrograph of an osmium stained bag cell. N, nucleus; G, golgi; RER, rough endoplasmic reticulum. Large arrow points to a member of the large bag cell DCV class, the medium arrow points to a representative granule from the second peak of the small DCV [see (B) and (C)], and the small arrow points to a DCV from the small class of (B) and (C). Bar = 300 nm. (B) Bag cell DCV sphere diameters. The bin sizes encompass 40 nm. Between 40 and 240 nm 640 vesicles were measured and 200 vesicles were measured with diameter greater than 240 nm. Large DCV are over represented so that their population distribution could be seen. On micrographs where both vesicle types were counted, large DCV (>240 nm) represented 5–10 per cent of the vesicles seen. (C) Bag cell small DCV sphere diameters, same data as 5B but the bin size has been reduced to 10 nm. Deviation from a normal distribution (ψ^2, $P < 0.001$). The two gaussians fit to the histogram represent two populations; 49 per cent of mean 114 nm and standard deviation of 16 nm, and 51 per cent of the population of mean 159 nm and standard deviation 20 nm. In the region corresponding to the S-shaped curve in Fig. 18.4D (90–170 nm) the two gaussians fit the curve extremely well (ψ^2, $P > 0.95$). (D) Q–Q plot of the data in Fig. 18.4C. Note the S-shaped section in the middle of the plot (90–170 nm).

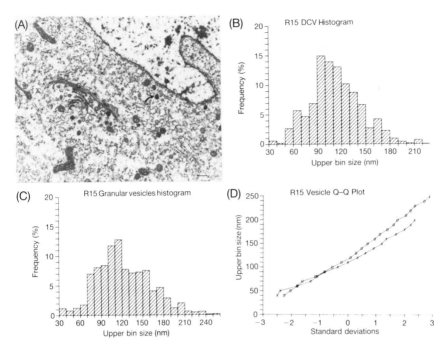

FIG. 18.6. R15 granular and dense core vesicles. (A) Electron micrograph of an osmium stained R15 neuron. N, nucleus; G, golgi. Dark arrow points to a dense cored vesicle and the light arrow points to a granular vesicle. Bar = 300 nm. (B) Histogram of granular sphere diameter. $n = 722$. Deviation from a normal distribution (ψ^2, $P < 0.001$). (C) Histogram of DCV sphere diameters, $n = 742$. Deviation from a normal distribution (ψ^2, $P < 0.001$). (E) Q–Q plot of data in Fig. 18.5B,C ($x \rightarrow$ DCV, $0 \rightarrow$ granular).

18.6D). The deviation from a straight line in both B1–B2 and R15 DCV and granular Q–Q plots indicate that the first half of the histogram rises faster than the second half decreases. This could be because of a non-normal distribution which leads to a tailing of vesicle sizes or could be indicative of more than one population. The single B1–B2 cells examined do not show the tailing as much as they show peaks, leading us to favour the multiple population hypothesis.

DISCUSSION

Vesicles change size during processing

An attractive hypothesis for differences in DCV size and density is that different populations of vesicles contain neuropeptides at different stages of processing. For R15, it may be that the granular vesicles contain less processed precursors which upon further cleavage condense into the DCV. There is evidence that the R15 peptide is processed intragranularly (Gainer *et al*. 1982), and it has been noted in other systems that vesicle size and morphology may change over time; these changes have been correlated with processing stages (Nordmann and Labourne 1981). Proinsulin and processed insulin can be localized to morphologically distinct vesicle classes (Tapia *et al*. 1983). The fact that the granular vesicles in R15 are larger than the dense core vesicles, and that the relative number and shape of the two distributions are similar suggests this may be the case in R15 as well. It is not mechanically clear how the vesicle size would change with a change in the state of the granule contents.

The very large DCV seen in the bag cell soma have not been reported in the axons or neurites (Frazier *et al*. 1967; Price and McAdoo 1979; T. Kreiner, unpublished) and therefore could represent a soma specific organelle; possible sites of precursor processing or peptide storage and/or turnover sites. It is possible that the smaller DCV arise from the larger soma specific granules prior to transport to the terminals. The ELH precursor was not found in bag cell granule preparations (Loh *et al*. 1977) and parts of the precursor protein appear to remain in the perikaryon (Arch *et al*. 1976). In these studies the very large granules would not have been included in the DCV isolation consistent with the possibility that the first cleavage occurs in these vesicles or the golgi complex. This would allow differential packaging and transport of peptides which are synthesized on a single precursor.

Differential transport

Neurons localize molecular components to distinct cellular locations such as axons *v*. dendrites. The nature of the specificity required for this process is unclear, however, it may be that a single neuron sends different chemical messengers to different terminals. Thus, the different DCV populations observed in this study could have different contents and, in the presence of the appropriate sorting apparatus, segregation to distinct terminals may occur. This is not a direct explanation for the different sizes, but one can imagine that the different contents of a vesicle would impose different size limitations on the granule. We are currently measuring sizes and exploring the contents of granules in the terminals of identified cells to address this issue.

Quantal vesicle sizes

There have been reports that the formation of vesicles is physically constrained to certain sizes (Bont *et al*. 1977; Bont 1978). In these studies, vesicles formed by ultrasonic disruption of membranes were shown to be formed in quantal sizes. These vesicle sizes agree fairly well with the peaks that we observe in this study (Table 18.1). The data from B1–B2 fit this model rather well; while the frequency of vesicles at a particular size changes, the peaks are seen in approximately the same place as would be predicted from the theory of Bont *et al*. (1977). These quantal sizes are presumably imposed by physical constraints; however the biological significance behind the quantal sizes remains unclear.

TABLE 18.1. *Comparison of predicted sphere diameters v. estimated sphere diameters.*

	Predicted vesicle diameters[*]			
	69	95	120	172
B1–B2 total histogram peaks	60–70	80–90	100–110	150–160
B1–B2 single cell 1 peaks		80–90	105–115	150–160
B1–B2 single cell 2 peaks	60–70	85–95	110–120	150–160
B1–B2 single cell 3 peaks	60–70	80–90	100–110	150–160
Bag cell small D.C.V. peaks			100–110	150–160
R15 D.C.V. peaks	60–70	90–100		160–170
R15 granular peaks	70–80		105–115	160–170

[*]From Bont *et al*. 1977; all values are in nm.

CONCLUSIONS

The preliminary data presented here are a further indication that vesiculation and processing of peptide messengers in the brain are processes with a great deal of complexity. Understanding the regulation of these steps in peptide biogenesis is clearly crucial in generating a complete picture of the function of peptide messengers. The potential for differential vesiculation, transport, and secretion further expands the range of activities which can be elicited by the already large number of peptide chemical messengers in the brain.

The large identifiable nerve cells of the mollusc *Aplysia* which have served the physiologist for many years are now yielding insight into cellular and molecular processes as well. With further studies of the type described here, we hope to merge the molecular, cellular, and physiological observations into a coherent understanding of the mechanisms utilized by the central nervous system to govern behavioural processes.

Acknowledgements

T.K. was supported by a predoctoral NSF fellowship. This work was supported by grants from the NIH and the NSF to RHS. RHS is a McKnight Foundation Scholar and an Alfred P. Sloan Fellow. The authors would like to thank Frances Thomas for technical assistance and Ruud Brands and Michael Bastiani for technical advice.

References

Abrams, T. W., Castellucci, V. F., Camardo, J. S., Kandel, E. R., and Lloyd, P. E. (1984). Two endogenous neuropeptides modulate the gill and siphon withdrawal reflex in *Aplysia* by presynaptic facilitation involving cAMP dependent closure of a serotonin-sensitive potassium channel. *Proc. Nat. Acad. Sci. USA*, **81**, 7956–60.

Arch, S., Smock, T., and Earley, P. (1976). Bag cell processing. *J. Gen. Physiol.* **68**, 211–9.

Bendayan, M. (1984). Protein A-gold electron microscopic immunocytochemistry: methods, applications, and limitations. *J. Electron Microsc. Techn.* **1**, 243–70.

Bont, W. S. (1978). The diameters of membrane vesicles fit in geometric series. *J. Theoret. Biol.* **74**, 361–75.

——, Boom, J., Hofs, H. P., and De Vries, M. (1977). Comparison between the sizes of granular vesicles in intact cells and vesicles obtained by fragmentation of biomembranes. *J. Memb. Biol.* **36**, 215–32.

Cruz-Orive, M. (1983). Distribution-free estimation of sphere size distribution from slabs showing overprojection and truncation, with a review of previous methods. *J. Microsc.* **131**, 265–90.

Fowlkes, E. B. (1979). Some methods for studying the mixture of two normal (lognormal) distributions. *J. Am. Statist. Ass.* **74**, 561–75.

Frazier, W. T., Kandel, E. R., Kupferman, I., Waziri, R., and Coggeshall, R. E. (1967). Morphological and functional properties of identified neurons in the abdominal ganglia of *Aplysia californica*. *J. Neurophysiol.* **30**, 1288–351.

Fried, G., Terenius, L., Hokfelt, T., and Goldstein, M. (1985). Evidence for differential localization of noradrenaline and neuropeptide Y in neuronal storage vesicles isolated from rat vas deferens. *J. Neurosci.* **5**, 450–8.

Gainer, H., Loh, Y. P., and Neale, E. A. (1982). The organization of post-translation precursor processing in peptidergic neurosecretory cells. In *Proteins in the Nervous System: Structure and Function* (eds B. Haber, J. R. Perez-Polo, and J. D. Coulter). p. 131. Alan R. Liss, New York.

——, Russel, J. T., and Loh, Y. P. (1985). The enzymology and intracellular organization of peptide precursor processing: the secretory vesicle hypothesis. *Prog. Neuroendocrinol.* **40**, 171–84.

Herbert, E. and Uhler, M. (1982). Biosynthesis of Polyprotein Precursors to Regulatory Peptides. *Cell*, **30**, 1–2.

Kaldany, R. R., Kreiner, T., Makk, G., Evans, C. J., and Scheller, R. H. (1986).

Proteolytic processing of a peptide precursor in *Aplysia* neuron R14. *J. Biol. Chem.* (In Press).

Kandel, E. R. (1976). *The Cellular Basis of Behavior*. W. H. Freeman and Co., San Francisco.

—— (1979). *Behavioral Biology of Aplysia*. W. H. Freeman and Co., San Francisco.

Klein, R. L., Wilson, D. J., Dzielak, W., Yang, H., and Viveros, O. H. (1982). Opioid peptides and Noradrenaline co-exist in large dense-cored vesicles from sympathetic nerves. *Neurosci.* **7**, 2255–61.

Krieger, D. T. (1983). Brain peptides: what, where and why? *Science*, **222**, 975–85.

Kupfermann, I. and Weiss, K. R. (1976). Water regulation by a presumptive hormone contained in identified neurosecretory cell R15 of *Aplysia*. *J. Gen. Physiol.* **67**, 113–23.

Lloyd, P. E. (1982). Cardioactive neuropeptides in gastropods. *Fed. Proc.* **41**, 2948–52.

——, Kupferman, I., and Weiss, K. R. (1984). Evidence for parallel actions of a molluscan neuropeptide and serotonin in mediating arousal in *Aplysia*. *Proc. Nat. Acad. Sci. USA*, **81**, 2934–7.

——, Mahon, A. C., Kupfermann, I., Cohen, J. L., Scheller, R. H., and Weiss, K. R. (1985). Biochemical and immunocytochemical localization of molluscan small cardioactive peptides (SCPS) in the nervous system of *Aplysia*. *J. Neurosci.* **5**, 1851–61.

Loh, Y. P., Brownstein, M. J., and Gainer, H. (1984). Proteolysis in Neuropeptide processing and other neural functions. *Ann. Rev. Neurosci.* **7**, 189–222.

——, Sarne, Y., Daniels, M. P., and Gainer, H. (1977). Subcellular fractionation studies related to the processing of neurosecretory proteins in *Aplysia* neurons. *J. Neurochem.* **29**, 135–9.

Mahon, A. C., Lloyd, P. E., Weiss, K. R., Kupfermann, I., and Scheller, R. H. (1985). The small cardioactive peptides A and B of *Aplysia* are derived from a common precursor molecule. *J. Neurosci.* **5**, 1872–80.

Nambu, J. R., Taussig, R., Mahon, A. C., and Scheller, R. H. (1983). Gene isolation with cDNA probes from identified *Aplysia* neurons: Neuropeptide modulators of cardiovascular physiology. *Cell*, **35**, 47–56.

Nordmann, J. J. and Labourne, J. (1981). Neurosecretory granules: evidence for an aging process within the neurohypophysis. *Science*, **211**, 595–7.

Price, C. H. and McAdoo, D. J. (1979). Anatomy and ultrastructure of the axons and terminals of neurons R3–R14 in *Aplysia*. *J. Comp. Neurol.* **188**, 647–78.

Probert, L., DeMey, J., and Polak, J. M. (1981). Distinct subpopulations of enteric p-type neurones contain substance P and vasoactive intestinal polypeptide. *Nature*, **294**, 470–1.

Roth, J., Bendyan, M., and Orci, L. (1978). Ultrastructural localization of interacellular antigens by the use of protein A-gold complex. *J. Biochem. Cytochem.* **26**, 1074–81.

Schaefer, M., Picciotto, M., Kreiner, T., Taussig, R., and Scheller, R. H. (1985). Identified neurons in *Aplysia* express a gene encoding multiple FMRFamide copies. *Cell* **41**, 457–67.

Scheller, R. H., Jackson, J. F., McAllister, L. B., Rothman, B. S., Mayeri, E., and Axel, R. (1983). A single gene encodes multiple neuropeptides mediating a stereotyped behavior. *Cell*, **35**, 7–22.

——, ——, ——, Schwartz, J. H., Kandel, E. R., and Axel, R. (1982). A family of genes that code for ELH, a neuropeptide eliciting a stereotyped pattern of behavior in *Aplysia*. *Cell*, **28**, 707–19.

Slot, J. W. and Geuze, H. J. (1981). Sizing of protein A-colloidal gold probes for immunoelectron microscopy. *J. Cell Biol.* **90**, 533–6.

Susswein, A. J. and Benny, M. (1985). Sexual behaviour in *Aplysia fasciata* induced by homogenates of the distal large hermaphroditic duct. *Neurosci. Letters* **59**, 325–30.

Tapia, F. S., Varndell, I. M., Probert, L., DeMey, J., and Polak, J. M. (1983). Double immunogold staining method for the simultaneous ultrastructural localization of regulatory peptides. *J. Histochem. Cytochem.* **31**, 977–81.

Taussig, R., Kaldany, R. R., and Scheller, R. H. (1984). A cDNA clone encoding neuropeptides isolated from *Aplysia* neuron L11. *Proc. Nat. Acad. Sci. USA*, **81**, 4988–92.

19

Molecular mechanisms for memory

D. E. KOSHLAND, Jr

The word 'memory' has many connotations. The most common one is the complex pattern of events or people which we can recall with appropriate stimulation such as the faces of our parents. Such memories are a sum of multitudinous modifications working in innumerable neurons. If one wishes to understand the mechanism of memory, two distinct problems must be resolved. One is the network of interactions among neurons which produce the associative memories. The second is the pattern of molecular events occurring in individual neurons, which form the 'building blocks' of the complex patterns. This paper focuses on this latter area.

Since billions of neurons are present in the human brain as well as in the brains of higher animals with similar memory patterns, there is a question of how one can even begin such an investigation. One approach is to pick a simple organism, demonstrate that it utilizes memory, and describe the observable molecular events occurring in its cells. This approach has been taken in our laboratory in studies on bacterial chemotaxis (Koshland, 1980a,b) and in the laboratories of Kandel and others (Kandel *et al*. 1983) in studies of *Aplysia*, and in the laboratories of Benzer, Quinn, and others in studies of *Drosophila* (Quinn *et al*. 1974). Another approach is to subject neurons in tissue culture to conditions such as repetitive stimuli which are likely to occur during memory, and then to investigate the molecular events occurring during these stimulations. This has been done by Lynch and Baudry (1984). Ultimately, of course, the molecular processes of isolated cells and of the simpler organisms must be related to the higher network interactions of more complex systems, which are already being studied at various levels to delineate the mechanisms of memory (Thompson *et al*. 1983; Squire and Cohen, 1984). Simpler organisms provide clues with respect to molecular events occurring in the much more complex systems. These clues may tell us what to look for in the chemistry of neurons which connected in an appropriate 'wiring diagram' will produce complex memories of our childhood or a home town.

To begin, we might consider some essential components required for a short-term memory process such as dialing a telephone number. As shown

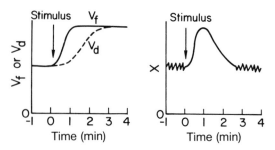

FIG. 19.1. A schematic illustration of short term memory. A potential neuron is shown schematically in which a stimulus such as reading a number in a telephone book occurs at time 0. If this activates two processes, leading to the formation and distruction of a molecule X, which controls the output of the neuron, under appropriate conditions a transient stimulus in which X rises and falls will occur. One potential set of conditions is shown in the figure, i.e. an activation of a forward reaction V_f which produces X, and its decomposition by a second reaction V_d. If the stimulus activates both V_f and V_d by the same amount but on different time scales, the actual level of X, the response regulator, will rise and fall as shown on the right hand side of the figure. During this interval all neurons so stimulated, therefore, will be changed and those changes can be recorded by other neurons. The memory trace must be long enough in the present case to allow dialling of the phone before it lapses back to the non-stimulus level.

in Fig. 19.1, the initial event or stimulus of reading a number in a telephone book must produce a change in one or more cells which persists for an interval long enough for us to start the dialing procedure. Once that occurs, the short-term memory chemical alteration can disappear. Such transient alterations of chemistry in individual cells have been observed and in the case of bacterial chemotaxis have been identified with a cell's memory (Macnab and Koshland, 1972). Some network pattern records the order of the numbers read. The arrangement of these neurons in order to provide proper recognition of the appropriate numbers is part of the networking problem referred to above. Our studies indicate that a relatively simple molecular process in individual neurons could explain a transient memory.

BACTERIAL MEMORY

The bacterium exhibits behavioural patterns essential to its survival. For instance, it shows a distinct ability to migrate towards nutrients which are good for it and to move away from toxic substances which will destroy it (Adler, 1969). We showed a number of years ago that these chemotactic functions depend upon a memory process (Macnab and Koshland, 1972) and we have been studying the molecular events which make this memory possible. Herein, we describe some of these molecular events and compare them with the memory capabilities of more complex species.

Why did bacteria develop a memory in the first place? Bacterial chemotaxis depends on the bacterium's ability to detect a gradient. If the bacterium is living in an environment with a uniform distribution of nutrients, there is no advantage in migrating and the bacterium swims around randomly. If, however, a gradient situation arises in which more nutrient could be obtained by swimming in one direction than another it would be advantageous to detect that gradient and to migrate appropriately. This is precisely what the bacterium does and to do it requires the development of the ability and capacity to compare past with present, as well as the capability to alter its swimming pattern accordingly. Not only can the bacterium perform these functions, but it does so with incredible efficiency: it can respond to a gradient in a concentration of 1 part in 10^4, measured over the distance from its head to its tail (Koshland, 1980a).

The molecular events occurring in bacterial chemotaxis have been described extensively, because this conveniently simple organism can be manipulated both genetically and biochemically in ways quite impossible to carry out in higher organisms. Delivery of a single stimulus to the organism produces a transient response very similar to that depicted for an individual neuron in Fig. 19.1. Genetic studies, biochemical studies, and recombinant DNA technology have been used to unravel many of the molecular events in the development of the signalling process. A schematic outline of our understanding of this process is shown in Fig. 19.2.

The molecular events occurring during bacterial chemotaxis can be divided into three conceptual domains. The first is the input process by which the stimulus interacts with the receptor, which then transmits the information to the interior of the cell. A typical receptor has been cloned, sequenced, purified, and mapped on the chromosome (Russo and Koshland, 1983). In process of transmitting the signal across the cell membrane, it is covalently modified by a transient process of methylation and demethylation (Springer *et al*. 1979; Koshland 1980c). Mutants in the receptor, the methylating enzyme or the demethylating enzyme, eliminate the chemotactic ability of the bacterium. Manipulation of the proteins produced by the individual genes by genetic engineering makes it possible to alter the timing of the bacterial response.

From these studies it has become clear that conformation of the receptor, and changes in conformation, together with the action of modifying enzyme are key elements in determining the endurance of the transient signal. For example, it has been possible by genetic engineering to overproduce the receptor such that it is in excess relative to the modifying enzymes (Russo and Koshland 1983). When that occurs, the level of methylation to terminate the signal requires a much larger interval because of the relative decrease of modifying enzymes, with the result that the response characteristics of the cell are changed dramatically. Mutants

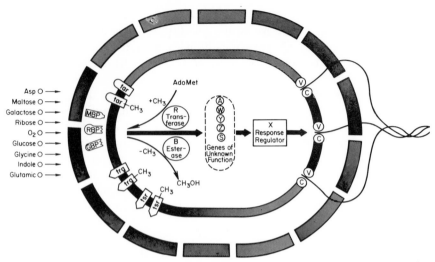

Fig. 19.2. Schematic illustration of the information processing system in bacterial sensing. The bacterial cell has an outer membrane where it is protected against injury, as our eyelids protect us from mechanical damage. Chemicals diffuse through holes in the outer membrane and bind to receptors in the periplasmic space or inner membrane. These receptors can be methylated and demethylated by two enzymes, the transferase and esterase which are coded for by the *cheR* and *cheB* genes, respectively. This information is further processed by a number of genes whose functions have not yet been totally delineated, illustrated as A, W, Y, Z, and S. The combined effect of these gene products produces a level of the response regulator which is responsible for the short term memory, as illustrated in Fig. 19.1. The level of the response regulator is detected by two detector genes, illustrated here as *cheV* and *cheC* in the membrane, which transmit a signal to the flagella, thereby controlling the migration of the bacterium.

which destroy either the methylating enzyme or the demethylating enzyme eliminate chemotaxis entirely (Springer and Koshland 1977; Stock and Koshland 1978). These experiments tell us a great deal about the response process but they have troubled those individuals who believed that an absolute concentration of some species within the cell was essential for memory. In that case, if damage to the methylating enzyme destroyed formation of that signal, then destruction of the demethylating enzyme should produce an even better concentration of the signal. The finding that mutants in either enzyme cause loss of function means that a dynamic system in which modification and demodification are occurring continuously is essential for the proper response. Thus, a simple summary of the input system suggests that binding of the stimular to the exterior of the receptor induces a conformational change which releases a signal into the cytoplasm. The intensity of the signal and the duration of the response are determined by the enzymes and by the kinetic characteristics of the cell in which these processes take place.

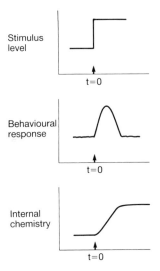

Stimulus
level

t=0

Behavioural
response

t=0

Internal
chemistry

t=0

FIG. 19.3. Illustration of a signal system that produces an adaptive response. When the stimulus shown as the top of the figure is suddenly increased from one level to another at time (t) = 0, a transient behavioural response occurs. In the case of an adapting cell, this behavioural response will return to its initial value, even though the stimulus is at a new and higher level. To achieve this adaptation in the face of a higher background level of stimulus, the internal chemistry of the cell must change and in the case of bacterial chemotaxis this is correlated with the methylation of the receptor. In the case of rhodopsin it is correlated with the phosphorylation of the receptor.

How these modifying enzymes react is shown schematically in Fig. 19.3. One finding in the characterization of the bacterial system is that bacteria adapt to a stimulus in a manner which is reminiscent of the adaptation of our eyes to light, of our taste system to persistent chemicals, and of the olfactory system to persistent odours. In the bacterial case this can be correlated with the methylation and demethylation of the receptors. By giving the bacterium a sudden stimulus of an attractant such as aspartic acid, the behavioural pattern of the cell is observed to change from one of random swimming to one of directed smooth swimming, which then reverts to random swimming. The additional methylation of the receptor apparently brings the receptor back to the conformation it had initially in the absence of stimulus, even though the stimulus is now bound to the receptor (Springer *et al.* 1979; Koshland 1980c).

At the molecular level, the simplest picture of this process shows that the initial stimulus binding to the receptor distorts its shape so that a signal is emitted in the interior of the cell. That change in shape alters the accessibility of glutamic acid residues on the cytoplasmic side of the

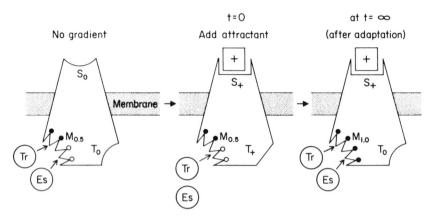

FIG. 19.4. A schematic illustration of the way in which adaptation is achieved at the molecular level. On the left hand side is shown a receptor which is not stimulated which has on the average two methyl groups per receptor and is sending no net signal through the cytoplasm. On binding an attractant, the conformation of the receptor is changed so that the signalling region changes from T_0 to T_+, thereby sending a signal to the cytoplasm. The conformation of the receptor is then changed to provide different access of transferase and esterase, leading as shown on the far right to a new higher level of methylation. This higher level of methylation alters the protein structure, so that the signalling region returns to the shape T_0, even though the stimulus is still binding to the receptor. Thus, the initial non-signalling condition is achieved in the presence of external stimuli by increased covalent modification of the receptor.

receptor, so that they are methylated and demethylated at new rates, leading to a net increase in methylation. That new level of methylation is able to alter the shape of the receptor so that the signalling part of the protein returns to its initial condition, even though the regions around the aspartate binding and the methylation sites are now altered from their original geometries. A schematic illustration of how this is achieved is shown in Fig. 19.4. An observer of this process might conclude that to achieve this new, higher level of methylation the aspartic acid stimulus was activating the methylase and inhibiting the demethylase. In fact, the aspartic acid never comes in contact with those enzymes but merely changes the shape of the receptor so that the kinetics of methylation and demethylation are altered by indirect action.

THE INTEGRATION DOMAIN

When the signal is released from the receptor, it undergoes the next step in processing. In this processing the system must be sensitive to increasing amounts of stimulus, to signals from different receptors, and to the length of time in which the receptors are stimulated. The bacterial cell is capable of carrying out all of these integration processes (Koshland 1980a; Tsang *et al*. 1973; Mesibov *et al*. 1973). It can integrate stimuli from a multitude of receptors and can even integrate information from changes of membrane potential with those of chemical gradients (Miller and Koshland 1977). Manipulation of the system by genetic engineering has shown that increase in the numbers per cell of a particular receptor will give a larger response to that particular stimulus (Russo and Koshland 1983). Deletion of a particular receptor will eliminate responses from that stimulus (Adler 1969). Positive and negative stimuli are integrated algebraically (Tsang *et al*. 1973). Two proteins which appear to be identified with this process are produced by the *cheA* and *cheW* genes and therefore the symbols A and W are shown as part of the integration phase, but this is only partially understood. That they are involved in transmitting the signal has been shown in experiments in which mutant strains lacking these proteins fail to transmit signals from receptor to flagella and in which overproduced genes alter the signalling (Parkinson 1977).

To explain the transient response and the integrative capacity of the system, the concept of a response regulator was introduced (Macnab and Koshland 1972; Koshland 1977). The response regulator can be viewed as either a single molecular function or as a combination of functions, i.e. a process wherein the changing levels of the parameter control behaviour much as the temperature in a room rises and falls and controls a thermostat. One can imagine enzymes producing one or more of the molecules of the response regulator system in exactly the same way which any chemical pool is created within the cytoplasm. Such a concept makes it easy to understand how a variety of signals, some from the increased numbers of a particular receptor and others from a multiplicity of different types of receptors, can all be integrated into a common pattern. On the exterior of the cell the specificity of the receptor determines which stimulus is to generate a signal in the information processing system. Inside the cell all these stimuli are converted so that increases in attractant gradient cause increases in the signal, decreases in attractant gradient cause decreases in the signal, increases in repellent gradient cause decreases in the signal, and decreases in repellent gradient cause increases in the signal. Thus, positive and negative stimuli can produce algebraically significant changes in the response regulator, which are all integrated into a common system. Recent studies in which the methylation system has been eliminated by genetic manipulation clarify some features of this part of the processing system (Stock *et al*. 1981, 1985; Block *et al*. 1982).

THE OUTPUT RESPONSE

In order to determine the output of the cell, there must be some detector of the response regulator level. In bacterial chemotaxis this seems to be centred in four proteins, the products of the *cheC*, *cheV*, *cheY*, and the *cheZ* genes. The *cheC* and the *cheV* proteins are part of the motor, apparently binding the response regulator which determines rotation direction. *CheY* and *cheZ* are involved in setting the level of the output detector. It had previously been shown that counter-clockwise rotation of the bacterial flagellum is identified with smooth swimming, whereas clockwise rotation is identified with tumbling (Larsen *et al*. 1974). This in turn provides an assay by which cells tethered through their flagella can be observed (Silverman and Simon 1974) and changes from clockwise to counter-clockwise behaviour can be assessed quantitatively (Spudich and Koshland 1976).

An experiment which revealed the role of the *Y* protein was performed introducing a plasmid containing *cheY* into mutant cells (Clegg and Koshland 1984). Essentially, all of the genes except those identified with the flagellar motor were removed from the cell. Then the protein on the *cheY* gene was introduced back into the cell to see whether it could influence the rotation of the bacterial flagella. It produced clockwise rotation. Removing the *cheY* gene only from a wildtype cell produced counter-clockwise behaviour. Thus, it was demonstrated that the *Y* gene product interacted with the motor in some way to produce clockwise rotation of the bacterial flagella (Clegg and Koshland 1974). Moreover, intergenic complementation suggested that the *cheY* protein interacts with the *cheC* protein in the flagella (Parkinson *et al*. 1983; Clegg and Koshland 1985).

Parallel experiments have also been performed with the Z protein (S. Kuo and D. E. Koshland, unpublished). It, in turn, has been shown to produce smooth swimming, or counter-clockwise behaviour when it is produced in excess, and tumbling or clockwise behaviour when it is removed from a wildtype cell. Cells which contain an excess of *cheZ* or *cheY*, but which have had the A, W, R, and B proteins removed, can influence the rotation of the flagellar motor by themselves, but there is no response of the bacterial system to signals from the receptors in the absence of those proteins. Thus, it seems clear that although the Y and Z proteins act on the flagella and can do so independently, they are not the agents by which the signal from the receptor is transmitted through the cytoplasm. Their influence on the detector may be altered by means of such a signal which comes from the integrating system, but they are also capable of affecting flagella rotation in the absence of these stimuli.

The two proteins, illustrated as V and C in figure 19.2, are part of the motor structure, and mutants in the *cheC* protein can have a variety of different behaviours. Some point mutations in this protein can produce a cell which tumbles all the time, others which swim smoothly all the time, and some which have intermediate behaviour patterns (Rubik and

Koshland 1978; Khan *et al*. 1978; Koshland 1983). These results are consistent with the model that this protein is part of the system which detects the level of response regulators. The detector is in essence like a thermostat on the wall of a room which can be set at various temperatures. The *cheC* gene product gives a variety of responses because its binding constant for the response regulator is varied either up or down from that of the wildtype cell. Similar results have been obtained with the *cheV* gene (Dean *et al*. 1983; Warrick *et al*. 1977). Moreover, the precise level at which the detector level is set is also controlled by the two proteins, *cheY* and *cheZ*, as described above. Thus, a complex detector system is obtained which controls the probability of tumbling of the bacterium at any one instant. A signal from the integration domain of the bacterium is received by the detector system which contains at least the proteins Y, Z, C, and V, to produce the proper output response.

RELATIONSHIP OF THE GEOMETRY OF THE CELL TO THE CONCEPTUAL DOMAIN

The above schematic separation of conceptual domains should not be construed to mean that these domains are totally separated within a particular cell. In fact, they are not. Moreover, there are geographical compartments and barriers which do separate some parts of the system from others and there are specificity properties of the individual proteins which allow for the temporal sequence of events. The stimuli on the outside of the cell interact with the exterior portion of the receptors. The receptor, either by itself or together with a transducer, transmits the information across the membrane barrier, to start the signal within the cell (cf Fig. 19.4). However, receptors are distributed throughout the membrane and not concentrated at any one locale, as might be inferred from the schematic outline of Fig. 19.2.

The signals received from the many receptors interact within the cytoplasm. The proteins indicated as R, B, A, W, Y, and Z are all cytoplasmic, soluble proteins, freely mixing in the intracellular milieu. Thus, the signal from the receptors and the integration system must operate in the cytoplasm. Which particular interactions occur depends on the specificity of the proteins and is not due to any topological compartmentalization.

The *cheC* and *cheV* proteins are placed within the motor apparatus itself. They are membrane associated, probably peripheral membrane proteins. They are the means by which the change in rotation of the flagella, the process which transmits information from inside the cell to outside, is achieved. The proteins are not localized in any one part of the cell, but are dispersed throughout the membrane. Thus, there is some separation of physical domains but not necessarily in a one-to-one correlation with the conceptual domains.

A revealing result of these studies has been that individual units can carry out their part of the information processing in the absence of the other components. Receptors can be purified and reconstituted into artificial membranes (Wang and Koshland 1980), or solubilized in detergents (Bogonez and Koshland 1985), and shown to have many of the properties of the intact receptors. For example, it is found that the binding of a stimulus to the receptor remains essentially the same in purified protein as in solubilized proteins. Moreover, methylation and demethylation of the receptor can occur in a solubilized system or in artificial liposomes constructed with pure enzymes (Bogonez and Koshland 1985). Furthermore, parts of the integration system can be studied in the absence of the methylating enzymes (Stock *et al*. 1981, in press). Thus, the chemical components of this system have the properties of a modulator system each unit of which has the capability for fairly extensive processing and can operate almost independently.

RELATIONSHIP OF BACTERIAL PROCESSING TO NEURONAL PROCESSING

The similarities between the processing system of the bacterium and what we know so far about neurons are striking. Neurons have receptors in the outer membranes of the cell which transmit information from the exterior to the interior. Although their distribution is somewhat different from the bacterial receptors, the neuron is similarly able to integrate signals from the receptors in different locations and from different types of receptors. There are inhibitory and excitatory signals to a neuronal cell and these varied signals can be integrated algebraically in the neuronal processing. The neuron, like the bacterium, can also integrate both electrical and chemical signals (Kandel 1976).

Receptors from neurons also show covalent modification. The light receptor rhodopsin is reversibly phosphorylated and the level of phosphorylation is proportional to the light intensity (Kuhn *et al*. 1977), just as the level of methylation of the bacterium is related to the chemical stimulus intensity (Koshland 1980c; Springer *et al*. 1979). The acetylcholine receptor is phosphorylated but it is too early to say exactly what role this plays in its function (Teichberg *et al*. 1977). It is interesting to note, however, that other receptors such as the EGF receptor and the insulin receptor are also covalently modified (Cohen *et al*. 1982; Kasuga *et al*. 1982).

The structure of these receptors also bears a striking similarity to the aspartate receptor. When the latter was identified and the structure deduced on the basis of the sequence (Russo and Koshland 1983), a thin transmembranal region in the middle of the molecule was postulated to play a major role in transmitting information from the N terminal end of the molecule (which bound aspartate) to the C terminal end of the molecule (which was involved in signalling, methylation, and de-methylation). Many individuals questioned whether such a structure could transmit a signal. In the interim, the sequences of the EGF receptor (Downward *et al.* 1984), the LDL receptor (Russell *et al.* 1984), and the insulin receptor (Ullrich *et al.* 1985; Ebina *et al.* 1985) have been determined and all have similar transmembrane regions. Thus, there appear to be similar features of a wide variety of receptors, devised as a means of transmitting information across the membrane. Studies are in process to understand the nature of that structural component.

Other systems show similar processing analogies to the bacterial model. *Drosophila* memory mutants have been obtained and deficiencies in the memory process are identified with both the loss of adenylate cyclase activity and the loss of phosphodiesterase activity (Quinn *et al.* 1974). As in the case of the bacterium, therefore, the memory process does not depend upon the absolute level of a particular chemical but rather must involve some dynamic system in which the constant formation and removal of a component is essential for the memory process.

The output system in the bacterial chemotaxis also has relations to higher species. It is now known that the release of neurotransmitters in a neuron is affected by calcium levels, by the formation of synaptic vesicles, and by a complex system of enzymes including some phosphorylated proteins. The components appear to be designed analogously to the bacterial system, except that the output response in this case is the release of a neural transmitter rather than a change in direction of a flagellar motor. Nevertheless, it seems probable that this system receives inputs, in some cases from a combination of electrical and chemical signals and in some cases from chemical signals alone, which then trigger the release of the neurotransmitter. Both systems work to produce an on-off response and might contain detector systems with analogous properties.

RELATIONSHIP OF BACTERIAL SYSTEMS TO FAST AND SLOW SIGNALS IN THE NERVOUS SYSTEM

The description of the bacterial information processing system has some relevant messages for the theme of this conference—fast and slow chemical signals—as described in the papers of Iversen and Schmitt (Iversen 1984; Schmitt 1984). The bacterial system shows that there are many ways in which the input of one stimulus can modify the responses to a second stimulus. A stimulus which inhibits the methylating enzymes, for example, would greatly increase the response time of a second stimulus which utilized a receptor which got methylated. A stimulus which caused complexing of the *cheY* protein would reset the detector system to either enhance or diminish signals which were received later. A stimulus which inhibited the *cheA* or the *cheW* genes might prevent transmission of a signal from the input receptor system to the output detector system. Moreover, the duration of the input stimuli would not necessarily be related to the length of the exposure of the neuron to a neurotransmitter or a neuromodulator. A short stimulus exposure such as from a neurotransmitter might give a prolonged response if the internal processing system had slow response characteristics. A long stimulus exposure could modify the neuron for a long time or it could initiate a covalent adaptation scheme so that the response inside the cell was of only brief duration. Because of the possibility for different receptors on the surface and the variety of components in the processing system, almost any permutation of responses to neurotransmitters, neuromodulators and informational substances could be devised. That the responses could be tailored to many purposes by simple modification of the levels of enzymes within the cell and the number of receptors on the cell surface provides enormous versatility to the neuronal system.

CONVERSION OF SHORT-TERM TO LONG-TERM MEMORY

In principle, there could be a continuum from short- to long-term memory. The analysis of the bacterial system indicates that the memory span of the bacterial cell can be altered in a variety of ways. It can be altered by increasing the number of receptors relative to the modifying enzymes. It can be altered by changing components in the processing system. Such modifications undoubtedly could be utilized by neurons to tailor their timing from milliseconds to hours, without much difficulty. Simple changes in the ratios of some of the enzymes involved in the dynamic system would achieve this, as described above. In fact, the internal processing systems of hormonal cells which respond over days or weeks do not appear to be extremely different from those of neurons which operate in milliseconds. Second messenger systems utilizing cyclic AMP, calcium, and inositol phosphates are common to all.

A more difficult problem arises, however, for long-term memory. Although a neuronal cell by itself exists for long intervals, the proteins within it do not. Thus, modification of an individual protein would not be sufficient to maintain a long-term memory response as exists in humans. Some intracellular event which is permanently changed in the cell is required. One tempting concept for such an event would be an alteration of the DNA, but most alterations of DNA occur at the time of DNA replication. A variant of those mechanisms, such as a methylation of the DNA or even some type of translocation mechanism of the type exhibited by antibodies in the absence of DNA duplication, might exist in neuronal cells. Nevertheless, it would be unusual and no evidence has yet been discovered. It is not necessary, however, to require a change in the DNA. Any change in an intracellular process which has some kind of amplification to maintain the modified state would suffice.

In the process of studying the receptors for bacterial chemotaxis, a clue of one type of mechanism was revealed. As discussed in Russo and Koshland (1983), the receptor for chemotaxis can be modified by removing the last 35 amino acids through genetic engineering. When that is done, the adaptation properties of the cells change completely. The wild type cell when exposed to aspartic acid has a response time which is over within a minute. A wild type cell in which the receptor has been overproduced has a response time which is changed by the order of two hours. Therefore, any mechanism which permanently turns on the receptor would permanently alter such a cell by continuously over-producing receptor proteins and make a permanent modification in the protein which would be the type of alteration needed for memory. Even more dramatic is the case in which 35 amino acids are removed from such an over-produced receptor. Such a cell shows little adaptation property because its receptor cannot be methylated or is methylated slowly. Aspartate turns on the cell but the cell cannot adapt, as does the normal wild type cell. In other words, the cell characteristics have been changed permanently. A feedback mechanism utilizing this principle could lead to a permanent alteration of the type needed for long term memory (Koshland *et al.* 1983). If, for example, some protease were released in such a cell which removed the 35 amino acids from the C-terminal end of the receptor, and that peptide triggered the synthesis of the protease, one would have a feedback loop which would continue through subsequent replications and syntheses of proteins. It would require no change in the DNA, but only a triggering of a protease activation in some kind of threshold event. Detailed mathematical calculations are being made on just such a system, and show that it is indeed possible that repetitive stimulation such as the repetition of a phone number or a multiplication table could have a threshold phenomenon which turned on such a system. It is furthermore intriguing to note that in the long-term potentiation studies of Lynch and Baudry (1984), there is a suggestion that the activation of a protease permanently alters such a cell and they have postulated that this protease may be involved in long-term memory.

It is too early of course to relate model systems to the much more complicated whole animal systems, but at least molecular mechanisms have been derived which could lead to the phenomenon of a permanent memory trace within a cell without any changes in our current knowledge of enzymes or phenomena. It will be intriguing therefore to see the development of these model studies and the clues they give in regard to the processing of neuronal cells in complex networks.

Acknowledgement

The author is grateful for financial support from the National Institutes of Health and the National Science Foundation.

References

Adler, J. (1969). Chemoreceptors in bacteria. *Science*, **166**, 1588–97.

Block, S. M., Segall, J. R., and Berg, H. C. (1982). Impulse responses in bacterial chemotaxis. *Cell*, **31**, 215–26.

Bogonez, E. and Koshland, D. E., Jr (1984). *Proc. Nat. Acad. Sci.* (in press).

Clegg, D. O. and Koshland, D. E. Jr (1984). The role of a signaling protein in bacterial sensing—behavioral effects of increased gene expression. *Proc. Nat. Acad. Sci.* **81**, 5056–60.

—— and —— (1985). Identification of a bacterial sensing protein and effects of its elevated expression. *J. Bacteriol.* **162**, 398–405.

Cohen, S., Ushiro, H., Stoscheck, C., and Chinkers, M. (1982). A native 170,000 epidermal growth-factor receptor-kinase complex from shed plasma-membrane vesicles. *J. Biol. Chem.* **257**, 1523–31.

Dean, G. E., Aizawa, S. I., and Macnab, R. M. (1983). FLAAI (MOTC, CHEV) of *Salmonella typhimurium* is a structural gene involved in energization and switching of the flagellar motor. *J. Bacteriol.* **154**, 84–91.

Downward, J., Yarden, Y., Mayes, E., Scrace, G., Totty, N., Stockwell, P., Ullrich, A., Schlessinger, J., and Waterfield, M. D. (1984). Close similarity of epidermal growth-factor receptor and V-ERB-B oncogene protein sequences. *Nature*, **307**, 521–7.

Ebina, Y., Ellis, L., Jarnagin, K., Edery, M., Graf, L., Clauser, E., Ou, J. H., Masiarz, F., KAn, Y. W., Goldfine, I. D., Roth, R. A., and Rutter, W. J. (1985). The human insulin-receptor cDNA—the structural basis for hormone activated transmembrane signaling. *Cell*, **40**, 747–58.

Iversen, L. L. (1984). Amino acids and peptides—fast and slow chemical signals in the nervous system? *Proc. Roy. Soc. Lond. Ser. B, Biol. Sci.* **221**, 245–60.

Kandel, E. R. (1976). *Cellular Basis of Behavior*. W. H. Freedman & Co., San Francisco.

——, Abrams, T., Bernier, L., Carew, T. J., Hawkins, R. D., and Schwartz, J. H. (1983). Classical conditioning and sensitization share aspects of the same molecular cascade in Aplysia. *Cold Spring Harbor Symp. Quant. Biol.* **48**, 821–30.

Kasuga, M., Karlsson, F. A., and Kahn, C. R. (1982). Insulin stimulates the phosphorylation of the 95,000-dalton subunit of its own receptor. *Science*, **215**, 185–7.

Khan, S., Macnab, R. M., Defranco, A. L., and Koshland, D. E. (1978). Inversion of a behavioral response in bacterial chemotaxis—explanation at molecular level. *Proc. Nat. Acad. Sci.* **75**, 4150–4.

Koshland, D. E., Jr (1977). Response regulator model in a simple sensory system. *Science*, **196**, 1055–63.

—— (1980a). Bacterial chemotaxis in relation to neurobiology. *Ann. Rev. Neurosci.* **3**, 43–75.

—— (1982b). *Bacterial Chemotaxis as a Model Behavioral System*. Raven Press, New York.

—— (1982c). Biochemistry of sensing and adaptation. *Trends Biochem. Sci.* **5**, 297–302.

—— (1983). The bacterium as a model Neuron. *Trends Neurosci.* **6**, 133–7.

——, Russo, A. F., and Gutterson, N. I. (1983). Information processing in a sensory system. *Cold Spring Harbor Symp. Quant. Biol.* **48**, 805–10.

Kuhn, H., McDowell, J. H., Leser, K. H., and Bader, S. (1977). Phosphorylation of rhodopsin as a possible mechanism of adaptation. *Biophys. Struct. Mech.* **3**, 175–80.

Larsen, S. H., Reader, R. W., Kort, E. N., Tso, W. W., and Adler, J. (1974). Change in direction in flagellar rotation is basis of chemotactic response in Escherichia coli. *Nature*, **249**, 74–7.

Lynch, G. and Baudry, M. (1984). The biochemistry of memory—a new and specific hypothesis. *Science*, **224**, 1057–63.

Macnab, R. M. and Koshland, D. E., Jr (1972). The gradient sensing mechanism in bacterial chemotaxis. *Proc. Nat. Acad. Sci.* **69**, 2509–12.

Mesibov, R., Ordal, G. W., and Adler, J. (1973). The range of attractant concentrations for bacterial chemotaxis and the threshold and size of response over this range. *J. Gen. Physiol.* **62**, 203–23.

Miller, J. B. and Koshland, D. E., Jr (1977). Sensory electrophysiology of bacteria—relationship of membrane potential to motility and chemotaxis in Bacillus subtilis. *Proc. Nat. Acad. Sci.* **74**, 4752–6.

Parkinson, J. S. (1977). Behavioral genetics in bacteria. *Ann. Rev. Genet.* **11**, 397–414.

——, Parker, S. R., Talbert, P. B., and Houts, S. E. (1983). Interactions between chemotaxis genes and flagellar genes in *Escherichia coli*. *J. Bacteriol.* **155**, 265–74.

Quinn, W. G., Harris, L., and Benzer, S. (1974). Conditioned behaviour in Drosophila melanogaster. *Proc. Nat. Acad. Sci.* **71**, 708–12.

Rubik, B. A. and Koshland, D. E., Jr (1978). Potentiation, desensitization, and inversion of response in bacterial sensing of chemical stimuli. *Proc. Nat. Acad. Sci.* **75**, 2820–4.

Russell, D. W., Schneider, W. J., Yamamoto, T., Luskey, K. L., Brown, M. S., and Goldstein, J. L. (1984). Domain map of the LDL receptor: sequence homology with the epidermal growth factor precursor. *Cell*, **37**, 577–85.

Russo, A. F. and Koshland, D. E., Jr (1983). Separation of signal transduction and adaptation functions of the aspartate receptor in bacterial sensing. *Science*, **220**, 1016–20.

Schmitt, F. O. (1984). Molecular regulators of brain function: a new view. *Neuroscience*, **13**, 991–1001.

Silverman, M. and Simon, M. (1974). Flagellar rotation and the mechanism of bacterial motility. *Nature*, **249**, 73–4.

Springer, M. S., Goy, M. F., and Adler, J. (1979). Protein methylation in behavioral control mechanisms and in signal transduction. *Nature*, **280**, 279–84.

Springer, W. R. and Koshland, D. E., Jr (1977). Identification of a protein methyltransferase as the *cheR* gene product in the bacterial sensing system. *Proc. Nat. Acad. Sci.* **74**, 533–7.

Spudich, J. L. and Koshland, D. E., Jr (1976). Non-genetic individuality—chance in single cell. *Nature*, **262**, 467–71.

Squire, L. R. and Cohen, N. J. (1984). In *The Neurobiology of Learning and Memory* (eds J. L. McGaugh *et al.*), pp. 3–64. Guildford Press, New York.

Stock, J., Kersulis, G., and Koshland, D. E., Jr *Cell* (in press).

Stock, J. B. and Koshland, D. E., Jr (1978). Protein methylesterase involved in bacterial sensing. *Proc. Nat. Acad. Sci.* **75**, 3659–63.

——, Madeiris, A. M., and Koshland, D. E., Jr (1981). Bacterial chemotaxis in the absence of receptor carboxylmethylation. *Cell*, **27**, 37–44.

Teichberg, V. I., Sobel, A., and Changeux, J.-P. (1977). *In vitro* phosphorylation of acetylcholine receptor. *Nature*, **267**, 540–2.

Thompson, R. F., Berger, T. W., and Madden, J. (1983). Cellular processes of learning and memory in the mammalian CNS. *Ann. Rev. Neurosci.* **6**, 447–91.

Tsang, N., Macnab, R., and Koshland, D. E., Jr (1973). Common mechanisms for repellents and attractants in bacterial chemotaxis. *Science*, **181**, 60–3.

Ullrich, A., Bell, J. R., Chen, E. Y., Herrera, R., Petruzzelli, L. M., Dull, T. J., Gray, A., Coussens, L., Liao, Y. C., Tsubokawa, M., Mason, A., Seeburg, P. H., Grunfeld, C., Rosen, O. M., and Ramachandran, J. (1985). Human insulin receptor and its relationship to the tyrosine kinase family of oncogenes. *Nature*, **313**, 756–61.

Wang, E. A. and Koshland, D. E. (1980). Receptor structure in the bacterial sensing system. *Proc. Nat. Acad. Sci.* **77**, 7157–61.

Warrick, H. M., Taylor, B. L., and Koshland, D. E. (1977). Chemotactic mechanism of *Salmonella typhimurium*—preliminary mapping and characterization of mutants. *J. Bacteriol.* **130**, 223–31.

20

Chemical signalling in the spatial, temporal continuum

FLOYD E. BLOOM

INTRODUCTION

The inauguration of this splendid research centre makes a strong statement about the opportunities for understanding and application that most of those gathered here recognize to exist within the neurosciences. As ever-more senior citizens within the growing populace of neuroscientists, we have witnessed marked changes in our expectations and anticipations of where the new insights will take us. By focussing our attention on the dynamic aspect of differing time spans for the action of different sets of interneuronal chemical signals, this book has immediately highlighted that interface between cellular and molecular neurobiology at which the most extensive progress has been made recently: the discovery of new neuronally-made substances satisfying many of the criteria of neurotransmitters, as I broadly define them (Bloom 1984a); the recognition of new anatomic connections linking neurons in circuits; and the recognition that the newly recognized chemicals in the newly recognized circuits can produce effects on the transmission process which expand considerably upon classical views of how transmitters act.

It seems important to try to define the standards by which the effects of transmitter substances can be classified and compared, not simply as 'factors' assessed *in vitro* for biochemical regulatory actions, but rather as physiological regulators operating within the context of their circuit anatomy and their overall signalling value, an occasional pastime of mine over the past dozen years (see Bloom 1973, 1979, 1984a,b). Looking back over those essays at 5-year intervals, one recognizes that the new data on cell structure and function and on the role of transmitter actions were impossible to predict. We have been blessed with an enormous richness of factual information in which the real puzzles continue to be how the data should be interpreted and what they may mean for understanding the operation of the intact brain.

From the vantage point of our present location in time and space, the points made by Schmitt (1984) in his recent thoughtful essay seem clearly to define the mainstreams of the new frameworks onto which the data are presently being conceptually placed: onto the classical, hard-wired nervous system of those who got us into this era of brain research has been superimposed a much looser nervous system in which neurons may release their signals to act at some distance from the release site, not strictly as primary communicators, but rather as signals that can modify the response of the intended target cells to the classical transmitter signals they are receiving simultaneously. In less than one decade, the nervous system ceased to be a rather dry place in which, to classical physiologists, transmitters meant relatively little, since they all worked either to excite or inhibit. In its stead we have a very juicy, flexible nervous system in which transmitters act on many different receptor transduction mechanisms to provide a very enriched repertoire of signalling capabilities across widely differing spatial domains and widely variant durations of action.

For this communication, I will concentrate my remarks on three accumulations of our data that seem relevant to these concerns of time and space in brain chemical signalling. Firstly, I shall update our group's recent efforts in defining the patterns of specificity within the widely divergent central noradrenergic nervous system to indicate that such systems are far from 'non-specific' in their innervation patterns; from these data I am inclined to conclude that such anatomic selectivity reflects target cell specificity regardless of one's views of the necessity, presence or absence of specialized synaptic contacts. Next, I shall examine one example of monoamine-peptide interaction, that between noradrenaline and vasoactive intestinal polypeptide (VIP) through their interactive regulation of neocortical adenylate cyclase. Finally, from these data, I will conclude that the actions of monoamines on their target cells, and the interactions between monoamines and peptides on their shared target cells may represent regulatory mechanisms that are 'conditional' and that such mechanisms may well set the scales on which other spatial and temporal signals are measured.

MONOAMINES IN SPATIAL AND TIME DOMAINS

Monoamines were among the first chemically defined transmitters to meet the more rigorous tests as central transmitters. This was due to two specific advantages: the broader armamentarium of drugs available to manipulate the central monoaminergic systems, and the detailed structural information on these systems, thanks to the early development of chemically specific and sensitive methods for their cytochemical localizations (see Morrison *et al*. 1984a,b). However, their unique cellular morphology—with a highly divergent axonal arborization connecting pontine nuclei with cortical regions by routes never visualized by the empirical methods of the metal-impregnation era—and their unique electrophysiological actions—altering membrane potential without increased ionic conductance (see Foote *et al.* 1983, and Siggins and Gruol 1985, for recent reviews) required considerable conceptual expansion of the concept of a neurotransmitter, actions that may well be outside the boundaries of conditions acceptable to some minds as 'real neurotransmitters'.

Time and space domains

The noradrenergic and other monoaminergic systems provide excellent illustrations of the features of time and space by which the possible functions of chemically-labelled neuronal systems may be abstracted (Bloom 1973, 1979, 1984a). Practically, the operations of all neurons can be factually charted on two domains, space and time, for comparative analysis. The spatial domain of a neuron is the total target cell area to which that neuron sends information. Similarly, the temporal domain is the time course of the neuron's effects on its targets especially the duration and pattern of its activity, and that of its synaptic targets. To appreciate the special properties of the monoaminergic neurons, requires a brief recapitulation to place them in a whole brain perspective.

Structural categories of discriminable transmitter systems

From my perspective as a cellular physiologist interested in broad classes of structural features which may serve to differentiate principles of neurotransmission, I find that most circuits can be lumped into three general categories (Bloom 1984b): (1) hierarchically arranged neurons in chained systematic, or throughout, connections; (2) divergent, single source, multi-targeted connections; and (3) local circuit neurons. The classical concept of the hierarchical or 'throughput' system offers highly precise information transfer, but suffers from the disadvantage that destruction of one link incapacitates the chain. The loss of function with destruction of defined links represents a time-honoured strategy in functional neuroanatomy. Nevertheless, the critical fact for the present context is that no primary transmitter molecules have been documented for any such throughput system, afferent or efferent, except for the final common output link from spinal or autonomic neurons to their effectors. Although not fully satisfying as rigorous identification, the prompt, vigorous, and classical actions of the excitatory amino acids Glu or Asp have been held as prototypes for throughput connections, as chapters in this book have attested.

The local circuit neurons are small, frequently unipolar, neurons whose efferent processes may bear the morphology of dendrites, and whose connections are established exclusively within the local vicinity in which the cell body is found. Although such small interneurons can exert significant control over information within their locale (presumably through conventionally coupled excitation-secretion mechanisms) they may also exhibit an unconventional 'dendritic' release of their transmitter, perhaps without action potentials (see Iversen 1984). Such morphological and chemical specializations transcend all known general transmitter categories. In some cases of inhibitory local circuits, the amino acids gamma-aminobutyrate (GABA) or glycine (Gly), and the neuropeptides somatostatin, enkephalin, dynorphin, and others have been incriminated as transmitters (see Bloom 1984b, for references).

By comparison with hierarchical neurons, or local circuit neurons, the monoaminergic and some peptidergic neurons operate over much larger spatial domains with much less constrained connectivities than the hierarchical systems. Most central monoamine systems (such as the noradrenergic locus coeruleus) fit the image of a single-source, divergent circuit (see Bloom 1984b; Foote *et al.* 1983; Morrison *et al.* 1984a).

Cortical organization of noradrenergic circuits

The continued analysis of the NA coeruleo-cortical system in the rat, has demonstrated two major characteristics: (1) there is a rich network of NA innervation throughout all layers and regions of the dorsal and lateral cortex which is characterized by a relatively uniform laminar pattern; and (2) the major NA fibres are orientated, and travel longitudinally through the grey matter and branch widely, furnishing the coeruleo-cortical system with the unique capacity to modulate neuronal activity synchronously throughout a vast expanse of neocortex (see Morrison *et al.* 1984a). Thus, the NA innervation of neocortex may be viewed as a tangential afferent system whose organization and pattern of termination is different from the highly localized, radial nature of the thalamo-cortical afferents. Although a tangential system of this design may largely reflect the lissencephalic, largely homogeneous structure of rat neocortex, more recent studies of monoamines in the far more highly differentiated, gyrencephalic neocortex of the primate shows the need for considerable refinement in concepts of monoamine circuit specificity.

We have used immunohistochemical methods to investigate these systems in three primates because these techniques in our hands are superior to direct amine fluorescence techniques in sensitivity and permanence. In the course of our studies, we have found that the specific density and pattern of NA (or 5-HT) innervation in a particular cortical locus varies systematically as a function of several factors: the cytoarchitectonic region, the cortical lamina, the species of animal, age of the animal, and the function of the region. Each of these factors influences monoaminergic innervation patterns specifically and have been described in detail in our recent publications (Morrison *et al.* 1984a; Foote and Morrison 1984).

The tangential intracortical trajectory that is characteristic of the rat brain is also a dominant feature of the NA innervation of the primate brain. However, the NA innervation of the primate brain exhibits a far greater degree of regional heterogeneity than the rat: most major cytoarchitectural regions exhibit distinctive patterns of innervation in terms of both density and laminar distribution of fibres (Fig. 20.1). These regional variations in primate NA innervation do not follow any simple geometric or functional organizational principles (i.e. preference for primary sensory cortices). By contrast, the 5-HT system does appear to innervate primary sensory cortices more densely than secondary sensory or association areas (see Morrison 1984a).

Laminar specificity

A detailed examination of primate primary visual cortex was undertaken in order to characterize possible laminar specialization of monoaminergic innervation patterns in the most obviously laminated and well-characterized region of neocortex. Initially, we characterized the NA and 5-HT innervation patterns within this cortical area in the squirrel monkey (*Saimiri sciureus*). The primary visual cortex has a relatively low density of NA innervation, whereas the 5-HT innervation, particularly in layer IV, is densest of all neocortical regions. These two fibre systems exhibited a high degree of laminar specialization, and were, in fact, distributed in a complementary fashion: layers V and VI receive a moderately dense NA projection and a sparse 5-HT projection, whereas layers IVa and IVc receive a very dense 5-HT projection and are largely devoid of NA fibres. These patterns of innervation imply that the two transmitter systems affect different stages of cortical information processing: the raphe-cortical 5-HT projections may preferentially innervate the spiny stellate cells of layers IVa and IVc, whereas the coeruleo-cortical NA projection may be directed predominantly at pyramidal cells.

We have recently compared these monoamine innervation patterns in the visual cortex of old and new world monkeys (Fig. 20.2). The old world monkeys receive an even denser 5-HT innervation of primary visual cortex than do the squirrel monkeys, whereas the overall density of the NA innervation is equivalent, or even slightly decreased in the old as compared

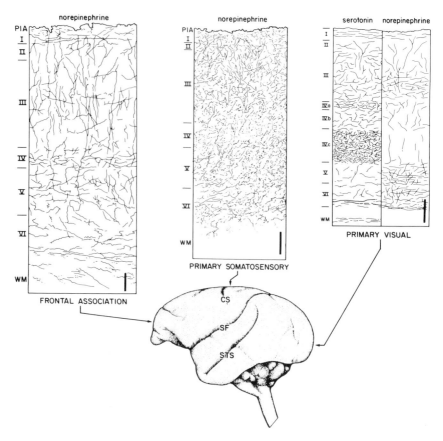

FIG. 20.1. Schematic depiction of the noradrenergic innervation of the squirrel monkey neocortex to illustrate the density and laminar variation differences across three primate neocortical regions: the dorsolateral frontal association cortex, the primary somatosensory cortex, and the primary visual cortex. The frontal and somatosensory cortex show substantial variation in noradrenergic fibre densities which are approximately equivalent across all six layers of cortex, but much denser in somatosensory cortex. In visual cortex, noradrenergic fibres are more sparse in general, but show some specific laminar omissions (layer IV) and some specific laminar concentrations (layers III, V, and VI). Also shown for visual cortex are the distribution of 5-HT fibre systems, showing the complementary laminar enrichment in those layers sparse in noradrenergic fibres, and the specific laminar sparseness in those layers most dense in noradrenergic fibres. Abbreviations: CS, central sulcus; SF, Sylvian fissure; STS, superior temporal sulcus. Calibration bars = 100 μm. Modified from Morrison and Magistretti (1983) with permission of the authors.

Squirrel monkey   Cynomolgus

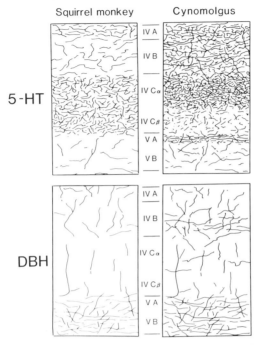

FIG. 20.2. Schematic illustration of the distributions within the sublaminae of Layer IV of 5-HT (upper row) and noradrenergic (lower row) dopamine-β-hydroxylase (DBH)-positive fibres in the primary visual cortex of New World (squirrel monkey; left column) and Old World (Cynomolgus; right column) Primates. Note that the complementarity between laminar density distributions for the two different monoamines holds across both primates for this most highly specialized laminar definition. Modified from Foote and Morrison (1984) with permission of the authors.

to new world monkey. These variations in 5HT patterns correlate with the further cortical specialization in the old world species.

These observations suggest that coincident with the extensive phylogenetic development and functional differentiation of neocortex in the primate there is a parallel elaboration and differentiation of the ascending NA and 5-HT projections. The remarkable heterogeneity of NA and 5-HT innervation observed in monkey neocortex is highly ordered and is systemically related to the cytoarchitectonic region, lamina, age, species, and functional nature of the particular area of neocortex being examined. In addition, there are important differences in the NA and 5-HT innervation in each of these 'dimensions'. In each case, the different innervation patterns point to functional distinctions between these two transmitter-specific systems and point towards a substantial and independent role for each of these monoamine systems in neocortical information processing in the primate.

In addition to there being substantially more presumptive specificity of anatomic innervation pattern for monoaminergic systems in primate cortex, recent data investigating the presumptive efferent trajectories of the locus coeruleus system in the rodent (Loughlin *et al.* 1985) support a far more precise system than earlier data had suggested. When retrograde transport of horseradish peroxidase is used to locate locus coeruleus neurons projecting to a given large terminal field, and the data are fitted to a computerized mean three-dimensional matrix of the perikarya coded for size and shape, it is possible to discriminate the noraderenergic neurons of origin for at least four terminal fields: spinal, hippocampal, hypothalamic, and neocortical. Such data suggest to us that even finer destinations of terminal field innervation areas might further subdivide the cluster of noradrenergic neurons into functionally separable sources, despite their likely high degree of intranuclear interaction.

Synaptic ultrastructure

One of the most controversial aspects of monoaminergic and peptidergic fibres, and the source for a contentious distinction between them and 'real' sympathetic transmitters, is the nature of their connections with potential target cells. The controversial interpretation is whether these fibres form 'true' synapses or end, instead, in a diffuse endocrine (paracrine or neurocrine) type of relationship with many unspecified target cells. In my view, the controversy stems from differences in the methods used to visualize the terminals and the region of the brain in which they are studied (see Foote *et al.* 1983, and Bloom 1984b for more extensive discussion).

As in all other cases of attempted ultrastructural-physiological correlation, the assumption that the specialized contact zone is the site of synaptic transmission (i.e. the actual release and response site) remains to be documented with certainty. It is clear that sympathetic fibres of the peripheral autonomic nervous system can transmit to smooth muscle without such specializations. An accurate determination of the precise cellular location of central adrenoreceptors or other specific receptors with reference to the specialized junctions of identified noradrenergic boutons (see Strader *et al.* 1983) may be ultimately required to settle the issue of post-synaptic specificity for these unconventional fibres. What remains wholly untested is the degree to which the varicose axons of a presumed conventional transmitter using neuron show non-specialized contacts as well. For vasopressin, it is said that non-specialized central boutons do not release (Buijs and Van Heerikhuize 1982). More detailed ultra-structural studies of immunocytochemically identified transmitter systems within defined brain regions are clearly necessary.

INTERACTIONS BETWEEN VASOACTIVE INTESTINAL POLYPEPTIDE AND NORADRENALINE IN RAT CEREBRAL CORTEX

Several lines of evidence support a role for vasoactive intestinal polypeptide (VIP) as a neuronal messenger in cerebral cortex. Biochemical data support its presence, release, binding, and at best one possible action (see Morrison and Magistretti 1983; Magistretti and Morrison 1985, for refs). Immunocytochemical data have localized VIP to cortical bipolar neurons (Morrison *et al.* 1984b). Single unit recording studies show that iontophoretically applied VIP excites some cortical neurons (Phillis *et al.* 1978), but in an inconsistent manner. Other cellular actions reported for VIP in cortex include the ability to stimulate cyclic AMP formation (Quik *et al.* 1979; Etgen and Browning 1983; Magistretti and Schorderet 1984, 1985; Magistretti *et al.* 1983) in cortical slices. It is somewhat more potent in these actions than noradrenaline (NA).

Cytochemically, VIP and NA containing circuits show a contrasting but complementary cortical anatomy (see Fig. 20.3; Morrison and Magistretti, 1983): VIP neurons are intrinsic, bipolar, radially-oriented, intracortical neurons (Morrison *et al.* 1984b;) the NA innervation arises only from locus coeruleus and innervates a broad expanse of cortex in a horizontal plane. The two fibre systems may have the same targets, the pyramidal cells. Identified cortical pyramidal neurons are depressed in spontaneous firing by iontophoresis of either NA or cyclic AMP (see Foote *et al.* 1983). The recent findings of Magistretti and Schorderet (1984, 1985) suggest that VIP and NA can act synergistically to increase cyclic AMP in cerebral cortex. Therefore, we tested VIP and NA on rat cortical neurons to evaluate this interaction at the cellular level using iontophoresis. Analysis of more than 100 cortical neurons suggests definitively that there are significant interactions: application of VIP during subthreshold NA administration causes pronounced inhibitors of cellular discharge regardless of the effect of VIP prior to NA (Ferron *et al.* 1985).

In our hands, the direct effects of VIP on the spontaneous firing rate of sensorimotor cortical neurons was not very impressive. VIP depressed 24 per cent, excited 20 per cent and had biphasic effects on 2 per cent; maximal currents (200 nA) passed through the VIP barrel had no effect on 54 per cent of these cells. The direction of the response and the proportion of responding neurons were similar for identified pyramidal neurons and for all cortical neurons. The direct effects of NA were predominantly depressant actions in the cortex, as previously reported by us (see Foote *et al.* 1983; Siggins and Gruol 1986). Therefore, in order to examine possible interactions between VIP and NA effects on single neurons, ejection currents of NA were reduced until they had little or no direct effect on

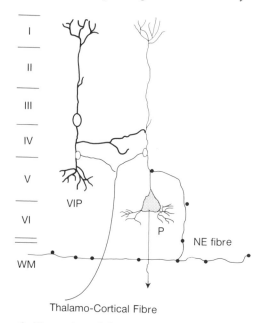

FIG. 20.3. Schematic illustration of the convergent, specialized afferent systems of VIP and noradrenergic neurons in the rodent neocortex. The intracortical, bipolar, VIP-immunoreactive neurons receive innervation from specific thalamic afferent systems, collaterals of which go to the primary cortical output neurons, the pyramidal cells (P). The output of the VIP neurons and the targets of the extracortical noradrenergic innervation are depicted as converging on the pyramidal neurons, as anatomical, biochemical and electrophysiological studies. The simultaneous period activity in the noradrenergic fibres, during a period of thalamic projection to neocortex could provide a mutually amplifying effect on the pyramidal neuron targets. Modified from Morrison and Magistretti (1983) with permission.

neuronal firing (1–5 μA). After testing a cell for direct effects (if any) of VIP, subthreshold currents of NA were then applied continuously for several minutes while VIP was repeatedly re-tested. Tests on approximately 50 neurons studied in this way showed that ejection of VIP during subthreshold NA administration now produced consistent inhibition of firing. Such synergistic inhibitions were seen in more than half the cells regardless of whether VIP alone had elicited excitatory, inhibitory, or negligible effects on the test neuron (Ferron *et al.* 1985).

When VIP alone had a depressant action, administration of subthreshold NA markedly enhanced this depressant effect. Even in cases where VIP alone gave excitatory effects, concurrent subthreshold NA treatment reversed the VIP effect from excitation to inhibition (six of nine cells).

Magistretti and Schorderet (1984, 1985) showed that the synergism of VIP by NA was blocked by phentolamine, an alpha-adrenergic receptor antagonist, and mimicked by phenylephrine, an alpha-receptor agonist. Therefore, we examined the effect of phenylephrine as well as NA pretreatment on neuronal responses to VIP. In 10 cells showing an interaction between VIP and NA, nine revealed equivalent interactive synergisms of depressant responses with phenylephrine (Ferron *et al.* 1985). Thus, the interaction of VIP and NA at the cellular level may also involve alpha receptor activation, although further testing is required.

Our electrophysiological indications of a VIP-NA interaction at the cellular level may arise from their biochemical effects *in vitro* on cyclic AMP generation. Such parallel findings strengthen the suggestions that cyclic AMP may mediate both NA and VIP evoked depressions of neuronal firing in cortex. The reported enhancement by NA of synaptic and other transmitter responses (including inhibitory ones) may be related phenomena (see Foote *et al.* 1983; Bloom 1984a). An apparently similar, cAMP-mediated, enhancement by beta receptors of noradrenergic target cell responsiveness to alpha adrenergic agonists has been reported for pineal (Klein *et al.* 1983; Sugden *et al.* 1985). Furthermore, in rat hepatocytes, increased cAMP levels induced by glucagon enhance binding of alpha adrenergic agonists to these cells (Morgan *et al.* 1984). Two modes of indirect amplification of post-synaptic target cell mechanisms may be involved in these interactions, namely a cAMP-mediated activation of protein kinases (as with beta-adrenergic and other responses) and a calcium activation of protein kinases (as with $alpha_1$-adrenergic, and other transmitter or hormone responses (see Nestler and Greengard 1984). However, it is not yet clear how these metabolic and electrophysiological events actually interact. If NA and VIP-containing fibres do indeed converge on the same cortical target cell, it is feasible that cyclic AMP is the intracellular mediator of their synergistic interaction.

CONCLUSIONS

While one would hesitate to rock the boat that in a conceptual ocean views monoamines and peptides as acting in the paracrine fashion rather than as conditional transmitters with their own special actions on a limited set of immediate target cells, the above summaries could certainly support such a view. As one who has spent considerable mental energy concerned with the utility of the micromethods of transmitter evaluation *in vivo* and their limitations (see Bloom 1974), I hesitate to place much emphasis on the relatively long time lag to onset of action reported for monoamines and peptides. It is only through evaluation of synaptic actions documented to be transmitted by specific agents that one can assess the minimum latency for onset of effects. Such data exist for some monoaminergic systems *in situ*, but not many (see Siggins and Gruol 1985, for review). Such *in vivo* data suggest that endogenous noradrenergic systems may have a synaptic latency that approaches 100–150 ms and that their own sensory activation requires 35–75 ms depending on the modality (see Foote *et al*. 1983). Such data could also reflect on the likely duration of those synaptic actions when applied in frequencies and train durations that would simulate the activity of the natural firing patterns of those monoaminergic single source divergent neurons. Such data also exist for a few studies of other rodent and primate monoamine neurons (see Foote *et al*. 1983, for review).

However, there are as yet no such data for peptide-containing systems in which a clustered group of neurons is susceptible to activation and recording under conditions that would define the possible time and space properties of the responses of the likely target cells. In fact, there are no instances of which I am aware in which the effects of a peptide containing system in the mammalian CNS have been identified as being peptidergic. Moreover, given the existing anatomic data, there are few places other than the hypothalamic paraventricular nucleus (Buijs and Van Heerikhuize 1982), where such stimulation paradigms might be effectively attempted. Until such data are at hand, it will fall upon general happy occasions as this to reminisce of how little we used to know, and how much smarter we all are now, especially when the hindsight allows us to focus our selective recollections of what we all predicted would be the 'truth'. Such retrospectoscopic visions also have the property of transcending time and space.

Acknowledgements

I would like to thank my colleagues, Drs Siggins, Foote, Morrison, and Ferron, for allowing me to quote from their recent studies in preparation, and Mrs Nancy Callahan for preparing this manuscript. Supported by grants from the Whittier Foundation.

References

Bloom, F. E. (1973). Dynamic synaptic communication: finding the vocabulary. *Brain Res.* **62**, 299–305.
—— (1974). To spritz or not to spritz: the doubtful value of aimless iontophoresis. *Life Sci.* **14**, 1819–34.
—— (1979). Chemical integrative processes in the central nervous system. In *Neurosciences—Fourth Intensive Study Program* (eds F. O. Schmitt and F. G. Worden), pp. 51–8. MIT Press, Cambridge.
—— (1984a). The functional significance of neurotransmitter diversity. *Am. J. Physiol.* **246**, C184–94.
—— (1984b). Chemical integrative processes in the central nervous system. In *Handbook of Chemical Neuroanatomy* (eds. A. Björklund, T. Hökfelt, M. J. Kuher), Vol. 2, pp. 51–8. Elsevier, Amsterdam.

Buijs, R. M. and Van Heerikhuize, J. J. (1982). Vasopressin and oxytocin release in the brain—a synaptic event. *Brain Res.* **252**, 71–6.

Etgen, A. M. and Browning, E. T. (1983). Activators of cyclic adenosine 3′:5′-monophosphate accumulation in rat hippocampal slices: action of vasoactive intestinal peptide (VIP). *J. Neurosci.* **3**, 2487–93.

Ferron, A., Siggins, G. R., and Bloom, F. E. (1985). Vasoactive intestinal polypeptide acts synergistically with noradrenaline to depress spontaneous discharge rates in cerebral cortical neurons. *Proc. Nat. Acad. Sci. USA*, (in press).

Foote, S. L., Bloom, F. E., and Aston-Jones, G. (1983). Nucleus locus ceruleus: new evidence of anatomical and physiological specificity. *Physiolog. Rev.* **63**, 844–914.

—— and Morrison, J. H. (1984). Postnatal development of laminar innervation patterns by monoaminergic fibers in monkey (*Macaca Fascicularis*) primary visual cortex. *J. Neurosci.* **4**, 2667–80.

Iversen, L. L. (1984). Amino acids and peptides: fast and slow chemical signals in the nervous system? *Proc. Roy. Soc. Lond.* **B221**, 245–60.

Klein, D. C., Sugden, D., and Weller, J. L. (1983). Postsynaptic α-adrenergic receptors potentiate the β-adrenergic stimulation of pineal serotonin N-acetyltransferase. *Proc. Nat. Acad. Sci. USA*, **80**, 599–603.

Loughlin, S. E., Foote, S. L., and Bloom, F. E. (1985). Efferent projections of nucleus locus coeruleus: I. Topographical organization of cells of origin demonstrated by three-dimensional reconstruction. *J. Neurosci.* (in press).

Magistretti, P. J. and Morrison, J. H. (1985). VIP neurons in the neocortex. *Trends Neurosci.* **8**, 7–8.

——, ——, Shoemaker, W. J., and Bloom, F. E. (1983). Effect of 6-hydroxydopamine lesions on norepinephrine-induced ³H-glycogen hydrolysis in mouse cortical slices. *Brain Res.* **261**, 159–62.

—— and Schorderet, M. (1984). VIP and noradrenaline act synergistically to increase cyclic AMP in cerebral cortex. *Nature*, **308**, 280–2.

—— and —— (1985). Norepinephine and histamine potentiate the increases in cyclic adenosine 3′:5′-monophosphate elicited by vasoactive intestinal polypeptide in mouse cerebral cortical slices: mediation by α₁-adrenergic and H₁-histaminergic receptors. *J. Neurosci.* **5**, 363–8.

Morgan, N. G., Charest, R., Blackmore, P. F., and Exton, J. H. (1984). Potentiation of alpha 1 adrenergic responses in rat liver by a cAMP-dependent mechanism. *Proc. Nat. Acad. Sci. USA*, **81**, 4208–12.

Morrison, J. H., Foote, S. L., and Bloom, F. E. (1984a). Laminar, regional, developmental and serotonergic innervation patterns in monkey cortex. In *Monoamine Innervation of Cerebral Cortex* (eds L. Descarries, T. A. Reader, and H. H. Jasper), pp. 61–75. Alan R. Liss, Inc. New York.

——, ——, Molliver, M. E., Bloom, F. E., and Lidov, H. G. W. (1982). Noradrenergic and serotonergic fibers innervate complementary layers in monkey primary visual cortex: an immunohistochemical study. *Proc. Nat. Acad. Sci.* **79**, 2401–5.

—— and Magistretti, P. J. (1983). Monoamines and peptides in cerebral cortex: contrasting principles of cortical organization. *Trends Neurosci.* **6**, 146–51.

——, ——, Benoit, R., and Bloom, F. E. (1984b). The distribution and morphological characteristics of the intracortical VIP-positive cell: an immunohistochemical analysis. *Brain Res.* **292**, 269–82.

Nestler, E. J. and Greengard, P. (1984). *Protein Phosphorylation in the Nervous System*. Rockefeller University Press, New York.

Quik, M., Emson, P. C., Fahrenkrug, J., and Iversen, L. L. (1979). Effect of kainic acid injections and other brain lesions on vasoactive intestinal polypeptide (VIP)-stimulated formation of cAMP in rat brain. *Naunyn-Schmiedeberg's Arch. Pharmacol.* **306**, 281–6.

Phillis, J. N., Kirkpatrick, J. R., and Said, S. I. (1978). Vasoactive intestinal polypeptide excitation of central neurons. *Can. J. Physiol. Pharmacol.* **56**, 337–40.

Schmitt, F. O. (1984). Molecular regulators of brain function: a new view. *Neurosci.* **4**, 994–1004.

Siggins, G. R. and Gruol, D. L. (1986). Synaptic mechanisms in the vertebrate central nervous system. In *Handbook of Physiology, Section 1. The Nervous System*. Vol. 4, *Intrinsic Regulatory Systems of the Brain* (ed. F. E. Bloom), The American Physiological Society, Bethesda, Maryland.

Strader, C. D., Pickel, V. M., Joh, T. H., Strohsacker, M. W., Shorr, R. G. L., Lefkowitz, R. J., and Caron, M. G. (1983). Antibodies to the β-adrenergic receptor: attenuation of catecholamine-sensitive adenylate cyclase and demonstration of postsynaptic receptor localization in brain. *Proc. Nat. Acad. Sci. USA*, **80**, 1840–4.

Sugden, D., Vamecek, J., Klein, D. C., Thomas, T. P., and Anderson, W. B. (1985). Activation of protein kinase C potentiates isoprenaline-induced cyclic AMP accumulation in rat pinealocytes. *Nature*, **314**, 359–61.

21

Chemically addressed neural communications: discussion

J. D. SALAMONE AND N. G. BOWERY

Injection of the amines octopamine and serotonin in the lobster causes the animal to adopt unusual postures for a long time, perhaps hours. This is much longer than the animals normally exhibit these postures and probably represents a non-physiological over stimulation of the system. The amine systems in the animal may normally act to set a general bias towards these types of behaviour, and then the postures themselves are phasically presented in specific situations. This is perhaps analogous to the effects of adrenaline in mammals, where the hormone sets a bias for the animal to engage in a 'fight or flight' reaction, but the particular response to a given situation is determined by a number of other factors. It was suggested that future research should try to approach a study of how the system normally operates. Interactions between serotonin and octopamine, and studies of what inputs activate these systems will be important.

The developmental and morphological studies of E. A. Kravitz indicate that aminergic neurones in crustacea appear to form groups which may act as behavioural units. The same may be true for mammals since there is little doubt that amine systems form groups within the mammalian brain. The noradrenergic system, however, appears to be less prone to grouping than other amine pathways. Nevertheless, grouping of nerve terminals within the noradrenergic system is emerging and this may reflect functional subunits for noradrenaline-mediated behaviours. The density of neuronal fibres in any brain region may not correlate with the underlying importance of the particular system. This is illustrated by the early work of F. E. Bloom on the noradrenergic innervation of the cerebellum. This work demonstrated an important input from the locus coeruleus which could be modified by pharmacological agents. However, the fibres responsible could hardly be detected. One wonders, therefore, whether the distribution of the noradrenergic fibres detected in the cerebral cortex really reflects their importance in the various cortical layers. Only physiology will answer this question and undoubtedly studies on locus coeruleus stimulation do provide us with clues about the noradrenergic system in the deep layers. In these regions conventional noradrenergic synapses can be detected. However, this is not so in the superficial layers of the cortex where fibres do not appear to make discrete synapses. This is true not just for noradrenergic fibres, but also for virtually all other systems so far studied. The purpose or function of these tangential fibres is unknown although it has been suggested that they regulate glial cell function since such cells are most numerous in the superficial layers of the cortex. Perhaps the fibres provide overall control of blood flow and oxygen transport into the cortex at the nexus between the external innervated and internal non-innervated blood vessels.

To categorize transmitters on a slow versus fast basis requires careful consideration for, despite the limited number of ways for an extracellular signal to be transmitted by the receptor, the combination of these responses could produce signals of different time courses. For example, α-adrenoceptor stimulation can accentuate the level of cAMP generated by conventional stimulants. It can even enhance the response to isoprenaline. So that α-adrenoceptor and β-adrenoceptor activation may act in concert under certain conditions. It is therefore possible that the noradrenergic system may act in the conventional single transmitter manner as well as in a slower paracrine fashion to influence other receptor mechanisms.

The adenylate cyclase system through which cAMP generation occurs forms a major enzyme mechanism within the brain and Professor D. E. Koshland pointed out that it may be prominent in memory processing. Like the phosphorylation/dephosphorylation mechanism proposed by Crick (1984) maintenance of the adenylate cyclase/phosphodiesterase steady state may be important for memory. If either adenylate cyclase activity or phosphodiesterase activity are removed memory is lost. Thus, it may not be the level of cAMP which is important but the regulation of the steady state. If this serves as an example for other enzyme activity patterns then conventional measurement of changes in second messenger levels may not be adequate for discerning changes related to memory mechanisms.

Reference

Crick, F. (1984). Neurobiology—memory and molecular turnover. *Nature* **312**, 101.

Index